CYTOSKELETAL MECHANICS

This book presents a full spectrum of views on current approaches to modeling cell mechanics. The authors of this book come from the biophysics, bioengineering, and physical chemistry communities and each joins the discussion with a unique perspective on biological systems. Consequently, the approaches range from finite element methods commonly used in continuum mechanics to models of the cytoskeleton as a cross-linked polymer network to models of glassy materials and gels. Studies reflect both the static, instantaneous nature of the structure, as well as its dynamic nature due to polymerization and the full array of biological processes. While it is unlikely that a single unifying approach will evolve from this diversity, it is our hope that a better appreciation of the various perspectives will lead to a highly coordinated approach to exploring the essential problems and better discussions among investigators with differing views.

Mohammad R. K. Mofrad is Assistant Professor of Bioengineering at the University of California, Berkeley, where he is also director of Berkeley Biomechanics Research Laboratory. After receiving his PhD from the University of Toronto he was a post-doctoral Fellow at Harvard Medical School and a principal research scientist at the Massachusetts Institute of Technology.

Roger D. Kamm is the Germeshausen Professor of Mechanical and Biological Engineering in the Department of Mechanical Engineering and the Biological Engineering Division at the Massachusetts Institute of Technology.

Cytoskeletal Mechanics

MODELS AND MEASUREMENTS

Edited by

MOHAMMAD R. K. MOFRAD
University of California, Berkeley

ROGER D. KAMM
Massachusetts Institute of Technology

CAMBRIDGE UNIVERSITY PRESS
Cambridge, New York, Melbourne, Madrid, Cape Town, Singapore, São Paulo

Cambridge University Press
32 Avenue of the Americas, New York, NY 10013-2473, USA

www.cambridge.org
Information on this title: www.cambridge.org/9780521846370

© Cambridge University Press 2006

This publication is in copyright. Subject to statutory exception
and to the provisions of relevant collective licensing agreements,
no reproduction of any part may take place without
the written permission of Cambridge University Press.

First published 2006

Printed in the United States of America

A catalog record for this publication is available from the British Library.

Library of Congress Cataloging in Publication Data

Cytoskeletal mechanics : models and measurements / Mohammad R. K. Mofrad
and Roger D. Kamm, Editors.
   p. cm.
Includes bibliographical references and index.
ISBN-13: 978-0-521-84637-0 (hardback)
ISBN-10: 0-521-84637-4 (hardback)
1. Cytoskeleton – Mechanical properties. I. Mofrad, Mohammad R. K.
II. Kamm, Roger D. III. Title.
QH603.C96C962 2006
571.6′54 – dc22   2006001153

ISBN-13 978-0-521-84637-0 hardback
ISBN-10 0-521-84637-4 hardback

Cambridge University Press has no responsibility for
the persistence or accuracy of URLs for external or
third-party Internet Web sites referred to in this publication
and does not guarantee that any content on such
Web sites is, or will remain, accurate or appropriate.

Contents

List of Contributors		page vii
Preface		ix
1	Introduction, with the biological basis for cell mechanics Roger D. Kamm and Mohammad R. K. Mofrad	1
2	Experimental measurements of intracellular mechanics Paul Janmey and Christoph Schmidt	18
3	The cytoskeleton as a soft glassy material Jeffrey Fredberg and Ben Fabry	50
4	Continuum elastic or viscoelastic models for the cell Mohammed R. K. Mofrad, Helene Karcher, and Roger D. Kamm	71
5	Multiphasic models of cell mechanics Farshid Guilak, Mansoor A. Haider, Lori A. Setton, Tod A. Laursen, and Frank P. T. Baaijens	84
6	Models of cytoskeletal mechanics based on tensegrity Dimitrije Stamenović	103
7	Cells, gels, and mechanics Gerald H. Pollack	129
8	Polymer-based models of cytoskeletal networks F. C. MacKintosh	152
9	Cell dynamics and the actin cytoskeleton James L. McGrath and C. Forbes Dewey, Jr.	170
10	Active cellular protrusion: continuum theories and models Marc Herant and Micah Dembo	204
11	Summary Mohammad R. K. Mofrad and Roger D. Kamm	225
Index		231

Contributors

FRANK P. T. BAAIJENS
Department of Biomedical Engineering
Eindhoven University of Technology

MICAH DEMBO
Department of Biomedical Engineering
Boston University

C. FORBES DEWEY, JR.
Department of Mechanical Engineering
and Biological Engineering Division
Massachusetts Institute of Technology

BEN FABRY
School of Public Health
Harvard University

JEFFREY FREDBERG
School of Public Health
Harvard University

FARSHID GUILAK
Department of Surgery
Duke University Medical Center

MANSOOR A. HAIDER
Department of Mathematics
North Carolina State University

MARC HERANT
Department of Biomedical Engineering
Boston University

PAUL JAMNEY
Institute for Medicine and Engineering
University of Pennsylvania

ROGER D. KAMM
Department of Mechanical Engineering
and Biological Engineering
Division
Massachusetts Institute of Technology

HELENE KARCHER
Biological Engineering Division
Massachusetts Institute of
Technology

TOD A. LAURSEN
Department of Civil and Environmental
Engineering
Duke University

F. C. MACKINTOSH
Division of Physics and Astronomy
Vrije Universiteit

JAMES L. MCGRATH
Department of Biomedical Engineering
University of Rochester

MOHAMMAD R. K. MOFRAD
Department of Bioengineering
University of California, Berkeley

Contributors

GERALD H. POLLACK
Department of Bioengineering
University of Washington

CHRISTOPH SCHMIDT
Institute for Medicine and Engineering
University of Pennsylvania

LORI A. SETTON
Department of Biomedical Engineering
Duke University

DIMITRIJE STAMENOVIĆ
Department of Biomedical Engineering
Boston University

Preface

Although the importance of the cytoskeleton in fundamental cellular processes such as migration, mechanotransduction, and shape stability have long been appreciated, no single theoretical or conceptual model has emerged to become universally accepted. Instead, a collection of structural models has been proposed, each backed by compelling experimental data and each with its own proponents. As a result, a consensus has not yet been reached on a single description, and the debate continues.

One reason for the diversity of opinion is that the cytoskeleton plays numerous roles and it has been examined from a variety of perspectives. Some biophysicists see the cytoskeleton as a cross-linked, branched polymer and have extended previous models for polymeric chains to describe the actin cytoskeleton. Structural engineers have drawn upon approaches that either treat the filamentous matrix as a continuum, above some critical length scale, or as a collection of struts or beams that resist deformation by the bending stiffness of each element. Others observe the similarity between the cell and large-scale structures whose mechanical integrity is derived from the balance between elements in tension and others in compression. And still others see the cytoskeleton as a gel, which utilizes the potential for phase transition to accomplish some of its dynamic processes. Underlying all of this complexity is the knowledge that the cell is alive and is constantly changing its properties, actively, as a consequence of many environmental factors. The ultimate truth, if indeed there is a single explanation for all the observed phenomena, likely lies somewhere among the existing theories.

As with the diversity of models, a variety of experimental approaches have been devised to probe the structural characteristics of a cell. And as with the models, different experimental approaches often lead to different findings, often due to the fact that interpretation of the data relies on use of one or another of the theories. But more than that, different experiments often probe the cell at very different length scales, and this is bound to lead to variations depending on whether the measurement is influenced by local structures such as the adhesion complexes that bind a bead to the cell.

We began this project with the intent of presenting in a single text the many and varied ways in which the cytoskeleton is viewed, in the hope that such a collection would spur on new experiments to test the theories, or the development of new theories themselves. We viewed this as an ongoing debate, where one of the leading proponents of each viewpoint could present their most compelling arguments in support of their model, so that members of the larger scientific community could form their own opinions.

As such, this was intended to be a monograph that captured the current state of a rapidly moving field. Since we began this project, however, it has been suggested that this book could fill a void in the area of cytoskeletal mechanics and might be useful as a text for courses taught specifically on the mechanics of a cell, or more broadly in courses that cover a range of topics in biomechanics. In either case, our hope is that this presentation might prove stimulating and educational to engineers, physicists, and biologists wishing to expand their understanding of the critical importance of mechanics in cell function, and the various ways in which it might be understood.

Finally, we wish to express our deepest gratitude to Peter Gordon and his colleagues at Cambridge University Press, who provided us with the encouragement, technical assistance, and overall guidance that were essential to the ultimate success of this endeavor. In addition, we would like to acknowledge Peter Katsirubas at Techbooks, who steered us through the final stages of editing.

1 Introduction, with the biological basis for cell mechanics

Roger D. Kamm and Mohammad R. K. Mofrad

Introduction

All living things, despite their profound diversity, share a common architectural building block: the cell. Cells are the basic functional units of life, yet are themselves comprised of numerous components with distinct mechanical characteristics. To perform their various functions, cells undergo or control a host of intra- and extracellular events, many of which involve mechanical phenomena or that may be guided by the forces experienced by the cell. The subject of *cell mechanics* encompasses a wide range of essential cellular processes, ranging from macroscopic events like the maintenance of cell shape, cell motility, adhesion, and deformation to microscopic events such as how cells sense mechanical signals and transduce them into a cascade of biochemical signals ultimately leading to a host of biological responses. One goal of the study of cell mechanics is to describe and evaluate mechanical properties of cells and cellular structures and the mechanical interactions between cells and their environment.

The field of cell mechanics recently has undergone rapid development with particular attention to the rheology of the cytoskeleton and the reconstituted gels of some of the major cytoskeletal components – actin filaments, intermediate filaments, microtubules, and their cross-linking proteins – that collectively are responsible for the main structural properties and motilities of the cell. Another area of intense investigation is the mechanical interaction of the cell with its surroundings and how this interaction causes changes in cell morphology and biological signaling that ultimately lead to functional adaptation or pathological conditions.

A wide range of computational models exists for cytoskeletal mechanics, ranging from finite element-based continuum models for cell deformation to actin filament-based models for cell motility. Numerous experimental techniques have also been developed to quantify cytoskeletal mechanics, typically involving a mechanical perturbation of the cell in the form of either an imposed deformation or force and observation of the static and dynamic responses of the cell. These experimental measurements, along with new computational approaches, have given rise to several theories for describing the mechanics of living cells, modeling the cytoskeleton as a simple mechanical elastic, viscoelastic, or poro-viscoelastic continuum, a porous gel or soft glassy material, or a tensegrity (tension integrity) network incorporating discrete structural elements that bear compression. With such remarkable disparity among these models, largely due to the relevant scales and biomechanical issues of interest, it may appear to the uninitiated that various authors are describing entirely different structures. Yet depending on the test conditions or length scale of the measurement, identical cells may be viewed quite differently: as either a continuum or a matrix with fine microstructure; as fluid-like or elastic; as a static structure; or as one with dynamically changing properties. This resembles the old Rumi tale about various people gathered in a dark room touching different parts of an elephant, each coming up with a different theory on what indeed that object was. Light reveals the whole object to prove the unity in diversity.

The objective of this book is to bring together diverse points of view regarding cell mechanics, to contrast and compare these models, and to attempt to offer a unified approach to the cell while addressing apparently irreconcilable differences. As with many rapidly evolving fields, there are conflicting points of view. We have sought in this book to capture the broad spectrum of opinions found in the literature and present them to you, the reader, so that you can draw your own conclusions.

In this Introduction we will lay the groundwork for subsequent chapters by providing some essential background information on the environment surrounding a cell, the molecular building blocks used to impart structural strength to the cell, and the importance of cell mechanics in biological function. As one would expect, diverse cell types exhibit diverse structure and nature has come up with a variety of ways in which to convey structural integrity.

The role of cell mechanics in biological function

This topic could constitute an entire book in itself, so it is necessary to place some constraints on our discussion. In this text, we focus primarily on eukaryotic cells of animals. One exception to this is the red blood cell, or erythrocyte, which contains no nucleus but which has been the prototypical cell for many mechanical studies over the years. Also, while many of the chapters are restricted to issues relating to the mechanics or dynamics of a cell as a material with properties that are time invariant, it is important to recognize that cells are living, changing entities with the capability to alter their mechanical properties in response to external stimuli. Many of the biological functions of cells for which mechanics is central are active processes for which the mechanics and biology are intrinsically linked. This is reflected in many of the examples that follow and it is the specific focus of Chapter 10.

Maintenance of cell shape

In many cases, the ability of a cell to perform its function depends on its shape, and shape is maintained through structural stiffness. In the circulation, erythrocytes exist in the form of biconcave disks that are easily deformed to help facilitate their flow through the microcirculation and have a relatively large surface-to-area ratio to enhance gas exchange. White cells, or leucocytes, are spherical, enabling them to roll

Introduction, with the biological basis for cell mechanics 3

(A) Neuron

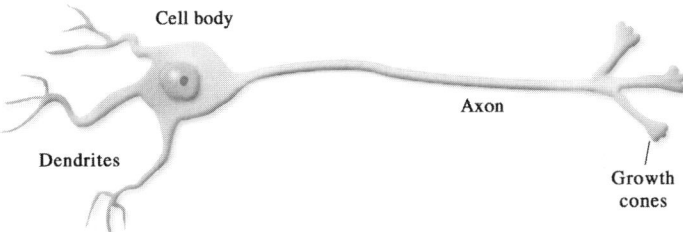

(B) Myocyte

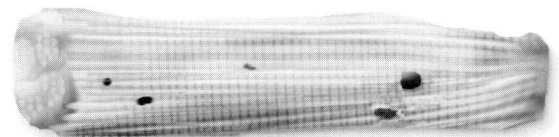

(C) Arterial wall cells

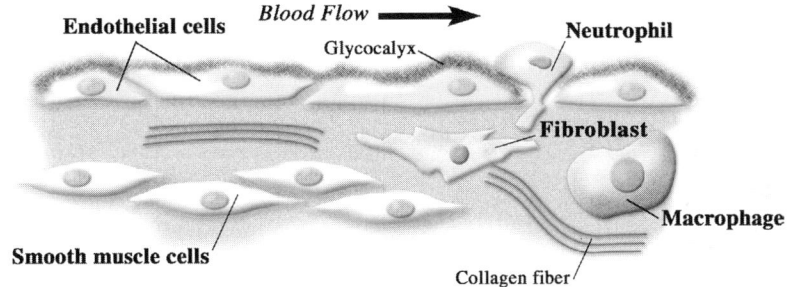

Fig. 1-1. Some selected examples of cell morphology. (A) Neuron, with long projections (dendrites and axons) that can extend a distance of 10 s of centimeters and form connections for communication with other cells. (B) Cardiac myocyte, showing the striations associated with the individual sarcomeres of the contractile apparatus. (C) Various cells found in the arterial wall. Endothelial cells line the vascular system, with a flattened, "pancake-like" morphology; neutrophils circulate in the blood until recruited by chemoattractants to transmigrate into the tissue and convert to macrophages; fibroblasts function as the "factories" for the extracellular matrix; and smooth muscle cells contribute to vessel contractility and flow control.

along the vascular endothelium before adhering and migrating into the tissue. Because their diameter is larger than some of the capillaries they pass through, leucocytes maintain excess membrane in the form of microvilli so they can elongate at constant volume and not obstruct the microcirculation. Neuronal cells extend long processes along which signals are conducted. Airway epithelial cells are covered with a bed of cilia, finger-like cell extensions that propel mucus along the airways of the lung. Some of the varieties of cell type are shown in Fig. 1-1. In each example, the internal

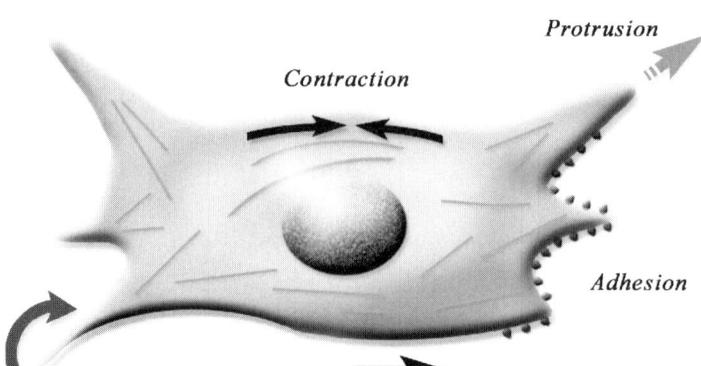

Fig. 1-2. The processes contributing to cell migration: protrusion, adhesion, contraction, and rear release. These steps can proceed in random order or simultaneously, but they all need to be operative for cell migration to take place.

structure of the cell, along with the cell membrane, provides the structural integrity that maintains the particular shape needed by the cell to accomplish its function, although the specific components of the structure are highly variable and diverse.

Cell migration

Many cells migrate, certainly during development (as the organism grows its various parts), but also at maturity for purposes of wound repair (when cells from the surrounding undamaged tissue migrate into the wound and renew the tissues) and in combating infection (when cells of the immune system transmigrate from the vascular system across the vessel wall and into the infected tissues). Migration is also an essential feature in cancer metastasis and during angiogenesis, the generation of new vessels.

Descriptions of cell migration depict a process that occurs in several stages: *protrusion*, the extension of the cell at the leading edge in the direction of movement; *adhesion* of the protrusion to the surrounding substrate or matrix; *contraction* of the cell that transmits a force from these protrusions at the leading edge to the cell body, pulling it forward; and *release* of the attachments at the rear, allowing net forward movement of the cell to occur (see, for example, DiMilla, Barbee et al., 1991; Horwitz and Webb, 2003; Friedl, Hegerfeldt et al., 2004; Christopher and Guan, 2000; and Fig. 1-2). These events might occur sequentially, with the cellular protrusions – called either filopodia ("finger-like") or lamellapodia ("sheet-like") projections – occurring as discrete events: suddenly reaching forward, extending from the main body of the cell, or more gradually and simultaneously, much like the progressive advance of a spreading pool of viscous syrup down an inclined surface. While it is well known that cells sense biochemical cues such as gradients in chemotactic agents, they can also apparently sense their physical environment, because their direction of migration can be influenced by variations in the stiffness of the substrate to which they adhere. Whatever the mode of migration, however, the central role of cell mechanics, both its passive stiffness and its active contractility, is obvious.

Introduction, with the biological basis for cell mechanics

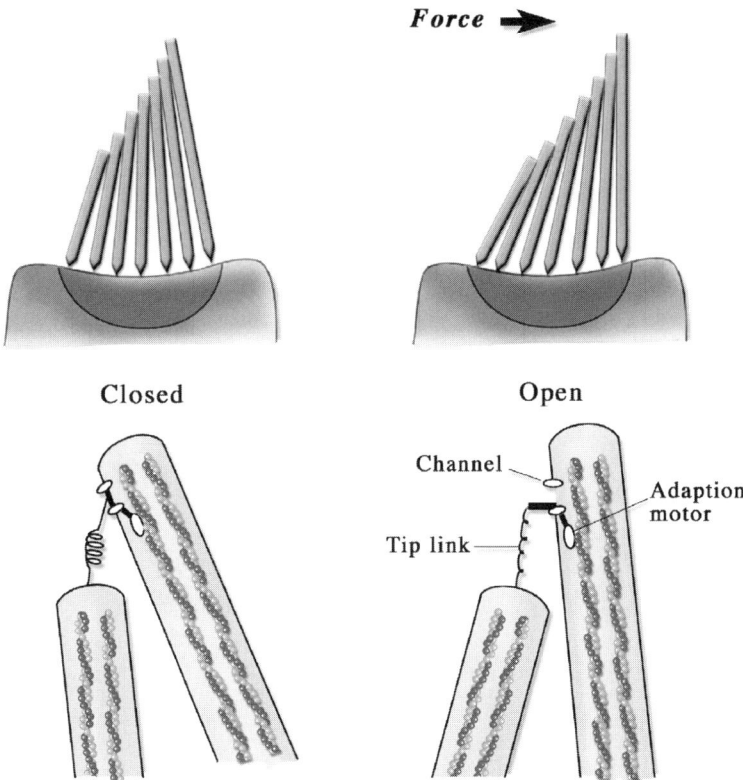

Fig. 1-3. Hair cells found in the inner ear transduce sound via the stereocilia that project from their apical surface. As the stereocilia bundle moves in response to fluid oscillations in the cochlea, tension in the tip link (a fine filament connecting the tip of one stereocilium to the side of another) increases, opening an ion channel to initiate the electrochemical response.

Mechanosensing

Nowhere is the importance of biology in cell mechanics more evident than in the ability of the cell to sense and respond to externally applied forces. Many – perhaps all – cells are able to sense when a physical force is applied to them. They respond through a variety of biological pathways that lead to such diverse consequences as changes in membrane channel activity, up- or down-regulation of gene expression, alterations in protein synthesis, or altered cell morphology. An elegant example of this process can be found in the sensory cells of the inner-ear, called hair cells, which transduce the mechanical vibration of the inner ear fluid into an electrical signal that propagates to the brain (Hamill and Martinac, 2001; Hudspeth, 2001; Hudspeth, Choe et al., 2000). By a remarkably clever design (Fig. 1-3), the stereocilia that extend from the apical surface of the cells form bundles. The individual stereocilia that comprise a bundle are able to slide relative to one another when the bundle is pushed one way or the other, but some are connected through what is termed a "tip link" – nothing more than a fine filament that connects the tip of one stereocilium to the side of another, the tension in which is modulated by an adaption motor that moves along the internal actin filaments and is tethered to the ion channel. As the neighboring filaments slide with respect to one another, tension is developed in the tip link, generating a force at the point where the filament connects to the side of the stereocilium. This force acts to change the conformation of a transmembrane protein that acts as an ion channel, causing it to open and allowing the transient entry of calcium ions. This flux of positive ions initiates the electrical signal that eventually reaches the brain and is perceived as sound.

Although the details of force transmission to the ion channel in the case of hair-cell excitation are not known, another mechanosensitive ion channel, the *M*echano*s*ensitive *c*hannel of *L*arge conductance (MscL) has been studied extensively (Chang, Spencer et al., 1998; Hamill and Martinac, 2001), and molecular dynamic simulation has been used to show how stresses in the cell membrane act directly on the channel and cause it to change its conductance (Gullingsrud, Kosztin et al., 2001).

This is but one example of the many ways a cell can physically "feel" its surroundings. Other mechanisms are only now being explored, but include: (1) conformational changes in intracellular proteins due to the transmission of external forces to the cell interior, leading to changes in reaction rates through a change in binding affinity; (2) changes in the viscosity of the cell membrane, altering the rate of diffusion of transmembrane proteins and consequently their reaction rates; and (3) direct transmission of force to the nucleus and to the chromatin contained inside, affecting expression of specific genes. These other mechanisms are less well understood than mechanosensitive channels, and it is likely that other mechanisms exist as well that have not yet been identified (for reviews of this topic, see Bao and Suresh, 2003; Chen, Tan et al., 2004; Huang, Kamm et al., 2004; Davies, 2002; Ingber, 1998; Shyy and Chien, 2002; Janmey and Weitz, 2004).

Although the detailed mechanisms remain ill-defined, the consequences of force applied to cells are well documented (see, for example, Dewey, Bussolari et al., 1981; Lehoux and Tedgui, 2003; Davies, 1995; McCormick, Frye et al., 2003; Gimbrone, Topper et al., 2000). Various forms of force application – whether transmitted via cell membrane adhesion proteins (such as the heterodimeric integrin family) or by the effects of fluid shear stress, transmitted either directly to the cell membrane or via the surface glycocalyx that coats the endothelial surface – elicit a biological response (see Fig. 1-4). Known responses to force can be observed in a matter of seconds, as in the case of channel activation, but can continue for hours after the initiating event, as for example changes in gene expression, protein synthesis, or morphological changes. Various signaling pathways that mediate these cellular responses have been identified and have been extensively reviewed (Davies, 2002; Hamill and Martinac, 2001; Malek and Izumo, 1994; Gimbrone, Topper et al., 2000).

Stress responses and the role of mechanical forces in disease

One reason for the strong interest in mechanosensation and the signaling pathways that become activated is that physical forces have been found to be instrumental in the process by which tissues remodel themselves in response to stress. Bone, for example, is known to respond to such changes in internal stress levels as occur following fracture or during prolonged exposure to microgravity. Many cells have shown that they can both sense and respond to a mechanical stimulus. While many of these responses appear designed to help the cell resist large deformations and possible structural

Introduction, with the biological basis for cell mechanics

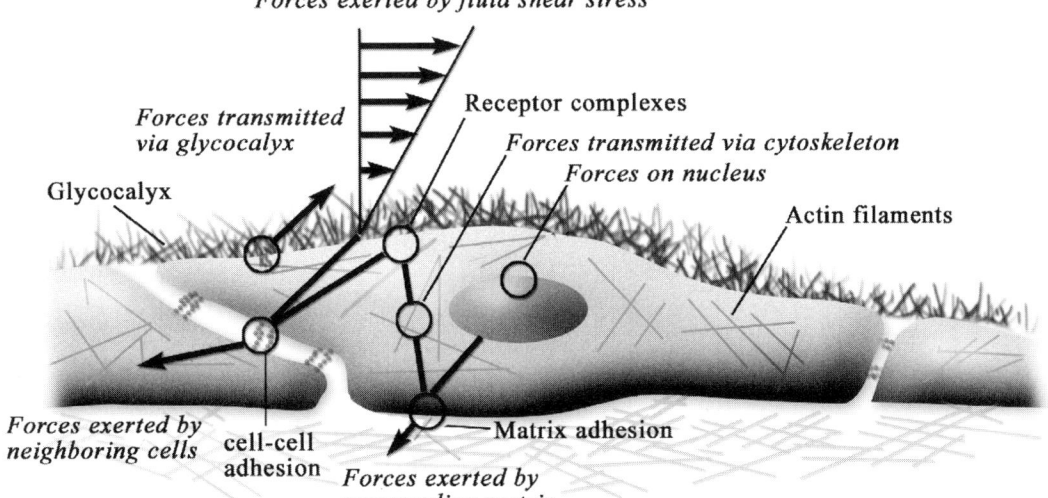

Fig. 1-4. Forces experienced by the endothelial lining of a blood vessel and the various pathways of force transmission, via receptor complexes, the glycocalyx, and the cytoskeleton even reaching the nucleus, cell-cell adhesions, and cell-matrix adhesions. Any of these locations is a potential site at which mechanical force can be transduced into a biochemical signal.

damage, others have an undesirable outcome, including atherosclerosis, arthritis, and pulmonary hypertension; there exists an extensive literature on each of these topics.

Active cell contraction

One important subset of cells primarily exists for the purpose of generating force. Cell types for which this is true include vascular smooth muscle cells, cardiac myocytes, and skeletal muscle cells. While the force-generating structures may differ in detail, the mechanisms of force generation have much in common. All muscle cells use the molecular motor comprised of actin and myosin to produce active contraction. These motor proteins are arranged in a well-defined structure, the sarcomere, and the regularity of the sarcomeres gives rise to the characteristic striated pattern seen clearly in skeletal muscle cells and cardiac myocytes (Fig. 1-5). Even nonmuscle cells contain contractile machinery, however, used for a variety of functions such as maintaining a resting level of cell tension, changing cell shape, or in cell migration. Most cells are capable of migration; in many, this capability only expresses itself when the cell is stimulated. For example, neutrophils are quiescent while in the circulation but become one of the most highly mobile migratory cells in the body when activated by signals emanating from a local infection.

Structural anatomy of a cell

Cells are biologically active, and their structure often reflects or responds to their physical environment. This is perhaps the primary distinction between traditional mechanics and the mechanics of biological materials. This is a fundamental difference

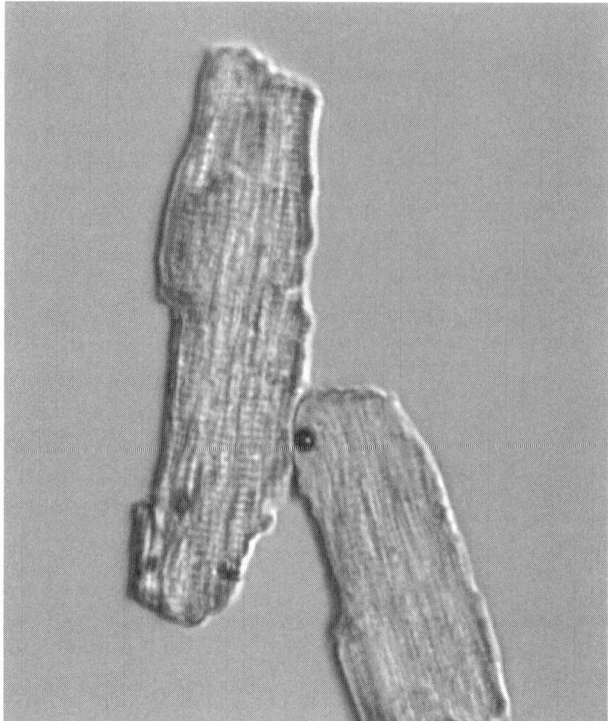

Fig. 1-5. Cardiac myocytes in culture showing the internal striations corresponding to the individual sarcomeres used for contraction. Courtesy of Jan Lammerding.

from inert materials and it must be kept in mind as we progress through the various descriptions found in this book. A second important distinction from most engineering materials is that thermal fluctuations often need to be considered, as these influence both the biochemical processes that lead to intracellular remodeling but also directly influence the elastic characteristics of the membrane and the biological filaments that comprise the cytoskeleton.

Cells often do not constitute the primary structural elements of the tissue in which they reside. For example, in either bone or cartilage, the mechanical stiffness of the resident cells are inconsequential in terms of their contribution to the modulus of the tissue, and their deformation is dictated almost entirely by that of the surrounding matrix – collagen, and hydroxyapatite in the case of bone, and a mix of collagen and proteoglycans with a high negative charge density in the case of cartilage. The role of cells in these tissues is not structural, yet through the mechanisms discussed above, cells are essential in regulating the composition and organization of the structures contained in the extracellular regions that determine the tissue's elasticity and strength through the cellular response to stress.

In other tissues, the structural role of the resident cells is much more direct and significant. Obviously, in muscle, the contractile force generated and the modulus of the tissue, either in the contracted or the relaxed state, are dominated by cellular activity. In other tissues, such as arterial wall or pulmonary airways, for example, collagen and elastin filaments in the extracellular matrix normally balance the bulk of

Introduction, with the biological basis for cell mechanics

Table 1-1. *Major families of adhesion molecules. (E)-extracellular; (I) intracellular*

Family	Location and/or function	Ligands recognized
Integrins	Focal adhesions, hemi-desmosomes, leukocyte ("spreading") adhesion, primarily focal adhesions to matrix but also in some cell-cell adhesions	(E) fibronectin, collagen, laminin, immunoglobulins, (I) actin filaments
Selectins	Circulating cells and endothelial cells, "rolling" adhesion	Carbohydrates
Ig superfamily (immunoglobulin)	Important in immune response	Integrins, homophillic
Cadherens	Adherens junctions, desmosomes	(E) homophillic, (I) actin filaments, intermediate filaments
Transmembrane proteoglycans	Fibroblasts, epithelial cells	(E) collagen, fibronectin (I) actin filaments, heterophillic

the stress. During activation of the smooth muscle, however, stress shifts from these extracellular constituents to the contractile cells, and the vessel constricts to a diameter much smaller than that associated with the passive wall stiffness. In the case of cardiac tissue, the contractile cells, or myocytes (Fig. 1-5), constitute a large fraction of the tissue volume and are primarily responsible for the stresses and deformations of the myocardium that are time varying through the cardiac cycle.

The extracellular matrix and its attachment to cells

Contrary to the situation in most cell mechanics experiments *in vitro*, where forces might be applied directly to the cells via tethered beads, a micropipette, an AFM probe, or fluid shear stress, forces *in vivo* are often transmitted to the cell via the extracellular matrix (ECM), which shares in the load-supporting function. Many cell membrane receptors contain extracellular domains that bind to the various proteins of the ECM. For example, members of the integrin family can bind to fibronectin, vitronectin, collagen, and laminin. Intracellular domains of these same proteins bind directly (or indirectly, through other membrane-associated proteins) to the cytoskeleton. The number and variety of linking proteins is quite remarkable, as described in detail in a recent review (Geiger and Bershadsky, 2002). Other adhesion molecules bind to the ECM, basement membrane, neighboring cells, or cells suspended in flowing blood. Adhesion molecules can be either homophillic (binding to other identical molecules) or heterophillic (Table 1-1). Of these transmembrane molecules (both proteins and proteoglycans) many attach directly to the cytoskeleton, which often exhibits a denser, more rigid structure in the vicinity of an adhesion site.

Transmission of force to the cytoskeleton and the role of the lipid bilayer

Cells are separated from the external environment by a thin lipid bilayer consisting of a rich mix of phospholipids, glycolipids, cholesterol, and a vast array of transmembrane proteins that constitute about 50 percent of the membrane by weight but only 1 to 2 percent of the total number of molecules residing in the membrane. Phospholipids, which are the most abundant, are amphipathic, having a hydrophilic part residing on the outside surface of the bilayer and a hydrophobic part on the inside. Some of the proteins serve as ion channels, others as a pathway for transmembrane signaling. Still others provide a structural bridge across the membrane, allowing for direct adhesion between the internal cytoskeleton and the extracellular matrix. Together, these are commonly referred to as *integral membrane proteins*. Roughly half of these integral proteins are able to freely diffuse within the membrane, while the rest are anchored to the cytoskeleton.

In addition to its role in communicating stress and biochemical signals into the cell, the membrane also serves a barrier function, isolating the cell interior from its extracellular environment and maintaining the appropriate biochemical conditions within for critical cell functions. By itself, the bilayer generally contributes little to the overall stiffness of the cell, except in situations in which the membrane becomes taut, as might occur due to osmotic swelling. In general, the bilayer can be thought of as a two-dimensional fluid within which the numerous integral membrane proteins diffuse, a concept first introduced in 1972 by Singer and Nicolson as the *fluid mosaic model* (Singer and Nicolson, 1972). The bilayer maintains a nearly constant thickness of about 6 nm under stress, and exhibits an area-expansion modulus, defined as the in-plane tension divided by the fractional area change, of about 0.1–1.0 N/m (for pure lipid bilayers) or 0.45 N/m (for a red blood cell) (Waugh and Evans, 1979). Rupture strength, in terms of the maximum tension that the membrane can withstand, lies in the range of 0.01–0.02 N/m, for a red blood cell and a lipid vesicle, respectively (Mohandas and Evans, 1994). Values for membrane and cortex bending stiffness reported in the literature (for example, $\sim 2-4 \times 10^{-19}$ N·m for the red blood cell membrane (Strey, Peterson et al., 1995; Scheffer, Bitler et al., 2001), and $1-2 \times 10^{-18}$ N·m for neutrophils (Zhelev, Needham et al., 1994), are not much larger than that for pure lipid bilayers (Evans and Rawicz, 1990), despite the fact that they include the effects of the membrane-associated cortex of cytoskeletal filaments, primarily spectrin in the case of erythrocytes and actin for leukocytes. When subjected to in-plane shear stresses, pure lipid bilayers exhibit a negligible shear modulus, whereas red blood cells have a shear modulus of about 10^{-6} N·s/m (Evans and Rawicz, 1990). Forces can be transmitted to the membrane via transmembrane proteins or proteins that extend only partially through the bilayer. When tethered to an external bead, for example, the latter can transmit normal forces; when forces are applied tangent to the bilayer, the protein can be dragged along, experiencing primarily a viscous resistance unless it is tethered to the cytoskeleton. Many proteins project some distance into the cell, so their motion is impeded even if they are not bound to the cytoskeleton due to steric interactions with the membrane-associated cytoskeleton.

Intracellular structures

In this text we primarily address the properties of a generic cell, without explicitly recognizing the distinctions, often quite marked, between different cell types. It is important, however, to recognize several different intracellular structures that influence the material properties of the cell that may, at times, need to be taken into account in modeling. Many cells (leucocytes, erythrocytes, and epithelial cells, for example) contain a relatively dense structure adjacent to the cell membrane called the *cortex*, with little by way of an internal network. In erythrocytes, this cortex contains another filamentous protein, spectrin, and largely accounts for the shape rigidity of the cell. Many epithelial cells, such as those found in the intestine or lining the pulmonary airways, also contain projections (called microvilli in the intestine and cilia in the lung) that extend from their apical surface. Cilia, in particular, are instrumental in the transport of mucus along the airway tree and have a well-defined internal structure, primarily due to microtubules, that imparts considerable rigidity.

Of the various internal structures, the nucleus is perhaps the most significant, from both a biological and a structural perspective. We know relatively little about the mechanical properties of the nucleus, but some recent studies have begun to probe nuclear mechanics, considering the separate contributions of the nuclear envelope, consisting of two lipid bilayers and a nuclear lamina, and the nucleoplasm, consisting largely of chromatin (Dahl, Kahn et al., 2004; Dahl, Engler et al., 2005).

Migrating cells have a rather unique structure, but again are quite variable from cell type to cell type. In general, the leading edge of the cell sends out protrusions, either lamellipodia or filopodia, that are rich in actin and highly cross-linked. The dynamics of actin polymerization and depolymerization is critical to migration and is the focus of much recent investigation (see, for example Chapter 9 and Bindschadler, Dewey, and McGrath, 2004). Active contraction of the network due to actin-myosin interactions also plays a central role and provides the necessary propulsive force.

Actin filaments form by polymerization of globular, monomeric actin (G-actin) into a twisted strand of filamentous actin (F-actin) 7–9 nm in diameter with structural polarity having a barbed end and a pointed end. Monomers consist of 375 amino acids with a molecular weight of 43 kDa. ATP can bind to the barbed end, which allows for monomer addition and filament growth, while depolymerization occurs preferentially at the pointed end (Fig. 1-6A). Filament growth and organization is regulated by many factors, including ionic concentrations and a variety of capping, binding, branching, and severing proteins. From actin filaments, tertiary structures such as fiber bundles, termed "stress fibers," or a three-dimensional lattice-like network can be formed through the action of various actin-binding proteins (ABPs). Some examples of ABPs are fimbrin and α-actinin, both instrumental in the formation of *stress fibers* or bundles of actin filaments, and filamin, which connects filaments into a three-dimensional space-filling matrix or gel with filaments joined at nearly a right angle. Recent rheological studies of reconstituted actin gels containing various concentrations of ABPs (see Chapter 2, or Tseng, An et al., 2004) have illustrated the rich complexities of even such simple systems and have also provided new insights into the nature of such matrices.

The importance of actin filaments is reflected in the fact that actin constitutes from 1 to 10 percent of all the protein in most cells, and is present at even higher concentrations in muscle cells. Actin is thought to be the primary structural component of most cells; it responds rapidly and dramatically to external forces and is also instrumental in the formation of leading-edge protrusions during cell migration. As the data in Table 1-2 illustrate, actin filaments measured by a variety of techniques (Yasuda,

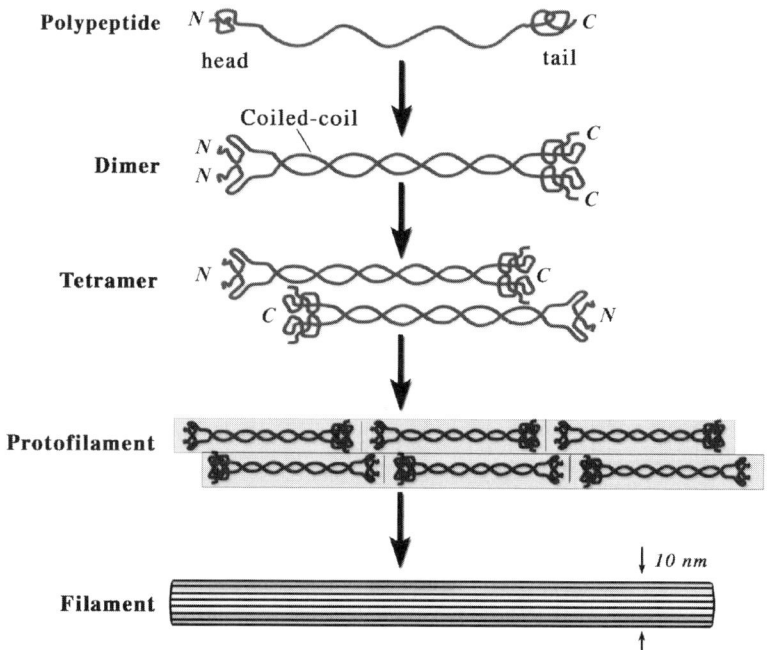

Fig. 1-6. Filaments that constitute the cytoskeleton. (A) Actin filaments. (B) Microtubules. (C) Intermediate filaments.

Miyata et al., 1996; Tsuda, Yasutake et al., 1996; Higuchi and Goldman, 1995) are stiff, having a persistence length of several microns, and an effective Young's modulus, determined from its bending stiffness and radius of $1 - 3 \times 10^9$ Pa, comparable to that of polystyrene (3×10^9 Pa) and nearly equal to that of bone (9×10^9 Pa).

Microtubules constitute a second major constituent of the cytoskeleton. These are polymerized filaments constructed from monomers of α- and β-tubulin in a helical

Table 1-2. *Elastic properties of cytoskeletal filaments*

	Diameter, $2a$ (nm)	Persistence length, l_p (μm)	Bending stiffness, K_B (Nm²)	Young's modulus, E (Pa)
Actin filament	6–8	15	7×10^{-26}	$1.3\text{–}2.5 \times 10^9$
Microtubule	25	6000	2.6×10^{-23}	1.9×10^9
Intermediate filament	10	~1	4×10^{-27}	1×10^9

The elastic properties of actin filaments and microtubules are approximately consistent with a prediction based on the force of van der Waals attraction between two surfaces (J. Howard, 2001). Persistence length (l_p) and bending stiffness (K_B) are related through the expression $l_p = K_B/k_B T$. Bending stiffness and Young's modulus (E) are related through the expression $K_B = EI = \frac{\pi}{4} a^4 E$ for a solid rod of circular cross-section with radius a, and $I = \frac{\pi}{4}(a_o^4 - a_i^4)$ for a hollow cylinder with inside and outside radii a_i and a_o, respectively.

arrangement, both 55 kDa polypeptides, that organize into a small, hollow cylinder (Fig. 1-6B). The filaments have an outer diameter of about 25 nm and exhibit a high bending stiffness, even greater than that of an actin filament (Table 1-2) with a persistence length of about 6 mm (Gittes, Mickey et al., 1993). Tubular structures tend to be more resistant to bending than solid cylinders with the same amount of material per unit length, and this combined with the larger radius accounts for the high bending stiffness of microtubules despite having an effective Young's modulus similar to that of actin. Because of their high bending stiffness, they are especially useful in the formation of long slender structures such as cilia and flagella. They also provide the network along which chromosomes are transported during cell division.

Microtubules are highly dynamic, even more so than actin, undergoing constant polymerization and depolymerization, so that the half-life of a microtubule is typically only a few minutes. (Mitchison and Kirschner, 1984). Growth is asymmetric, as with actin, with polymerization typically occurring rapidly at one end and more slowly at the other, and turnover is generally quite rapid; the half-life of a microtubule is typically on the order of minutes.

Intermediate filaments (IFs) constitute a superfamily of proteins containing more than fifty different members. They have in common a structure consisting of a central α-helical domain of over 300 residues that forms a coiled coil. The dimers then assemble into a staggered array to form tetramers that connect end-to-end, forming protofilaments (Fig. 1-6C). These in turn bundle into ropelike structures, each containing about eight protofilaments with a persistence length of about 1 μm (Mucke, Kreplak et al., 2004). Aside from these differences in structure, intermediate filaments differ from microfilaments and microtubules in terms of their long-term stability and high resistance to solubility in salts. Also, unlike polymerization of other cytoskeletal filaments, intermediate filaments form without the need for GTP or ATP hydrolysis.

In recent experiments, intermediate filaments have been labeled with a fluorescent marker and used to map the strain field within the cell (Helmke, Thakker et al., 2001). This is facilitated by the tendency for IFs to be present throughout the entire cell at a sufficiently high concentration that they can serve as fiducial markers.

Of course these are but a few of the numerous proteins that contribute to the mechanical properties of a cell. The ones mentioned above – actin filaments, microtubules,

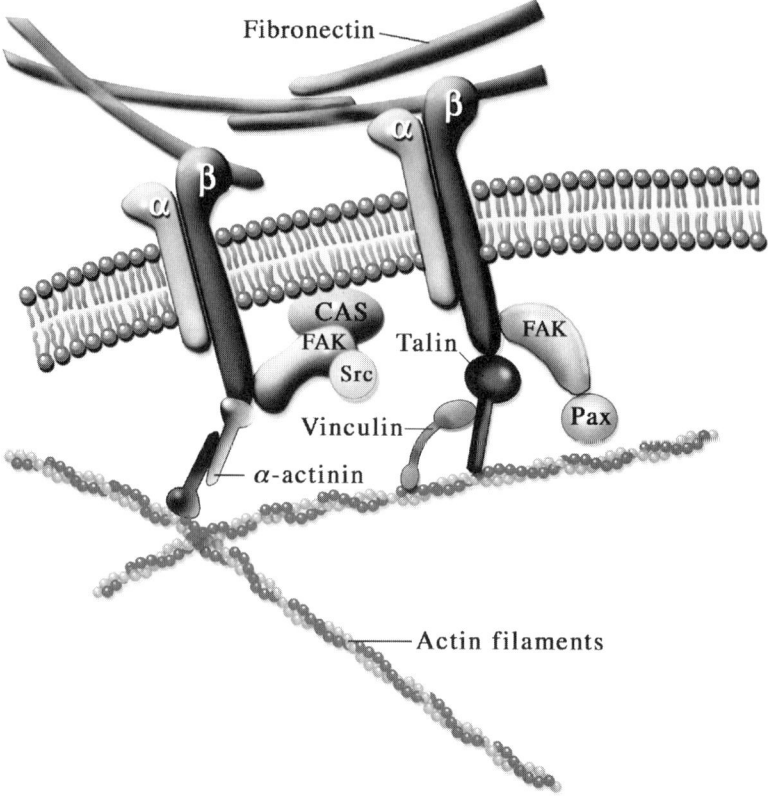

Fig. 1-7. A small sampling of the proteins found in a focal adhesion complex (FAC). Forces are typically transmitted from the extracellular matrix (for example, fibronectin), via the integral membrane adhesion receptors (α- and β-integrins), various membrane-associated proteins (focal adhesion kinase (FAK), paxillin (Pax), talin, Crk-associated substrate (CAS)), to actin-binding proteins (α-actinin) that link the FAC to the cytoskeleton. Adapted from Geiger and Bershadsky 2002.

and intermediate filaments – are primarily associated with the cytoskeleton, but even within the cytoskeletal network are found numerous linking proteins (ABPs constituting one family) that influence the strength and integrity of the resulting matrix. In addition to these are the molecular constituents of the cell membrane, nuclear membrane, and all the organelles and other intracellular bodies that influence the overall mechanical response of a cell. In fact, intracellular structure should be noted for its complexity, as can be seen in Fig. 1-7, which shows just a small subset of the numerous proteins that link the extracellular matrix and the cytoskeleton. Any of these constitutes a pathway for transmitting force across the cell membrane, between the proteins found in the adhesion complexes, and through the cytoskeletal network. To the extent that a particular protein is located along the force transmission pathway, not only does it play a role in transmitting stress, but it also represents a candidate for mechanosensing due to the conformational changes that arise from the transmission of force.

Active contraction is another fundamental feature of the cytoskeleton that influences its structural properties. While this is an obvious characteristic of the various types of muscle cell, most cells contain contractile machinery, and even in their resting state can exert a force on their surroundings. Forces have been measured in resting fibroblasts, for example, where intracellular tension gives rise to stresses in the focal adhesions of the cell adherent to a flexible two-dimensional substrate of about 5 nN/μm^2, or 5 kPa (Balaban, Schwarz et al., 2001). In experiments with various cell types grown in a three-dimensional gel such as collagen, the cells actively contract the matrix by more than 50 percent (Sieminski, Hebbel et al., 2004). These contractile forces are associated with intracellular molecular motors such as those in the myosin family.

Overview

This book presents a full spectrum of views on current approaches to modeling cell mechanics. In part, this diversity of opinion stems from the different backgrounds of contributors to the field. Indeed, the authors of this book come from the biophysics, bioengineering, and physical chemistry communities, and each joins the discussion with a unique perspective on biological systems. Consequently, the approaches range from finite element methods commonly used in continuum mechanics to models of the cytoskeleton as a cross-linked polymer network to models of soft glassy materials and gels. Studies reflect both the static, instantaneous nature of the structure as well as its dynamic nature due to polymerization and the full array of biological processes. It is unlikely that a single unifying approach will evolve from this diversity, in part because of the complexity of the phenomena underlying the mechanical properties of the cell. It is our hope, however, that a better appreciation of the various perspectives will lead to a more highly coordinated approach to the essential problems and might facilitate discussions among investigators with differing views.

Perhaps the most important purpose of this monograph is to stimulate new ideas and approaches. Because no single method has emerged as clearly superior, this might reflect the need for approaches not yet envisaged. That much of the work presented here derives from publications over the past several years reinforces the notion that cell mechanics is a rapidly evolving field. The next decade will likely yield further advances not yet foreseen.

References

Balaban, N. Q., U. S. Schwarz, et al. (2001). "Force and focal adhesion assembly: a close relationship studied using elastic micropatterned substrates." *Nat. Cell Biol.*, **3**(5): 466–72.

Bao, G. and S. Suresh (2003). "Cell and molecular mechanics of biological materials." *Nat. Mater.*, **2**(11): 715–25.

Bindschadler M., O. E., Dewey C. F. Jr, McGrath, J. L. (2004). "A mechanistic model of the actin cycle." *Biophys. J.*, **86**: 2720–2739.

Chang, G., R. H. Spencer, et al. (1998). "Structure of the MscL homolog from Mycobacterium tuberculosis: a gated mechanosensitive ion channel." *Science*, **282**(5397): 2220–6.

Chen, C. S., J. Tan, et al. (2004). "Mechanotransduction at cell-matrix and cell-cell contacts." *Annu. Rev. Biomed. Eng.*, **6**: 275–302.

Christopher, R. A. and J. L. Guan (2000). "To move or not: how a cell responds (Review)." *Int. J. Mol. Med.*, **5**(6): 575–81.

Dahl, K. N., A. J. Engler, et al. (2005). "Power-law rheology of isolated nuclei with deformation mapping of nuclear substructures." *Biophys. J.*, **89**(4): 2855–64.

Dahl, K. N., S. M. Kahn, et al. (2004). "The nuclear envelope lamina network has elasticity and a compressibility limit suggestive of a molecular shock absorber." *J. Cell Sci.*, **117**(Pt 20): 4779–86.

Davies, P. F. (1995). "Flow-mediated endothelial mechanotransduction." *Physiol. Rev.*, **75**(3): 519–60.

Davies, P. F. (2002). "Multiple signaling pathways in flow-mediated endothelial mechanotransduction: PYK-ing the right location." *Arterioscler Thromb Vasc. Biol.*, **22**(11): 1755–7.

Dewey, C. F., Jr., S. R. Bussolari, et al. (1981). "The dynamic response of vascular endothelial cells to fluid shear stress." *J. Biomech. Eng.*, **103**(3): 177–85.

DiMilla, P. A., K. Barbee, et al. (1991). "Mathematical model for the effects of adhesion and mechanics on cell migration speed." *Biophys. J.*, **60**(1): 15–37.

Evans, E. and W. Rawicz (1990). "Entropy-driven tension and bending elasticity in condensed-fluid membranes." *Phys. Rev. Lett.*, **64**(17): 2094–2097.

Friedl, P., Y. Hegerfeldt, et al. (2004). "Collective cell migration in morphogenesis and cancer." *Int. J. Dev. Biol.*, **48**(5-6): 441–9.

Geiger, B. and A. Bershadsky (2002). "Exploring the neighborhood: adhesion-coupled cell mechanosensors." *Cell*, **110**(2): 139–42.

Gimbrone, M. A., Jr., J. N. Topper, et al. (2000). "Endothelial dysfunction, hemodynamic forces, and atherogenesis." *Ann. N Y Acad. Sci.*, **902**: 230-9; discussion 239–40.

Gittes, F., B. Mickey, et al. (1993). "Flexural rigidity of microtubules and actin filaments measured from thermal fluctuations in shape." *J. Cell Biol.*, **120**(4): 923–34.

Gullingsrud, J., D. Kosztin, et al. (2001). "Structural determinants of MscL gating studied by molecular dynamics simulations." *Biophys. J.*, **80**(5): 2074–81.

Hamill, O. P. and B. Martinac (2001). "Molecular basis of mechanotransduction in living cells." *Physiol. Rev.*, **81**(2): 685–740.

Helmke, B. P., D. B. Thakker, et al. (2001). "Spatiotemporal analysis of flow-induced intermediate filament displacement in living endothelial cells." *Biophys. J.*, **80**(1): 184–94.

Higuchi, H. and Y. E. Goldman (1995). "Sliding distance per ATP molecule hydrolyzed by myosin heads during isotonic shortening of skinned muscle fibers." *Biophys. J.*, **69**(4): 1491–507.

Horwitz, R. and D. Webb (2003). "Cell migration." *Curr. Biol.*, **13**(19): R756–9.

Howard, J., (2001) Mechanics of Motor Proteins and the Cytoskeleton, Sinauer Associates, Inc., pp. 288–289.

Huang, H., R. D. Kamm, et al. (2004). "Cell mechanics and mechanotransduction: pathways, probes, and physiology." *Am. J. Physiol. Cell Physiol.*, **287**(1): C1–11.

Hudspeth, A. J. (2001). "How the ear's works work: mechanoelectrical transduction and amplification by hair cells of the internal ear." *Harvey Lect.*, **97**: 41–54.

Hudspeth, A. J., Y. Choe, et al. (2000). "Putting ion channels to work: mechanoelectrical transduction, adaptation, and amplification by hair cells." *Proc. Natl. Acad. Sci. USA*, **97**(22): 11765–72.

Ingber, D. E. (1998). "Cellular basis of mechanotransduction." *Biol. Bull.*, **194**(3): 323–5; discussion 325–7.

Janmey, P. A. and D. A. Weitz (2004). "Dealing with mechanics: mechanisms of force transduction in cells." *Trends Biochem. Sci.*, **29**(7): 364–70.

Lehoux, S. and A. Tedgui (2003). "Cellular mechanics and gene expression in blood vessels." *J. Biomech.*, **36**(5): 631–43.

Malek, A. M. and S. Izumo (1994). "Molecular aspects of signal transduction of shear stress in the endothelial cell." *J. Hypertens.*, **12**(9): 989–99.

McCormick, S. M., S. R. Frye, et al. (2003). "Microarray analysis of shear stressed endothelial cells." *Biorheology*, **40**(1–3): 5–11.

Mitchison, T. and M. Kirschner (1984). "Dynamic instability of microtubule growth." *Nature*, **312**(5991): 237–42.

Mohandas, N. and E. Evans (1994). "Mechanical properties of the red cell membrane in relation to molecular structure and genetic defects." *Annu. Rev. Biophys. Biomol. Struct.*, **23**: 787–818.

Mucke, N., L. Kreplak, et al. (2004). "Assessing the flexibility of intermediate filaments by atomic force microscopy." *J. Mol. Biol.*, **335**(5): 1241–50.

Scheffer, L., A. Bitler, et al. (2001). "Atomic force pulling: probing the local elasticity of the cell membrane." *Eur. Biophys. J.*, **30**(2): 83–90.

Shyy, J. Y. and S. Chien (2002). "Role of integrins in endothelial mechanosensing of shear stress." *Circ. Res.*, **91**(9): 769–75.

Sieminski, A. L., R. P. Hebbel, et al. (2004). "The relative magnitudes of endothelial force generation and matrix stiffness modulate capillary morphogenesis in vitro." *Exp. Cell Res.*, **297**(2): 574–84.

Singer, S. J. and G. L. Nicolson (1972). "The fluid mosaic model of the structure of cell membranes." *Science*, **175**(23): 720–31.

Strey, H., M. Peterson, et al. (1995). "Measurement of erythrocyte membrane elasticity by flicker eigenmode decomposition." *Biophys. J.*, **69**(2): 478–88.

Tseng, Y., K. M. An, et al. (2004). "The bimodal role of filamin in controlling the architecture and mechanics of F-actin networks." *J. Biol. Chem.*, **279**(3): 1819–26.

Tsuda, Y., H. Yasutake, et al. (1996). "Torsional rigidity of single actin filaments and actin-actin bond breaking force under torsion measured directly by in vitro micromanipulation." *Proc. Natl. Acad. Sci. USA*, **93**(23): 12937–42.

Waugh, R. and E. A. Evans (1979). "Thermoelasticity of red blood cell membrane." *Biophys. J.*, **26**(1): 115–31.

Yasuda, R., H. Miyata, et al. (1996). "Direct measurement of the torsional rigidity of single actin filaments." *J. Mol. Biol.*, **263**(2): 227–36.

Zhelev, D. V., D. Needham, et al. (1994). "Role of the membrane cortex in neutrophil deformation in small pipets." *Biophys. J.*, **67**(2): 696–705.

2 Experimental measurements of intracellular mechanics

Paul Janmey and Christoph Schmidt

ABSTRACT: Novel methods to measure the viscoelasticity of soft materials and new theories relating these measurements to the underlying molecular structures have the potential to revolutionize our understanding of complex viscoelastic materials like cytoplasm. Much of the progress in this field has been in methods to apply piconewton forces and to detect motions over distances of nanometers, thus performing mechanical manipulations on the scale of single macromolecules and measuring the viscoelastic properties of volumes as small as fractions of a cell. Exogenous forces ranging from pN to nN are applied by optical traps, magnetic beads, glass needles, and atomic force microscope cantilevers, while deformations on a scale of nanometers to microns are measured by deflection of lasers onto optical detectors or by high resolution light microscopy.

Complementary to the use of external forces to probe material properties of the cell are analyses of the thermal motion of refractile particles such as internal vesicles or submicron-sized beads imbedded within the cell. Measurements of local viscoelastic parameters are essential for mapping the properties of small but heterogeneous materials like cytoplasm; some methods, most notably atomic force microscopy and optical tracking methods, enable high-resolution mapping of the cell's viscoelasticity.

A significant challenge in this field is to relate experimental and theoretical results derived from systems on a molecular scale to similar measurements on a macroscopic scale, for example from tissues, cell extracts, or purified polymer systems, and thus provide a self-consistent set of experimental methods that span many decades in time and length scales. At present, the new methods of nanoscale rheology often yield results that differ from bulk measurements by an order of magnitude. Such discrepancies are not a trivial result of experimental inaccuracy, but result from physical effects that only currently are being recognized and solved. This chapter will summarize some recent advances in methodology and provide examples where experimental results may motivate new theoretical insights into both cell biology and material science.

Introduction

The mechanical properties of cells have been matters of study and debate for centuries. Because cells perform a variety of mechanical processes, such as locomotion, secretion, and cell division, mechanical properties are relevant for biological function. Certain cells, such as plant cells and bacteria, have a hard cell wall that dominates the mechanics, whereas most other cells have a soft membrane and their mechanical properties are determined largely by an internal protein polymer network, the cytoskeleton. Early observations of single cells by microscopy showed regions of cytoplasm that were devoid of particles undergoing Brownian motion, and therefore were presumed to be "glassy" (see Chapter 3) or in some sense solid (Stossel, 1990). The interior of the cell, variously called the protoplasm, the ectoplasm, or more generally the cytoplasm, was shown to have both viscous and elastic features. A variety of methods were designed to measure these properties quantitatively.

Forces to which cells are exposed in a biological context

The range of stresses (force per area) to which different tissues are naturally exposed is large. Cytoskeletal structures have evolved accordingly and are not only responsible for passively providing material strength. They are also intimately involved in the sensing of external forces and the cellular responses to those forces. How cells respond to mechanical stress depends not only on specific molecular sensors and signaling pathways but also on their internal mechanical properties or rheologic parameters, because these material properties determine how the cell deforms when subjected to force (Janmey and Weitz, 2004).

It is likely that different structures and mechanisms are responsible for different forms of mechanical sensing. For example, cartilage typically experiences stresses on the order of 20 MPa, and the individual chondrocytes within it alter their expression of glycosaminoglycans and other constituents as they deform in response to such large forces (Grodzinsky et al., 2000). Bone and the osteocytes within it respond to similarly large stresses (Ehrlich and Lanyon, 2002), although the stress to which a cell imbedded within the bone matrix is directly exposed is not always clear. At the other extreme, endothelial cells undergo a wide range of morphological and transcriptional changes in response to shear stresses less than 1 Pa (Dewey et al., 1981), and neutrophils activate in response to similar or even smaller shear stresses (Fukuda and Schmid-Schonbein, 2003). Not only the magnitude but the geometry and time course of mechanical perturbations are critical to elicit specific cellular effects. Some tissues like tendons or skeletal muscle experience or generate mainly uniaxial forces and deformations, while others, such as the cells lining blood vessels, normally experience shear stresses due to fluid flow. These cells often respond to changes in stress or to oscillatory stress patterns rather than to a specific magnitude of stress (Bacabac et al., 2004; Davies et al., 1986; Florian et al., 2003; Ohura et al., 2003; Turner et al., 1995). Many cells, including the cells lining blood vessels and epithelial cells in the lung, experience large-area-dilation forces, and in these settings both the magnitude and the temporal characteristics of the force are critical to cell response (Waters et al., 2002).

Methods to measure intracellular rheology by macrorheology, diffusion, and sedimentation

The experimental designs to measure cytoplasmic (micro)rheology have to overcome three major challenges: the small size of the cell; the heterogeneous structure of the cell interior; and the active remodeling of the cytoplasm that occurs both constitutively, as part of the resting metabolic state, and directly, in response to the application of forces necessary to perform the rheologic measurement. The more strongly the cell is perturbed in an effort to measure its mechanical state, the more it reacts biochemically to change that state (Glogauer et al., 1997). Furthermore it is important to distinguish linear response to small strains from nonlinear response to larger strains. Structural cellular materials typically have a very small range of linear response (on the order of 10 percent) and beyond that react nonlinearly, for example by strain hardening or shear thinning or both in sequence. To overcome these problems a number of experimental methods have been devised.

Whole cell aggregates

The simplest and in some sense crudest method to measure intracellular mechanics is to use standard rheologic instruments to obtain stress/strain relations on a macroscopic sample containing many cells, but in which a single cell type is arranged in a regular pattern. Perhaps the most successful application of this method has been the study of muscle fibers, in which actin-myosin-based fibers are arranged in parallel and attached longitudinally, allowing an inference of single cell quantities directly from the properties of the macroscopic sample. One example of the validity of the assumptions that go into such measurements is the excellent agreement of single molecule measurements of the force-elongation relation for titin molecules with macroscopic compliance measurements of muscle fibers where the restoring force derives mainly from a large number of such molecules working in series and in parallel (Kellermayer et al., 1997). Another simple application of this method is the measurement of close-packed sedimented samples of a single cell type, with the assumption that during measurement, the deformation is related to the deformation of the cell interior rather than to the sliding of cells past each other. Such measurement have, for example, shown the effects of single actin-binding protein mutations in Dictyostelium (Eichinger et al., 1996) and melanoma cells (Cunningham et al., 1992). These simple measurements have the serious disadvantage that properties of a single cell require assumptions or verification of how the cells attach to each other, and in most cases the contributions of membrane deformation cannot be separated from those of the cell interior or the extracellular matrix.

Sedimentation of particles

To overcome the problems inherent in the measurement of macroscopic samples, a variety of elegant solutions have been devised. Generally, in order to resolve varying viscoelastic properties within a system, one has to use probes of a size comparable to or smaller than the inhomogeneities. Such microscopic probes can be fashioned in different ways. One of the simplest and oldest methods to measure cytoplasmic viscosity relies on observations of diffusion or sedimentation of intracellular granules with higher specific gravity through the cytoplasmic continuum. Generally these measurements were performed on relatively large cells containing colored or retractile particles easily visible in the microscope. Such measurements (reviewed in Heilbrunn, 1952; Heilbrunn, 1956) are among the earliest to obtain values similar to those measured currently, but they are limited to specialized cells and cannot measure elasticity in addition to viscosity.

Sedimentation measurements are done by a variety of elegant methods. One of the earliest such studies (Heilbronn, 1914) observed the rate of falling of starch grains within a bean cell and compared the rate of sedimentation of the same starch particles purified from the cells in fluids of known densities and viscosities measured by conventional viscometers, to obtain a value of 8 mPa.s for cytoplasmic viscosity. These measurements were an early application of a falling-sphere method commonly used in macroscopic rheometry (Rockwell et al., 1984). The viscosity of the cytoplasm in this application was determined by relation to calibrated liquids; because the starch particles are relatively uniform and could be purified from the cell, inaccuracies associated with measurement of their small size were avoided. More generally, any gravity-dependent velocity V of a particle of radius r and density σ in the cytoplasm of density ρ could be used to measure cytoplasmic viscosity η by use of the relation

$$V = \frac{2g(\sigma - \rho)r^2}{9\eta}, \quad (2.1)$$

in which g is the gravitational acceleration. Without centrifugation internal organelles rarely sediment, but a sufficiently large density difference between an internal particle and the surrounding cytoplasm could be created by injecting a small droplet of inert oil into a large cell, like a muscle fiber (Reiser, 1949) to obtain values of 29 mPa.s from the rate at which the drop rose in the cytoplasm. Alternatively, internal organelles could be made to sediment by known gravitational forces in a centrifuge, and in various cells – including oocytes, amoebas, and slime molds – cytoplasmic viscosities between 2 and 20 mPa.s have been commonly reported, although some much higher values greater than 1 Pa.s were also observed (reviewed in Heilbrunn, 1956). The large differences in viscosity were presumed to arise from experimental differences in the sedimentation rates, because these early studies also showed that cytoplasm was a highly non-Newtonian fluid and that the apparent viscosity strongly decreased with increasing shear rate.

Diffusion

Measurements of viscosity by diffusion (Heilbrunn, 1956) have been done by first centrifuging a large cell, such as a sea urchin egg or an amoeba, with sedimentation forces typically between 100 and 5000 times that of gravity, sufficient for internal organelles to get concentrated at the bottom while the cell remains intact. Then the displacement of a single particle of radius r in one direction $x(t)$ is monitored, and the cytoplasmic viscosity is measured from the Stokes-Einstein relation:

$$\langle x^2(t) \rangle = \frac{k_B T}{3\pi \eta r}, \quad (2.2)$$

where the brackets denote ensemble averaging, k_B is the Boltzmann constant, and T is temperature. Such measurements, dating from at least the 1920s (decades before video microscopy and image processing) showed that the viscosity within sea urchin egg cytoplasm was 4 mPa.s (4 cP), only four times higher than that of water, but that other cells exhibited much higher internal viscosities. Three other important features of intracellular material properties were evident from these studies. First, it was shown that the apparent viscosity of the protoplasm depended strongly on flow rates, as varied, for example, by changing the sedimentation force in the centrifuge. Second, viscous flow of internal organelles could generally be measured only deeper inside the cell, away from the periphery, where an elastic cortical layer could be distinguished from the more liquid cell interior. Third, the cellular viscosity was often strongly temperature dependent.

Mechanical indentation of the cell surface

Glass microneedles

Glass needles can be made thin enough to apply to a cell measured forces large enough to deform it but small enough that the cell is not damaged. An early use of such needles was to pull on individual cultured neurons (Bray, 1984); these studies showed how such point forces could be used to initiate neurite extension. Improved instrumentation and methods allowed an accurate estimate of the forces needed to initiate these changes. The method (Heidemann and Buxbaum, 1994; Heidemann et al., 1999) begins with calibration of the bending constant of a wire needle essentially by hanging a weight from the end of a thin metal wire and determining its spring constant from deflection of the loaded end by the relation:

$$y(L) = \frac{FL^3}{3EI} \tag{2.3}$$

where $y(L)$ is the displacement of the end of a wire of length L, F is the force due to the weight, E is the material's Young's modulus of elasticity, and $I = \pi r^4/4$ is the second moment of inertia of the rod of radius r.

The product EI is a constant for each rod; in practice the first calibrated rod is used to provide a known (smaller) force to a thinner, usually glass, rod, to calibrate that rod, and repeat the process until a rod is calibrated that can provide nN or smaller forces depending on its length and radius.

Cell poker

A pioneering effort to apply forces locally to the surface of live cells was the development of the cell poker (Daily et al., 1984; Petersen et al., 1982). In this device, shown schematically in Fig. 2-1, a cell is suspended in fluid from a glass coverslip on an upright microscope, below which is a vertical glass needle attached at its opposite end to a wire needle that is in turn coupled to a piezoelectric actuator that moves the wire/needle assembly up and down. The vertical displacements of both ends of the wire are measured optically, and a difference in the displacements of these points, x, occurs because of resistance to moving the glass needle tip as it makes contact with the cell. The force exerted by the tip F on the cell surface is determined by Hooke's law $F = kx$ from the stiffness of the wire k, which can be calibrated by macroscopic means such as the hanging of known weights from a specified length of wire. Using this

Experimental measurements of intracellular mechanics

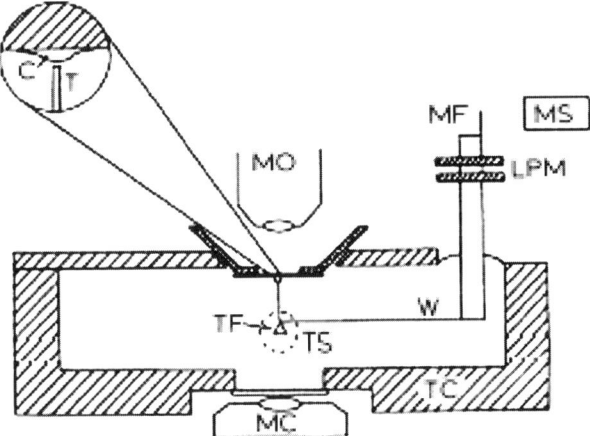

Fig. 2-1. Schematic representation of the cell-poking apparatus. Positioning of the cell (C) relative to the poker tip (T) is achieved by translating the top of the temperature control unit (TC) or by rotating the holder on which the coverslip is mounted. The motor assembly can be translated to ensure the tip is positioned in the field of view. W, steel wire; LPM, linear piezoelectric motor; MS and TS, optical sensors; MF, motor flag; TF, tip flag; MO, modulation contrast objective; MC, matching condenser.

instrument, displacements less than 100 nm can be resolved corresponding to forces less than 10 nN. A typical force vs. displacement curve from this instrument as shown in Fig. 2-2 reveals a significant degree of both elasticity and unrecoverable deformation from plasticity or flow of the cytoplasm. Such measurements have demonstrated both a significant elastic response as well as a plastic deformation of the cell, and the time course and magnitudes of these processes can be probed by varying the rate at which the forces are applied. Because the tip is considerably smaller than the cross-sectional area of the cell, local viscoelasticity could be probed at different regions of the cell or as active motion or other responses are triggered. The earliest such measurements revealed a large difference in relative stiffness over different areas of the cell and a high degree of softening when actin-filament-disorganizing drugs like

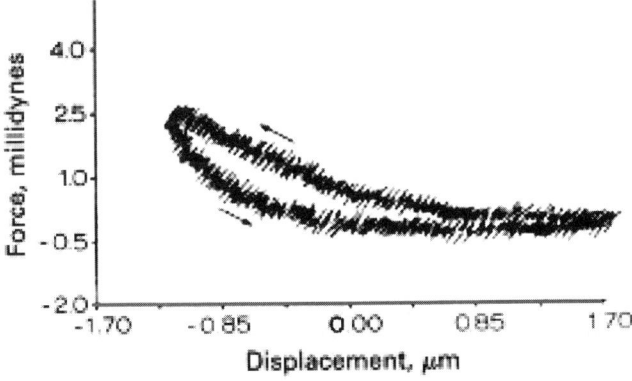

Fig. 2-2. Force displacement curve as the cell poker tip first indents the cell (upper curve) and then is lowered away from the cell contact (lower curve).

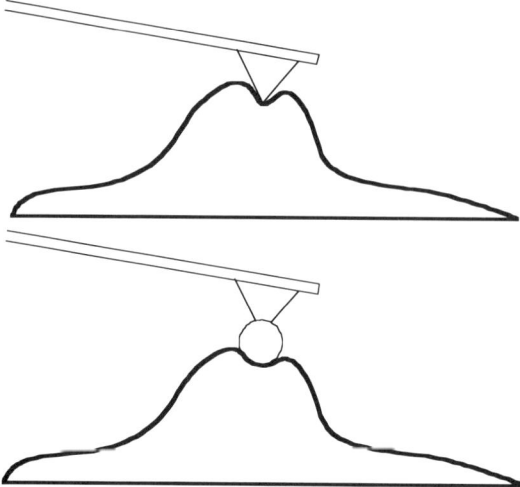

Fig. 2-3. Cell poking with the tip of an atomic force microscope. Upper image: If a regular sharp tip is used, inhomogeneities encountered on the nm scale of the tip radius are likely to make the result difficult to interpret. Lower image: Using a micrometer-sized bead attached to the tip, force sensitivity is maintained while the cell response is averaged over a micrometer scale.

cytochalasin were applied. The measurements also showed that the apparent stiffness of the cell increased as the amplitude of indentation increased. How this nonlinear elastic response is related to the material properties of the cell is, however, not straightforward to deduce, because of a number of complicating effects, as the earliest such studies pointed out.

For a homogeneous, semi-infinite elastic solid, given the geometry of the glass needle tip and the force of indentation, the force-displacement curves are determined by two material properties, the Young's modulus and Poisson's ratio, in a way that is described by the Hertz relation. For a sphere the result for the force as a function of indentation depth δ is (Hertz, 1882; Landau and Lifshitz, 1970):

$$F_{sphere} = \frac{4}{3} \frac{E}{(1-\nu^2)} R^{1/2} \delta^{3/2} \qquad (2.4)$$

with the Young's modulus E, Poisson ratio ν, and sphere radius R. For indentation with a conical object, the result is:

$$F_{cone} = \frac{\pi}{2} \frac{E}{(1-\nu^2)} \tan(\alpha) \delta^2 \qquad (2.5)$$

with the cone opening angle α.

The application of the Hertz model in relation to cell-poking measurements is, however, often not meaningful for at least three reasons. First, the Hertz relation is not valid if the cell thickness is not much greater than the degree of indentation. Second, the cell cytoskeleton is in most cases far from being an isotropic homogeneous material. And third, forces exerted on a cell typically initiate biochemical as well as other active reactions. These issues have been extensively discussed both in terms of the cell poker (Daily et al., 1984), and more recently in applications of the scanning force microscope that operates on the same principle.

Atomic force microscopy

A very sensitive local mechanical probe is provided by atomic force microscopy (AFM). An AFM in an imaging mode works by scanning a sharp microfabricated tip over a surface while simultaneously recording tip deflection. The deflection time course is then converted into an image of the surface profile (Binnig et al., 1986). Imaging can be done in different modes – contact mode (Dufrene, 2003), tapping mode (Hansma et al., 1994), jumping mode (de Pablo et al., 1998) or others – which are usually designed to minimize damage to the sample or distortions of the surface by the imaging method. When one wants to probe the mechanical properties of a material surface, however, one can also use an AFM tip to exert precisely controlled forces in selected locations and record the corresponding sample displacements. In many ways this method is related to cell poking with larger probes, but it holds the potential of better spatial and force resolution. The obvious limitation of the technique is that manipulation can only occur through the accessible surface of a cell, that is one cannot measure elastic moduli well inside the cell without an influence of boundary conditions. One can both indent cells or pull on cells when the tips are attached strongly enough to the cell surface. The indentation approach has been used to test the elastic properties of various types of cell. Initial studies have used conventional sharp (radius of 10s of nm) tips and applied the Hertz model as described above (reviewed in MacKintosh and Schmidt, 1999). The same caveats hold in this case as in the discussion of other cell-poking experiments: the cell is not a homogeneous, isotropic, passive elastic solid. The thin parts of cells, at the cell periphery in surface-attached cells, are particularly interesting to study because they are crucial for cell motility but are usually too thin to apply the standard Hertz model. When using an AFM with a sharp tip, the spatial inhomogeneity of cells – for example the presence of bundles of actin (stress fibres), microtubules, and more – is likely more of a problem, because spatial averaging in the case of a larger probe tends to make the material look more homogeneous. Results of initial experiments were thus rather qualitative, but differences between the cell center and its periphery could be detected (Dvorak and Nagao, 1998). A problem with quasi-static or low-frequency measurements is that the cell will react to forces exerted on it and the response measured will not only reflect passive material properties, but also active cellular responses. AFM has also been used on cells in a high-frequency mode, namely the tapping mode. It was observed that cells dynamically stiffened when they were probed with a rapidly oscillating tip, as one would expect (Putman et al., 1994).

A more quantitative technique has been developed more recently, using polystyrene beads of carefully controlled radius attached to the AFM tips to contact cells (Mahaffy et al., 2004; Mahaffy et al., 2000). This creates a well-defined probe geometry and provides another parameter, namely bead radius, to control for inhomogeneities. Values for zero-frequency shear moduli were between 1 and 2 kPa for the fibroblast cells studied. The probing was in this case also done with an oscillating tip, to measure frequency-dependent viscoelastic response with a bandwidth of 50–300 Hz and data were evaluated with an extended Hertz model valid for oscillating probes (Mahaffy et al., 2000). A problem for determining the viscous part of the response is the hydrodynamic drag on the rest of the cantilever that dominates and changes with decreasing distance from the surface and with tip-sample contact and is not easy to correct for.

The Hertz model has been further modified to account for finite sample thickness and boundary conditions on the substrate (Mahaffy et al., 2004), which makes it possible to estimate elastic constants also for the thin lamellipodia of cells, which were found again to be between 1 and 2 kPa in fibroblasts.

Mechanical tension applied to the cell membrane

Pulling on a cell membrane by controlled suction within a micropipette has been an important tool to measure the viscosity and elastic response of cells to controlled forces. The initial report of a cell elastimeter based on micropipette aspiration (Mitchison and Swann, 1954) has guided many studies that have employed this method to deform the membranes of a variety of cells, especially red blood cells, which lack a three-dimensional cytoskeleton but have a continuous viscoelastic protein network lining their outer membrane (Discher et al., 1994; Evans and Hochmuth, 1976). One important advantage of this method is that the cell can either be suspended in solution while bound to the micropipette or attached to a surface as the micropipette applies negative pressure from the top. The ability to probe nonadherent cells has made micropipette aspiration a powerful method to probe the viscoelasticity of blood cells including erythrocytes, leukocytes, and monocytes (Chien et al., 1984; Dong et al., 1988; Richelme et al., 2000).

A typical micropipette aspiration system is shown in Fig. 2-4. Images of two red blood cells partly pulled into a micropipette are shown in Fig. 2-5. Micropipette aspiration provides measures of three quantities: the cortical tension in the cell membrane; the cytoplasmic viscosity; and the cell elasticity. If the cell can be modeled as a liquid drop with a cortical tension, as appears suitable to leukocytes under some conditions, the cortical tension t is calculated from the pressure at which the aspirated part of the cell forms a hemispherical cap within the pipette.

For a cell modeled as an elastic body, its Young's modulus E is determined by the relation

$$\Delta P = \frac{2\pi}{3} E \frac{L_p}{R_p} \phi, \quad (2.6)$$

where ΔP is the pressure difference inside and outside the pipette, L_p is the length of the cell pulled into the pipette with radius R_p, and ϕ is a geometric constant with a value around 2.1 (Evans and Yeung, 1989).

For liquid-like flow of cells at pressures exceeding the cortical tension, the cytoplasmic viscosity is calculated from the relation

$$\eta = \frac{R_p \Delta p}{\left(\frac{dL_p}{dt}\right) m (1 - R_p/R)}, \quad (2.7)$$

where η is the viscosity, R is the diameter of the cell outside the pipette, and m is a constant with a value around 9 (Evans and Yeung, 1989).

Experimental measurements of intracellular mechanics

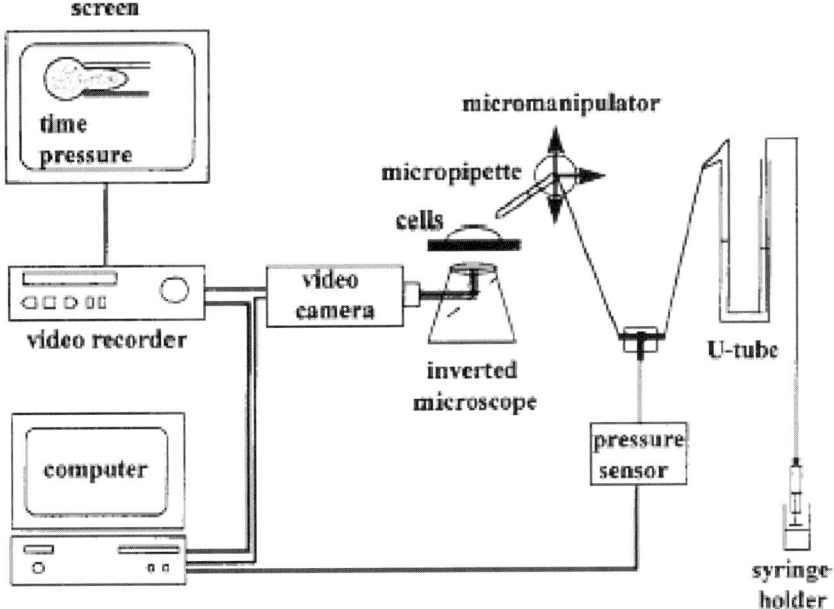

Fig. 2-4. Experimental study of cell response to mechanical forces. Cells are deposited on the stage of an inverted microscope equipped with a video camera. The video output is connected to a digitizer mounted on a desk computer. Cells are aspirated into micropipettes connected to a syringe mounted on a syringe holder. Pressure is monitored with a sensor connected to the computer. Pressure and time values are superimposed on live cell images before recording on videotapes for delayed analysis. From Richelme et al., 2000.

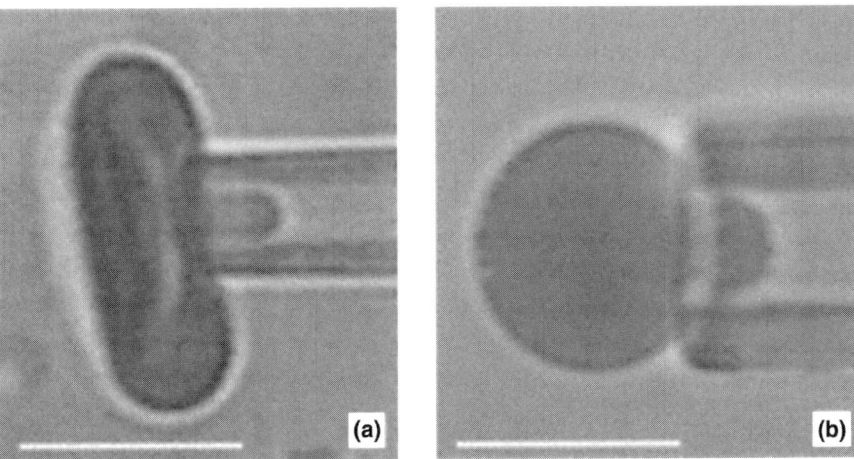

Fig. 2-5. Aspiration of a flaccid (a) and swollen (b) red blood cell into a pipette. The diameter of the flaccid cell is approximately 8 µm and that of the swollen cell is about 6 µm. The scale bars indicate 5 µm. From Hochmuth, 2000.

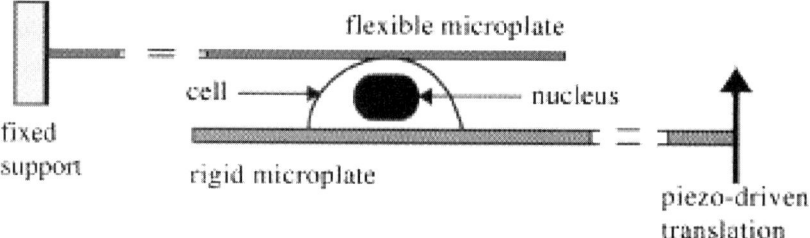

Fig. 2-6. Diagram for a device for compression of a cell between microplates. Variations of this design also allow for imposition of shear deformation. From Caille et al., 2002.

Shearing and compression between microplates

For cells that normally adhere to surfaces, an elegant but technically challenging method to measure viscoelasticity is by attaching them at both top and bottom to glass surfaces that can be moved with respect to each other in compression, extension, or shear (Thoumine et al., 1999). A schematic diagram of such a system is shown in Fig. 2-6.

In this method a cell such as a fibroblast that adheres tightly to glass surfaces coated with adhesion proteins such as fibronectin is grown on a relatively rigid plate; a second, flexible plate is then placed on the top surface. Piezo-driven motors displace the rigid plate a known distance to determine the strain, and the deflection of the flexible microplate provides a measure of the stress imposed on the cell surface. Use of this device to provide well-defined strains with simultaneous imaging of internal structures such as the nucleus provides a measure of the elastic modulus of fibroblasts around 1000 Pa, consistent with measurements by AFM, and has shown that the stiffness of the nucleus is approximately ten times greater than that of the cytoplasmic protein networks (Caille et al., 2002; Thoumine and Ott, 1997). A recent refinement of the microcantilever apparatus allows a cell in suspension to be captured by both upper and lower plates nearly simultaneously and to measure the forces exerted by the cell as it begins to spread on the glass surfaces (Desprat et al., 2005).

Fluid flow

Cells have to withstand direct mechanical deformations through contact with other cells or the environment, but some cells are also regularly exposed to fluid stresses, such as vascular endothelial cells in the circulating system or certain bone cells (osteocytes) within the bone matrix. Cells sense these stresses and their responses are crucial for many regulatory processes. For example, in vascular endothelial cells, mechanosensing is believed to control the production of protective extracellular matrix (Barbee et al., 1995; Weinbaum et al., 2003); whereas in bone, mechanosensing is at the basis of bone repair and adaptive restructuring processes (Burger and Klein-Nulend, 1999; Wolff, 1986). Osteocytes have been studied *in vitro* after extraction from the bone matrix in parallel plate flow chambers (Fig. 2-7). Monolayers of osteocytes coated onto one of the chamber surfaces were exposed to shear stress while the

Experimental measurements of intracellular mechanics

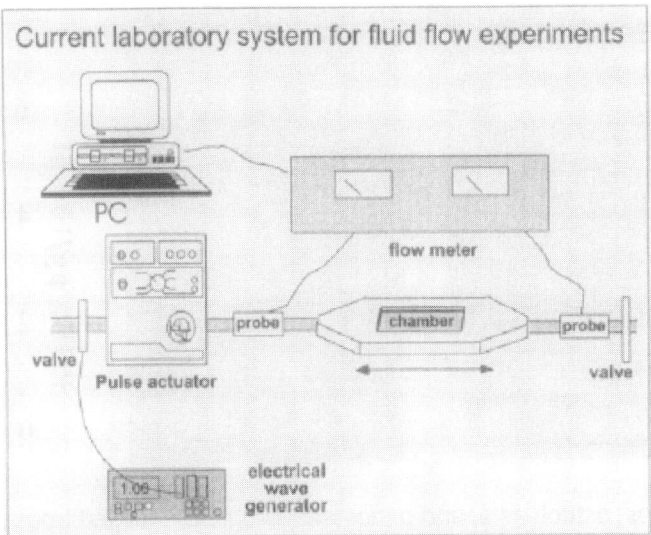

Fig. 2-7. Fluid flow system to stimulate mechanosensitive bone cells, consisting of a culture chamber containing the cells, a pulse generator controlling the fluid flow, and flow meters. The response of the cells is either biochemically measured from the cells after the application of flow (for example prostaglandin release) or measured in the medium after flowing over the cells (for example nitric oxide). From Klein-Nulend et al., 2003.

response was measured by detecting the amount of nitric oxide produced as a function of fluid flow rate (Bacabac et al., 2002; Rubin and Lanyon, 1984).

The strain field within individual surface-attached cells in response to shear flow has been mapped in bovine vascular endothelial cells with the help of endogenous fluorescent vimentin (Helmke et al., 2003; Helmke et al., 2001). It was found that the spatial distribution of strain is rather inhomogeneous, and that strain is focused to localized areas within the cells. The method can only measure strain and not stress. The sites for mechanosensing might be those where strain is large if some large distortion of the sensing element is required to create a signal, in other words, if the sensor is "soft." On the other hand, the sites for sensing might also be those where stress is focused and where little strain occurs if the sensing element requires a small distortion, or is "hard," and functions by having a relatively high force threshold.

Numerical simulations can be applied to both the cell and the fluid passing over it. A combination of finite element analysis and computational fluid dynamics has been used to model the flow across the surface of an adhering cell and to calculate the shear stresses in different spots on the cell (Barbee et al., 1995; Charras and Horton, 2002). This analysis provides a distribution of stress given a real (to some resolution) cell shape, but without knowing the material inhomogeneities inside, the material had to be assumed to be linear elastic and isotropic. The method was also applied to model stress and strain distributions inside cells that were manipulated by AFM, magnetic bead pulling or twisting, and substrate stretching, and proved useful to compare the effects of the various ways of mechanical distortion.

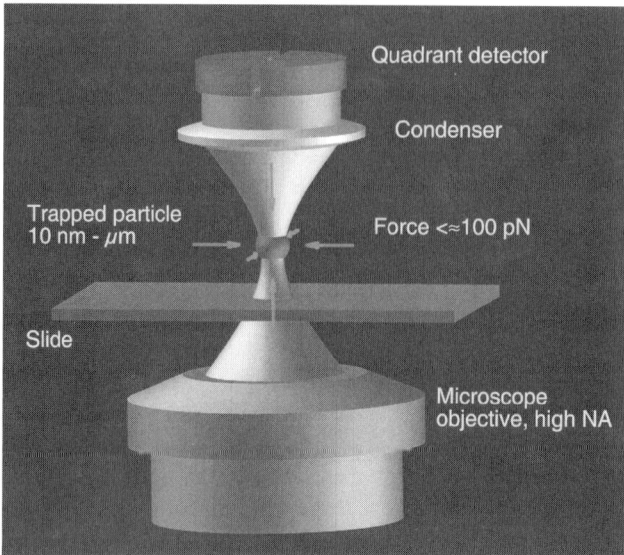

Fig. 2-8. Schematic diagram of an optical trap.

Optical traps

Optical traps (see Fig. 2-8) use a laser beam focused through a high-numerical aperture microscope objective lens to three-dimensionally trap micron-sized refractile particles, usually silica or latex beads (Ashkin, 1997; Svoboda and Block, 1994). The force acting on the bead at a certain distance from the laser focus is in general very difficult to calculate because (1) a high-NA laser focus is not well approximated by a Gaussian, and (2) a micron-sized refractive particle will substantially affect the light field. Approximations are possible for both small particles (Rayleigh limit) and large particles (ray optics limit) with respect to the laser wave length. For a small particle, the force can be subdivided into a "gradient force" pulling the particle towards the laser focus and a scattering force pushing it along the propagation direction of the laser (Ashkin, 1992). Assuming a Gaussian focus and a particle much smaller than the laser wavelength, the gradient forces in radial and axial direction are (Agayan et al., 2002):

$$F_g^{axial} \propto -\alpha' I_0 z \frac{w_0^4}{z_0^2} \left[\frac{1}{w^4(z)} - \frac{2r^2}{w^6(z)} \right] \cdot \exp\left(-\frac{2r^2}{w^2(z)}\right) \quad (2.8)$$

$$F_g^{radial} \propto -\alpha' I_0 r \frac{w_0^2}{w^4(z)} \cdot \exp\left(-\frac{2r^2}{w^2(z)}\right) \quad (2.9)$$

and the scattering forces:

$$F_s^{axial} \propto \alpha'' I_0 \frac{w_0^2}{w^2(z)} \left[k_m \left(1 - \frac{r^2}{2} \frac{(z^2 - z_0^2)}{(z^2 + z_0^2)^2}\right) - \frac{w_0^2}{z_0 w^2(z)} \right] \cdot \exp\left(-\frac{2r^2}{w^2(z)}\right) \quad (2.10)$$

$$F_s^{radial} \propto \alpha'' I_0 \frac{w_0^2}{w^2(z)} \frac{k_m r}{R(z)} \exp\left(-\frac{2r^2}{w^2(z)}\right) \quad (2.11)$$

with complex polarizability $\alpha = \alpha' + i\alpha''$, laser intensity I_0, w_0 the beam radius in the focus, and $w(z) = w_0\sqrt{1 + (z/z_0)^2}$ the beam radius near the focus; $z_0 = \pi w_0^2/\lambda_m$ the Rayleigh range, $k_m = 2\pi/\lambda_m$ the wave vector, with λ_m the wavelength in the medium with refractive index n_m. (For details and prefactors see Agayan et al., 2002).

Stable trapping will only occur if the gradient force wins over the trapping force all around the focus. Trap stability thus depends on the geometry of the applied field and on properties of the trapped particle and the surrounding medium. The forces generally depend on particle size and the relative index of refraction $n = n_p/n_m$, where n_p and n_m are the indices of the particle and the medium, respectively, which is hidden in the polarizability α in Eqs. 2.8–2.11. In the geometrical optics regime, maximal trap strength is particle-size-independent, but increases with n over some intermediate range until, at larger values of n, the scattering force exceeds the gradient force. The scattering force on a nonabsorbing Rayleigh particle of diameter d is proportional to its scattering cross-section, thus the scattering force scales with the square of the polarizability (volume) (Jackson, 1975), or as d^6. The gradient force scales linearly with polarizability (volume), that is, it has a d^3-dependence (Ashkin et al., 1986; Harada and Asakura, 1996).

A trappable bead can then be attached to the surface of a cell and can be used to deform the cell membrane locally. The method has the advantage that no mechanical access to the cells is necessary. Using beads of micron size furthermore makes it possible to choose the site to be probed on the cell with relatively high resolution. A disadvantage is that the forces that can be exerted are difficult to increase beyond about 100 pN, orders of magnitude smaller than can be achieved with micropipettes or AFM tips. At high laser powers, local heating may not be negligible (Peterman et al., 2003a). Force and displacement can be detected, however, with great accuracy, sub-nm for the displacement and sub-pN for the force, using interferometric methods (Gittes and Schmidt, 1998; Pralle et al., 1999). This makes the method well suited to study linear response parameters of cells. Interferometric detection can also be as fast as 10 µs, opening up another dimension in the study of cell viscoelasticity. Focusing on different frequency regimes should make it, for example, possible to differentiate between active, motor-driven responses and passive viscoelasticity. Such an application of optical tweezers is closely related to laser-based microrheology, which can also be applied inside the cells (see **Passive Microrheology**). We will here focus on experiments that have used the optical manipulation of externally attached beads.

Optical tweezers have been used by several groups to manipulate human red blood cells (see Fig. 2-9), which have no space-filling cytoskeleton but only a membrane-associated 2D protein polymer network (spectrin network). The 2D shear modulus measured for the cell membrane plus spectrin network varies between 2.5 µN/m (Henon et al., 1999; Lenormand et al., 2001) and 200 µN/m (Sleep et al., 1999), possibly due to different modeling approaches in estimating the modulus. The nonlinear part of the response of red blood cells has been explored by using large beads and high laser power achieving a force of up to 600 pN. The shear modulus of the cells levels off at intermediate forces before rising again at the highest forces, which was simulated in finite element models of the cells under tension (Dao et al., 2003; Lim et al., 2004).

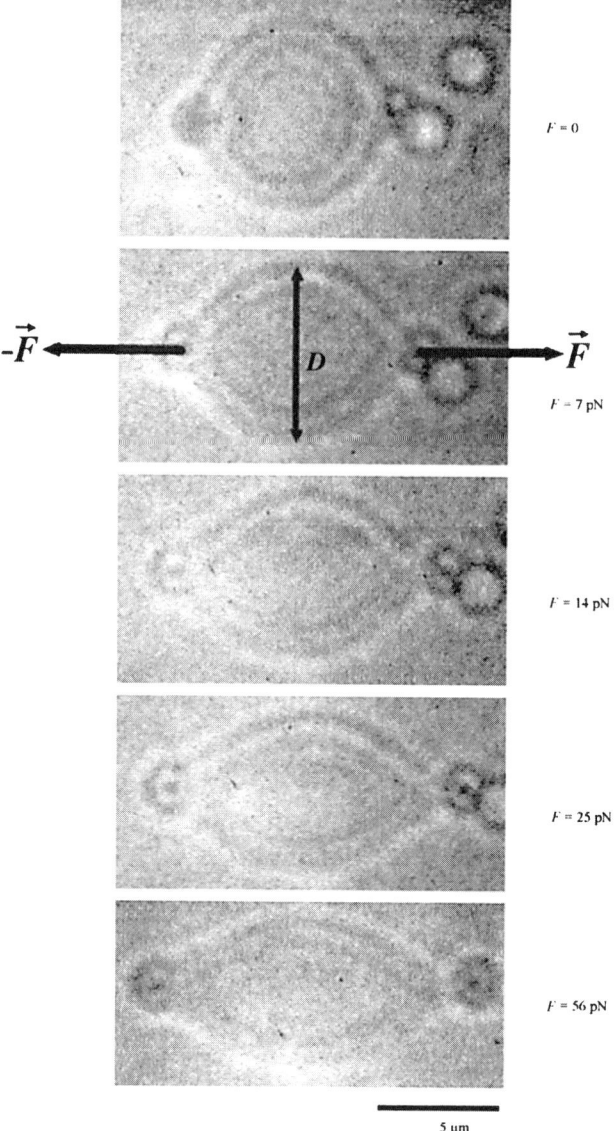

Fig. 2-9. Stretching of red blood cells by optical tweezers, using a pair of beads attached to diametrically opposed ends of the cell. Forces are given next to the panels. From Henon et al., 1999.

Magnetic methods

Using magnetic particles has the advantage that large forces (comparable to AFM) can be exerted, while no open surface is required. One can use magnetic fields to apply forces and/or torques to the particles. Ferromagnetic particles are needed to apply torques; paramagnetic particles are sufficient to apply force only. A disadvantage of the magnetic force method is that it is difficult to establish homogeneous field gradients (only gradients exert a force on a magnetic dipole), and the dipole moments of microscopic particles typically scatter strongly. Furthermore, one is often limited

Experimental measurements of intracellular mechanics

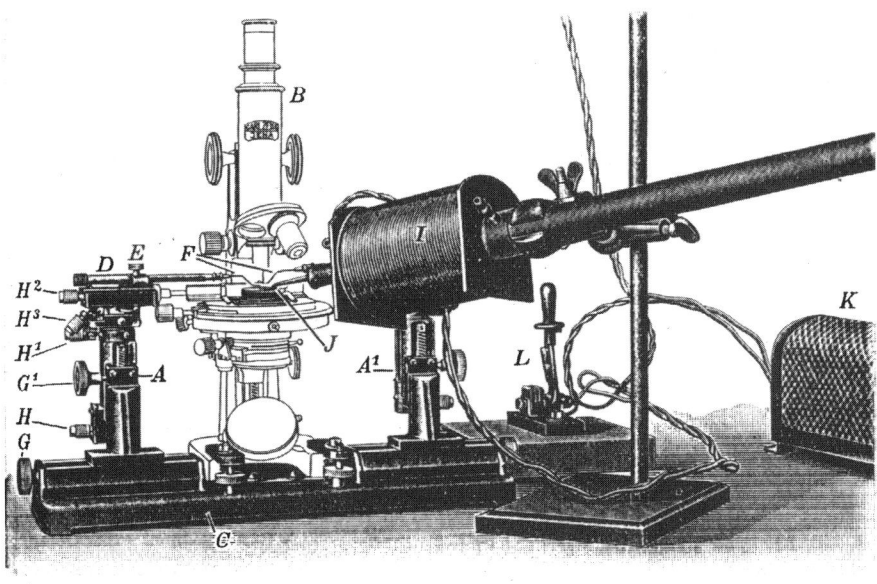

Fig. 2-10. A magnetic manipulation system to measure viscoelasticity in a single cell. From Freundlich and Seifriz, 1922.

to video rates for displacement detection when using the force method. Rotations can be detected by induction for ensembles of particles.

One of the first reports of an apparatus to measure intracellular viscoelasticity was from Freundlich and Seifriz (1922). A diagram of the instrument is shown in Fig. 2-10.

In this instrument, a micromanipulator mounted next to the microscope objective was used to insert a magnetic particle, made of nickel or magnetite, into a relatively large cell like a sand dollar egg. Then a magnetic field gradient, produced by an electromagnet placed as close as possible to the cell, was used to impose a force on the bead, whose displacement was measured by the microscope. The strength of the force on the bead could be calibrated by measuring the rate of its movement through a calibration fluid of viscosity that could be measured by conventional rheometers. This magnetic manipulation instrument was the precursor of current magnet-based microrheology systems, and was further enhanced by work of Crick and Hughes (1950), who made two important modifications of the experimental design. One was to first magnetize the particle with a large magnetic field, and then use a smaller probing magnetic field directed at a different angle to twist particles on or within the cell. The second change was to use phagocytic cells that would engulf the magnetic particle, thereby avoiding possible damage to the cell when magnetic particles were forced through its membrane. These early studies were done before the cytoskeleton was visualized by electron or fluorescence microscopy and before the phospholipid bilayer forming the cell membrane was characterized, so a critical evaluation could not be done of how disruptive either way of introducing the beads was. Further pioneering work was done on amoebae (Yagi, 1961) and on squid axoplasm (Sato et al., 1984).

In principle, the motion of embedded probes will depend on the probe size. Small particles can diffuse through the meshes, and this has been used to determine effective mesh sizes in model systems (Jones and Luby-Phelps, 1996; Schmidt et al., 1989; Schnurr et al., 1997) and in cells (Jones et al., 1997; Luby-Phelps, 1994; Valentine et al., 2001). On the other hand, the beads might interact with and stick to the cytoskeleton, possibly mediated by an enveloping lipid membrane and by motor proteins, which would cause active motion. How micron-sized beads are coupled to the network in which they are imbedded is still a major issue in evaluating microrheology measurements, and no optimal method to control or prevent interactions yet exists. Entry of a particle through phagocytosis certainly places it in a compartment distinct from the proteins forming the cytoskeleton, and how such phagosomes are bound to other cytoplasmic structures is unclear. Likewise, both the mechanical and chemical effects of placing micron-sized metal beads in the cell raises issues about alignment and reorganization of the cytoskeleton. These issues will be further considered in the following sections.

Pulling by magnetic field gradients

Magnetic particles can either be inserted into cells or bound – possibly via specific attachments – to cell surfaces. Both superparamagnetic particles (Bausch et al., 1998; Keller et al., 2001) and ferro- as well as ferrimagnetic particles (Bausch et al., 1999; Trepat et al., 2003; Valberg and Butler, 1987) have been used. Paramagnetic particles will experience a translational force in a field gradient, but no torque. With sharpened iron cores reaching close to the cells, forces of up to 10 nN have been generated (Vonna et al., 2003). Ferromagnetic (as well as ferrimagnetic) particles have larger magnetic moments and therefore need less-strong gradients, which can be produced by electromagnetic coils without iron cores and can therefore be much more rapidly modulated. The particles have to be magnetized initially with a strong homogeneous field. Depending on the directions of the fields, particles will experience both torque and translational forces in a field gradient (see Fig. 2-11). Forces reported are on the order of pN (Trepat et al., 2003).

Forces exerted by the cell in response to an imposed particle movement, both on the membrane and inside the cell, are mostly dominated by the cytoskeleton. Exceptions are cases where the particle size is smaller than the cytoskeletal mesh size; specialized cells, such as mammalian red blood cells without a three-dimensional cytoskeleton; or cells with a disrupted cytoskeleton after treatment with drugs such as nocodazole or cytochalasin. The interpretation of measured responses needs to start from a knowledge of the exact geometry of the surroundings of the probe and it is difficult, even when the particle is inserted deeply into the cell. This is due to the highly inhomogeneous character of the cytoplasm, consisting of different types of protein fibers, bundles, organelles, and membranes. Because a living cell is an active material that is slowly and continuously changing shape, responses are in general time dependent and contain passive and active components. Passive responses to low forces are often hidden under the active motions of the cell, while responses to large forces do not probe linear response parameters, but rather nonlinear behavior and rupture of the networks. Given all the restrictions mentioned, a window to measure

Experimental measurements of intracellular mechanics

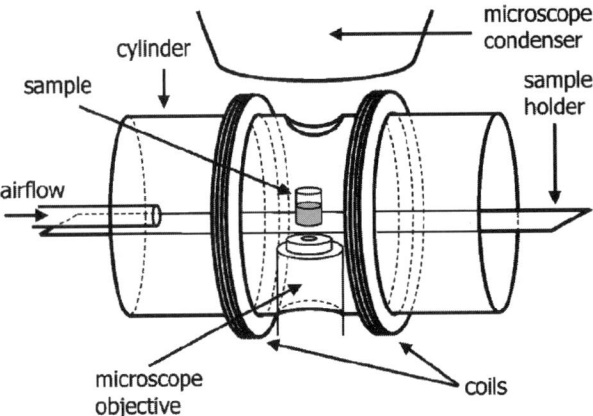

Fig. 2-11. Schematic diagram of a magnetic tweezers device using a magnetic field gradient. From Trepat et al., 2003.

the passive mechanical properties of cells appears to be to apply large strains, or to apply relatively high-frequency oscillatory strain at small amplitudes, while active responses can best be measured at low frequencies.

A number of experiments performed inside cells have observed the creep response to the instantaneous application of large or small forces (Bausch et al., 1999; Bausch et al., 1998; Feneberg et al., 2001). Bulk shear moduli were found to vary from ∼20 Pa in the cytoplasm of Dictyostelium to ∼300 Pa inside macrophages. At higher forces and strains, differences in rupture forces were found between mutant Dictyostelium cells and wild-type controls, highlighting the roles of regulatory proteins for the properties of the cytoskeleton (Feneberg et al., 2001).

With particles attached to the surface of cells, a shear modulus between 20 and 40 kPa was measured in the cortex of fibroblasts (Bausch et al., 1998), qualitative differences were measured between unstimulated and stimulated (stiffening) vascular endothelial cells (Bausch et al., 2001), and an absolute value of about 400 Pa was estimated from subsequent work (Feneberg et al., 2004). Active responses of macrophages and the formation of cell protrusion under varying forces were also tested with externally attached magnetic beads (Vonna et al., 2003).

Twisting of magnetized particles on the cell surface and interior

Applying a pure torque to magnetic particles avoids the difficulties of constructing a well-controlled field gradient. Homogeneous fields can be created rather easily. The method most widely used was pioneered by Valberg and colleagues (Valberg and Butler, 1987; Valberg and Feldman, 1987; Wang et al., 1993) and consists of using a strong magnetic field pulse to magnetize a large number of ferromagnetic particles that were previously attached to an ensemble (20,000–40,000) of cells. A weaker probe field oriented at 90° to the induced dipoles then causes rotation, which is measured in a lock-in mode with a magnetometer. In an homogeneous infinite medium, an effective shear modulus can be determined simply from the angle α rotated in response to an applied torque T: $G = T/\alpha$. On the surface of cells, however, the boundary conditions are highly complex, and a substantial polydispersity within the bead ensemble is expected (Fabry et al., 1999). Therefore the method has been mainly used for determining qualitative behavior, for comparative studies of different cell types, and for studies of relative changes in a given cell population. Frequency dependence of the viscoelastic response was measured with smooth muscle cells between 0.05 and 0.4 Hz (Maksym et al., 2000) and with bronchial epithelial cells up to 16 Hz (Puig-de-Morales et al., 2001) (see also Chapter 3). The shear elastic modulus was found to be around 50 Pa with a weak frequency dependence in both cases.

Rotation in response to torque can also be detected on individual particles by video tracking when beads are attached to the outsides of cells. In that case the torque causes center-of-mass displacement, which can be tracked with nm accuracy (Fabry et al., 2001). Tracking individual particles makes it possible to study the heterogeneity of response between different cells and in different locations on cells. In conjunction with fluorescent labeling it is possible to explore the strains caused by locally imposed stresses; initial studies reveal that the strain field is surprisingly long-range (Hu et al., 2003). Using oscillatory torque and phase-locking techniques, the bandwidth of this technique was extended to 1 kHz (Fabry et al., 2001). While absolute shear moduli were still not easy to determine because of unknown geometrical factors such as depth of embedding, the bandwidth was wide enough to study the scaling behavior of the complex shear modulus more extensively. The observed weak power-laws (exponent between 0.1 and 0.3) appear to be rather typical for cells in that frequency window and were interpreted in terms of a soft glass model. Finite element numerical modeling has been applied to analyze the deformations of cells when attached magnetic beads are rotated (Mijailovich et al., 2002) to test the limits of linearity in the response as well as the effect of finite cell thickness and surface attachment.

Passive microrheology

To measure the viscoelastic properties of a system, it is not necessary to apply external forces when one employs microscopic probes. In a soft-enough medium, thermal fluctuations will be measurable and these fluctuations precisely report the linear-response viscoelastic parameters of the medium surrounding the probe. This connection is formalized in the fluctuation-dissipation (FD) theorem of linear-response theory (Landau et al., 1980). When possible, that is when the medium is soft enough, this method even elegantly circumvents the need to extrapolate to zero-force amplitude, which is usually necessary in active methods to obtain linear response parameters. Particularly in networks of semiflexible polymers such as the cytoskeleton, nonlinear response occurs typically for rather small strains on the order of a few percent (Storm et al., 2005).

A microscopic probe offers both the possibility to study inhomogeneities directly in the elastic properties of the cytoskeleton, and to measure viscoelasticity at higher frequencies, above 1 kHz or even up to MHz, because inertia of both probe and embedding medium can be neglected at such small length scales (Levine and Lubensky, 2001; Peterman et al., 2003b). The possibility to observe thermal fluctuations instead of actively applying force or torque in principle exists for all the techniques using microscopic probes described above. It has, however, mainly been used in several related and recently developed techniques, collectively referred to as passive microrheology, employing beads of micron size embedded in the sample (Addas et al., 2004; Lau et al., 2003; MacKintosh and Schmidt, 1999; Mason et al., 1997; Schmidt et al., 2000; Schnurr et al., 1997). Passive microrheology has been used to probe, on microscopic scales, the material properties of systems ranging from simple polymer solutions to the interior of living cells.

Optically detected individual probes

The simplest method in terms of instrumentation uses video microscopy to record the Brownian motion of the embedded particles. Advantages are the use of standard equipment coupled with well-established image processing and particle tracking (Crocker and Grier, 1996), and the fact that massively parallel processing can be done (100 s of particles at the same time). Disadvantages are the relatively low spatial and temporal resolution, although limits can be pushed to nm in spatial displacement resolution and kHz temporal resolution with specialized cameras. Much higher spatial and temporal resolution still can be achieved using laser interferometry with laser beams focused on individual probe particles (Denk and Webb, 1990; Gittes and Schmidt, 1998; Pralle et al., 1999). Due to high light intensities focused on the particles, high spatial resolution (sub-nm) can be reached. Because the detection involves no video imaging, 100 kHz bandwidth can be reached routinely.

One-particle method

Once particle positions as a function of time are recorded by either method, the complex shear modulus $G(\omega)$ of the viscoelastic particle environment has to be calculated. This can be done by calculating the mean square displacement $\langle x_\omega^2 \rangle$ of the Brownian motion by Fourier transformation. The complex compliance $\alpha(\omega)$ of the probe particle with respect to a force exerted on it is defined by:

$$x_\omega = (\alpha' + i\alpha'')f_\omega \qquad (2.12)$$

The FD theorem relates the imaginary part of the complex compliance to the mean square displacement:

$$\langle x_\omega^2 \rangle = \frac{4k_B T \alpha''}{\omega}. \qquad (2.13)$$

Using a Kramers-Kronig relation (Landau et al., 1980):

$$\alpha'(\omega) = \frac{2}{\pi} P \int_0^\infty \frac{\zeta \alpha''(\zeta)}{\zeta^2 - \omega^2} d\zeta, \qquad (2.14)$$

where ζ is the frequency variable to integrate over and P denotes a principal value integral, one can then calculate the real part of the compliance. Knowing both real and imaginary parts of the compliance, one then finds the complex shear modulus via a generalized Stokes law:

$$\alpha(\omega) = \frac{1}{6\pi G(\omega) R}, \quad (2.15)$$

where R is the probe bead radius. This procedure is explained in detail in Schnurr et al. (1997).

The shear modulus can also be derived from position fluctuation data in a different way. After first calculating the mean square displacement as a function of time, one can obtain (using equipartition) a viscoelastic memory function by Laplace transformation, The shear modulus follows by again using the generalized Stokes law (Mason et al., 1997).

Two-particle methods

Large discrepancies between macroscopic viscoelastic parameters and those determined by one-particle microrheology can arise if the presence of the probe particle itself influences the viscoelastic medium in its vicinity or if active particle movement occurs and is interpreted as thermal motion. The shear strain field coupled to the motion of a probe particle extends into the medium a distance that is similar to the particle radius. Any perturbation of the medium caused by the presence of the particle will decay over a distance that is the shorter of the particle radius or characteristic length scales in the medium itself, such as mesh size of a network or persistence length of a polymer. Thus it follows that if any characteristic length scales in the system exceed the probe size, the simple interpretation of data with the generalized Stokes law is not valid. This is probably always the case when micron-sized beads are used to study the cytoskeleton, because the persistence length of actin is already 17 μm (Howard, 2001), while that of actin bundles or microtubules is much larger still. A perturbation of the medium could be caused by a chemical interaction with the probe surface, which can be prevented by appropriate surface coating. It is unavoidable, though, that the probe bead locally dilutes the medium by entropic depletion. To circumvent these pitfalls, two-particle microrheology has been developed (Crocker et al., 2000; Levine and Lubensky, 2000). In this variant, the cross-correlation of the displacement fluctuations of two particles, located at a given distance from each other, is measured (Fig. 2-12). The distance between the probes takes over as relevant length scale and probe size or shape become of secondary importance.

Instead of the one-particle compliance, a mutual compliance is defined by:

$$x_\omega^{mi} = \alpha_{ij}^{mn}(\omega) f_\omega^{nj}, \quad (2.16)$$

with particle indices n, m and coordinate indices $i, j = x, y$. If the particles are separated by a distance r along the x-axis, two cases are relevant, namely α_{xx}^{12}, which will be denoted as $\alpha_\parallel^{12}$, and α_{yy}^{12}, which will be denoted as $\alpha_\perp^{12}$ (the other combinations are second order). The Fourier transform of the cross-correlation function is related to the imaginary part of the corresponding compliance:

$$S_{ij}^{12}(\omega) = X_i^1(\omega) X_j^2(\omega)^\dagger = \frac{4 k_B T}{\omega} \alpha_{ij}^{12''}(\omega), \quad (2.17)$$

where † denotes the complex conjugate.

Experimental measurements of intracellular mechanics

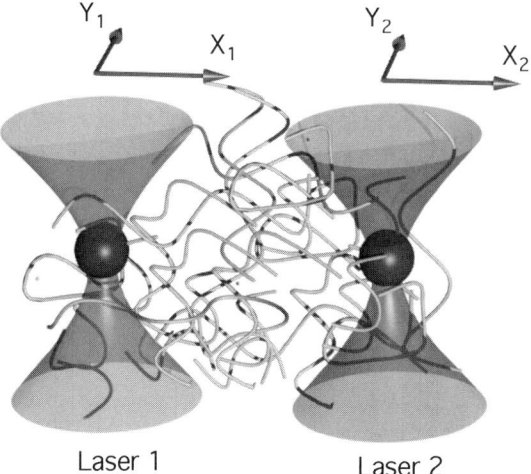

Fig. 2-12. Sketch of 1-particle and 2-particle microrheology using lasers for trapping and detection. Either one laser beam is focused on one particle at a time, or the two beams, displaced by some distance, are each focused on seperate particles. In both cases the motion of the particle in the two directions normal to the laser propagation direction is measured by projecting the laser light onto quadrant photodiodes downstream from the sample.

A Kramers-Kronig integral can again be used to calculate the real parts, and elastic moduli can be derived according to Levine and Lubensky (2002) from:

$$\alpha_{\parallel}^{12}(\omega) = \frac{1}{4\pi r \mu_0(\omega)} \qquad (2.18)$$

$$\alpha_{\perp}^{12}(\omega) = \frac{1}{8\pi r \mu_0(\omega)} \left[\frac{\lambda_0(\omega) + 3\mu_0(\omega)}{\lambda_0(\omega) + 2\mu_0(\omega)} \right], \qquad (2.19)$$

written here (following Levine and Lubensky (2002)) with the Lamé coefficients $\lambda_0(\omega)$ and $\mu_0(\omega)$, where $\mu_0(\omega) = G(\omega)$. One can thus measure directly the compressional modulus and the shear modulus in the sample.

The technique can again be implemented using video recording and particle tracking (Crocker et al., 2000) or laser interferometry. In cells, so far only a video-based variant has been used (Lau et al., 2003), exploring the low-frequency regime of the cellular dynamics in mouse macrophages and mouse carcinoma cells. It was found that the low-frequency passive microrheology results were strongly influenced by active transport in the cells, so that the fluctuation-dissipation theorem could not be used for calculating viscoelastic parameters.

Dynamic light scattering and diffusing wave spectroscopy

A well-established method to study the dynamics of large ensembles of particles in solutions is dynamic light scattering (DLS) (Berne and Pecora, 1990). To obtain smooth data and good statistics, it is obviously advantageous to average over a large number of particles. In DLS, a collimated laser beam is typically sent through a sample of milliliter volume, and scattered light is collected under a well-defined angle with a photomultiplier or other sensitive detector. The intensity autocorrelation function:

$$g_2(\tau) = \frac{\langle I(t)I(t+\tau)\rangle}{\langle I(t)\rangle^2} = 1 + \beta e^{-q^2\langle \Delta r^2(\tau)\rangle/3} \tag{2.20}$$

can be used to calculate the average mean square displacement $\langle \Delta r^2(\tau)\rangle$ of particles, with the scattering vector $q = 4\pi n \sin(\theta/2)/\lambda$, wavelength λ, scattering angle θ and index of the solvent n, and a coherence factor β. This relationship assumes that all particles are identical and that the solution is homogeneous across the scattering volume. It also assumes that the particles dominate the scattering intensity compared to the scattering from the embedding medium itself. DLS has also been used extensively to study polymer solutions without added probe particles (Berne and Pecora, 1990). In that case the medium has to be modeled to extract material properties from the observed intensity autocorrelation function. This has been done for example for pure actin networks as models for the cytoskeleton of cells (Isambert et al., 1995; Liverpool and Maggs, 2001; Schmidt et al., 1989). Unfortunately this technique is not well applicable to study the interior of cells, because the cellular environment is highly inhomogeneous and it is not well defined which structures scatter the light in any given location. Larger objects dominate the scattered intensity (Berne and Pecora, 1990). DLS has been applied to red blood cells (Peetermans et al., 1987a; Peetermans et al., 1987b), but results have been qualitative. It is also difficult to introduce external probe particles that scatter light strongly enough in sufficient concentrations without harming the cells.

A related light-scattering technique is diffusing wave spectroscopy (DWS) (Pine et al., 1988; Weitz et al., 1993), which measures again intensity correlation functions of scattered light, but now in very dense opaque media where light is scattered many times before it is detected so that the path of a photon becomes a random walk and resembles diffusion. There is no more scattering-vector dependence in the field correlation function, which is directly related to the average mean square displacement $\langle \Delta r^2(\tau)\rangle$ of the scattering particles (Weitz and Pine, 1993):

$$g_1(\tau) \propto \int_0^\infty P(s) \exp\left[-\frac{k_0^2 s \langle \Delta r^2(\tau)\rangle}{3l^*}\right] ds. \tag{2.21}$$

$P(s)$ is the probability that the light travels a path length s, $k_0 = 2\pi/\lambda$ is the wave vector, and l^* is the transport mean free path. The final steps to extract a complex shear modulus are the same as described above, either using the power spectral density method (Schnurr et al., 1997) or the Laplace transform method (Mason et al., 1997).

The advantage of the technique is that it is sensitive to very small motions (of less than nm) because the path of an individual photon reflects the sum of the motions of the all particles by which it is scattered. The bandwidth of the technique is also high (typically 10 Hz to 1 MHz), so that ensemble-averaged mean-square displacements, and from that viscoelastic response functions, can be measured over many decades in frequency. The technique has been used to study polymer solutions, colloidal systems, and cytoskeletal protein solutions (actin) (Mason et al., 1997; Mason et al., 2000) and

Experimental measurements of intracellular mechanics

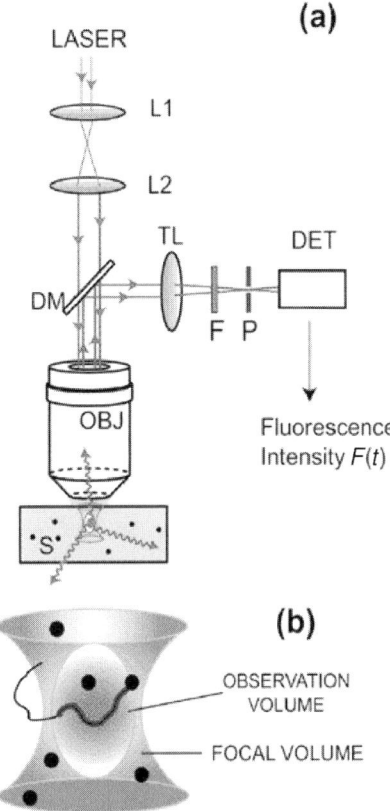

Fig. 2-13. (a) Experimental set-up for fluorescence correlation spectroscopy. A laser beam is expanded (L1, L2) and focussed through a microscope objective into a fluorescent sample. The fluorescence light is collected through the same objective and split out with a dichroic mirror toward the confocal pinhole (P) and then the detector. (b) Magnified focal volume with the fluorescent particles (spheres) and the diffusive path of one particle highlighted. From Hess et al., 2002.

results agree with those from conventional methods in the time/frequency regimes where they overlap. The application to cells is hindered, just as in the case of DLS, by the inhomogeneity of the cellular environment. Furthermore, typical cells are more or less transparent, in other words one would need to introduce high concentrations of scattering particles, which likely would disturb the cell's integrity.

Fluorescence correlation spectroscopy

Many complications can be avoided if specific particles or molecules of interest in a cell can be selected from other structures. A way to avoid collecting signals from unknown cellular structures is to use fluorescent labeling of particular molecules or structures within the cell. This method is extensively used in cell biology to study the localization of certain proteins in the cell. Fluorescence can also be used to measure dynamic processes in video microscopy, but due to low emission intensities of fluorescent molecules and due to their fast Brownian motion when they are not fixed to larger structures, it is difficult to use such data to extract diffusion coefficients or viscoelastic parameters inside cells. A method that is related to dynamic light scattering and is a nonimaging method that can reach much faster time scales is fluorescence correlation spectroscopy (Hess et al., 2002; Webb, 2001), where a laser is focused to a small volume and the fluctuating fluorescence originating from molecules entering and leaving this volume is recorded with fast detectors (Fig. 2-13).

From the fluorescence intensity fluctuations $\delta F(t) = F(t) - \langle F(t) \rangle$ one calculates the normalized autocorrelation function:

$$G(\tau) = \frac{\langle \delta F(t) \delta F(t+\tau) \rangle}{\langle F(t) \rangle^2}, \quad (2.22)$$

from which one can calculate in the simplest case, in the absence of chemical reactions involving the fluorescent species, the characteristic time τ_D a diffusing molecule spends in the focal volume:

$$G_D(\tau) = \frac{1}{N(1 + \tau/\tau_D)(1 + \tau/\omega^2 \tau_D)^{1/2}} \quad (2.23)$$

with an axial-to-lateral-dimension ratio ω and the mean number of fluorescent molecules in the focus N. Eq. 2.23 is valid for a molecule diffusing in 3D. The method can also be used in other cases, for example 2D diffusion in a membrane.

With some knowledge of the geometry of the situation – for example 2D membrane-bound diffusion – one can again extract diffusion coefficients. This has been done on cell surfaces and even inside cells (see references in Hess et al., (2002)). Data typically have been interpreted as diffusion in a purely viscous environment or as diffusion in an inhomogeneous environment with obstacles. Compared with the rheology methods described above, fluorescence correlation spectroscopy is particularly good when studying small particles such as single-enzyme molecules. For the motion of such particles it is not appropriate to model the environment inside a cell as a viscoelastic continuum, because characteristic length scales of the cytoskeleton are as large or larger than the particles.

Optical stretcher

A novel optical method related to optical traps employs two opposing nonfocused laser beams to both immobilize and stretch a suspended cell (Guck et al., 2001; Guck et al., 2000). Viscoelastic properties are determined from the time-dependent change in cell dimensions as a function of optical pressures. This method has the significant advantage over other optical trapping methods that it can be scaled up and automated to allow measurement, and potentially sorting, of many cells within a complex mixture for use in diagnosing abnormal cells and sorting cells on the basis of their rigidity (Lincoln et al., 2004).

Acoustic microscopy

Ultrasound transmission and attenuation through cells and biological tissues can also provide measurements of viscoelasticity, and acoustic microscopy has the potential to provide high-resolution imaging of live cells in a minimally invasive manner (Viola and Walker, 2003). Studies of purified systems such as F-actin (Wagner et al., 2001; Wagner et al., 1999), and alginate capsules (Klemenz et al., 2003), suggest that acoustic signals can be related to changes in material properties of these biopolymer gels, but there are numerous challenges related to interpreting the data and relating them to viscoelastic parameters before the potential of this method for quantitative high-resolution elastic imaging on cells is realized.

Outstanding issues and future directions

The survey of methods used to study the rheology of cells presented here shows the wide range of methods that various groups have designed and employed. At present there appears to be no one ideal method suitable for most cell types. In many cases, measurements of similar cell types by different methods have yielded highly different values for elastic and viscous parameters. For example, micropipette aspiration of leukocytes can variably be interpreted as showing that these cells are liquid droplets with a cortical tension or soft viscoelastic fluids, while atomic force microscopy measures elastic moduli on the order of 1000 Pa. In part, differences in measurements stem from differences in the time scale or frequency and in the strains at which the measurements are done. Also, it is almost certain that cells respond actively to the forces needed to measure their rheology, and the material properties of the cell often cannot be interpreted as those of passive material. Combining rheological measurements with simultaneous monitoring or manipulation of intracellular signals and cytoskeletal structures can go a long way toward resolving such challenges.

Currently a different and equally serious challenge is presented by the finding that even when studying purified systems like F-actin networks, micro- and macrorheology methods sometimes give very different results, for reasons that are not completely clear. In part there are likely to be methodological problems that need to be resolved, but it also appears that there are interesting physical differences in probing very small displacements of parts of a network not much larger than the network mesh size and the macroscopic deformations that occur as the whole network deforms in macrorheologic measurements. Here a combination of more experimentation and new theories is likely to be important.

The physical properties of cells have been of great interest to biologists and physiologists from the earliest studies that suggested that cells may be able to convert from solid to liquid states as they move or perform other functions. More recently, unraveling the immense complexity of the molecular biology regulating cell biology and high-resolution imaging of intracellular structures have provided molecular models to suggest how the dynamic viscoelasticity of the cell may be achieved. Now the renewed interest in cell mechanics together with technological advances allowing unprecedented precision and sensitivity in force application and imaging can combine with molecular information to increase our understanding of the mechanisms by which cells maintain and change their mechanical properties.

References

Addas KM, Schmidt CF, Tang JX. (2004). "Microrheology of solutions of semiflexible biopolymer filaments using laser tweezers interferometry." *Phys. Rev. E.*, **70**(2):Art. No. 021503.

Agayan RR, Gittes F, Kopelman R, Schmidt CF. (2002). "Optical trapping near resonance absorption." *Appl. Opt.*, **41**(12):2318–27.

Ashkin A. (1992). "Forces of a single-beam gradient laser trap on a dielectric sphere in the ray optics regime." *Biophys. J.*, **61**(2):569–82.

Ashkin A. (1997). "Optical trapping and manipulation of neutral particles using lasers." *Proc. Natl. Acad. Sci. USA*, **94**(10):4853–60.

Ashkin A, Dziedzic JM, Bjorkholm JE, Chu S. (1986). "Observation of a single-beam gradient force optical trap for dielectric particles." *Opt. Lett.*, **11**(5):288–90.

Bacabac RG, Smit TH, Heethaar RM, van Loon JJWA, Pourquie MJMB, Nieuwstadt FTM, Klein-Nulend J. (2002). "Characteristics of the parallel-plate flow chamber for mechanical stimulation of bone cells under microgravity." *J. Gravitat. Physiol.*, **9**:P181–2.

Bacabac RG, Smit TH, Mullender MG, Dijcks SJ, Van Loon J, Klein-Nulend J. (2004). "Nitric oxide production by bone cells is fluid shear stress rate dependent." *Biochem. Biophys. Res. Commun.*, **315**(4):823–9.

Barbee KA, Mundel T, Lal R, Davies PF. (1995). "Subcellular distribution of shear stress at the surface of flow-aligned and nonaligned endothelial monolayers." *Am. J. Physiol.*, **268**(4 Pt 2): H1765–72.

Bausch AR, Hellerer U, Essler M, Aepfelbacher M, Sackmann E. (2001). "Rapid stiffening of integrin receptor-actin linkages in endothelial cells stimulated with thrombin: A magnetic bead microrheology study." *Biophys. J.*, **80**(6):2649 57.

Bausch AR, Moller W, Sackmann E. (1999). "Measurement of local viscoelasticity and forces in living cells by magnetic tweezers." *Biophys. J.*, **76**(1):573–9.

Bausch AR, Ziemann F, Boulbitch AA, Jacobson K, Sackmann E. (1998). "Local measurements of viscoelastic parameters of adherent cell surfaces by magnetic bead microrheometry." *Biophys. J.*, **75**(4):2038–49.

Berne BJ, Pecora R. 1990. *Dynamic light scattering*. Malabar: Robert E. Krieger Publishing Co.

Binnig G, Quate CF, Gerber C. (1986). "Atomic Force Microscope." *Phys., Rev., Lett.*, **56**(9):930–3.

Bray D. (1984). "Axonal growth in response to experimentally applied mechanical tension." *Dev. Biol.*, **102**(2):379–89.

Burger EH, Klein-Nulend J. (1999). "Mechanotransduction in bone – role of the lacuno-canalicular network." *Faseb. J.*, **13**:S101–S112.

Caille N, Thoumine O, Tardy Y, Meister JJ. (2002). "Contribution of the nucleus to the mechanical properties of endothelial cells." *J. Biomech.*, **35**(2):177–87.

Charras GT, Horton MA. (2002). "Determination of cellular strains by combined atomic force microscopy and finite element modeling." *Biophys. J.*, **83**(2):858–79.

Chien S, Schmid-Schonbein GW, Sung KL, Schmalzer EA, Skalak R. (1984). "Viscoelastic properties of leukocytes." *Kroc. Found Ser.*, **16**:19–51.

Crick F, Hughes A. (1950). "The physical properties of cytoplasm." *Exp. Cell Res.*, 1:37–80.

Crocker JC, Grier DG. (1996). "Methods of digital video microscopy for colloidal studies." *J. Colloid. Interface Sci.*, **179**(1):298–310.

Crocker JC, Valentine MT, Weeks ER, Gisler T, Kaplan PD, Yodh AG, Weitz DA. (2000). "Two-point microrheology of inhomogeneous soft materials." *Phys. Rev. Lett.*, **85**(4):888–91.

Cunningham CC, Gorlin JB, Kwiatkowski DJ, Hartwig JH, Janmey PA, Byers HR, Stossel TP. (1992). "Actin-binding protein requirement for cortical stability and efficient locomotion." *Science*, **255**(5042):325–7.

Daily B, Elson EL, Zahalak GI. (1984). "Cell poking. Determination of the elastic area compressibility modulus of the erythrocyte membrane." *Biophys. J.*, **45**(4):671–82.

Dao M, Lim CT, Suresh S. (2003). "Mechanics of the human red blood cell deformed by optical tweezers." *J. Mechan. Phys. Solids*, **51**(11–12):2259–80.

Davies PF, Remuzzi A, Gordon EJ, Dewey CF, Jr., Gimbrone MA, Jr. (1986). "Turbulent fluid shear stress induces vascular endothelial cell turnover in vitro." *Proc. Natl. Acad. Sci. USA*, **83**(7): 2114–17.

de Pablo PJ, Colchero J, Gomez-Herrero J, Baro AM. (1998). "Jumping mode scanning force microscopy." *Appl. Phys. Lett.*, **73**(22):3300–2.

Denk W, Webb WW. (1990). "Optical measurement of picometer displacements of transparent microscopic objects." *Appl. Opt.*, **29**(16):2382–91.

Desprat N, Richert A, Simeon J, Asnacios A. (2005). "Creep function of a single living cell." *Biophys. J.*, **88**(3):2224–33.

Dewey CF, Jr., Bussolari SR, Gimbrone MA, Jr., Davies PF. (1981). "The dynamic response of vascular endothelial cells to fluid shear stress." *J. Biomech. Eng.*, **103**(3):177–85.

Discher DE, Mohandas N, Evans EA. (1994). "Molecular maps of red cell deformation: hidden elasticity and in situ connectivity." *Science*, **266**(5187):1032–5.

Dong C, Skalak R, Sung KL, Schmid SG, Chien S. (1988). "Passive deformation analysis of human leukocytes." *J. Biomech. Eng.*, **110**(1):27–36.

Dufrene YF. (2003). "Recent progress in the application of atomic force microscopy imaging and force spectroscopy to microbiology." *Curr. Op. Microbiol.*, **6**(3):317–23.

Dvorak JA, Nagao E. (1998). "Kinetic analysis of the mitotic cycle of living vertebrate cells by atomic force microscopy" *Exp. Cell Res.*, **242**:69–74.

Ehrlich PJ, Lanyon LE. (2002). "Mechanical strain and bone cell function: a review." *Osteoporos. Int.*, **13**(9):688–700.

Eichinger L, Koppel B, Noegel AA, Schleicher M, Schliwa M, Weijer K, Witke W, Janmey PA. (1996). "Mechanical perturbation elicits a phenotypic difference between Dictyostelium wild-type cells and cytoskeletal mutants." *Biophys. J.*, **70**(2):1054–60.

Evans E, Yeung A. (1989). "Apparent viscosity and cortical tension of blood granulocytes determined by micropipet aspiration." *Biophys. J.*, **56**(1):151–60.

Evans EA, Hochmuth RM. (1976). "Membrane viscoelasticity." *Biophys. J.*, **16**(1):1–11.

Fabry B, Maksym GN, Butler JP, Glogauer M, Navajas D, Fredberg JJ. (2001). "Scaling the microrheology of living cells." *Phys. Rev. Lett.*, **8714**(14):Art. No.148102.

Fabry B, Maksym GN, Hubmayr RD, Butler JP, Fredberg JJ. (1999). "Implications of heterogeneous bead behavior on cell mechanical properties measured with magnetic twisting cytometry." *J. Magn. and Magn. Mater.*, **194**(1–3):120–5.

Feneberg W, Aepfelbacher M, Sackmann E. (2004). "Microviscoelasticity of the apical cell surface of human umbilical vein endothelial cells (HUVEC) within confluent monolayers." *Biophys. J.*, **87**(2):1338–50.

Feneberg W, Westphal M, Sackmann E. (2001). "Dictyostelium cells' cytoplasm as an active viscoplastic body." *Eur. Biophys. J.*, **30**(4):284–94.

Florian JA, Kosky JR, Ainslie K, Pang ZY, Dull RO, Tarbell JM. (2003). "Heparan sulfate proteoglycan is a mechanosensor on endothelial cells." *Circul. Res.*, **93**(10):E136–42.

Freundlich H, Seifriz W. (1922). "Über die Elastizität von Sollen und Gelen." *Z. Phys. Chem.*, **104**:233–61.

Fukuda S, Schmid-Schonbein GW. (2003). "Regulation of CD18 expression on neutrophils in response to fluid shear stress." *Proc. Natl. Acad. Sci. USA*, **100**(23):13152–7.

Gittes F, Schmidt CF. (1998). "Interference model for back-focal-plane displacement detection in optical tweezers." *Opt. Lett.*, **23**(1):7–9.

Glogauer M, Arora P, Yao G, Sokholov I, Ferrier J, McCulloch CA. (1997). "Calcium ions and tyrosine phosphorylation interact coordinately with actin to regulate cytoprotective responses to stretching." *J. Cell Sci.*, **110** (Pt 1):11–21.

Grodzinsky AJ, Levenston ME, Jin M, Frank EH. (2000). "Cartilage tissue remodeling in response to mechanical forces." *Annu. Rev. Biomed. Eng.*, **2**:691–713.

Guck J, Ananthakrishnan R, Mahmood H, Moon TJ, Cunningham CC, Kas J. (2001). "The optical stretcher: a novel laser tool to micromanipulate cells." *Biophys. J.*, **81**(2):767–84.

Guck J, Ananthakrishnan R, Moon TJ, Cunningham CC, Kas J. (2000). "Optical deformability of soft biological dielectrics." *Phys. Rev. Lett.*, **84**(23):5451–4.

Hansma PK, Cleveland JP, Radmacher M, Walters DA, Hillner PE, Bezanilla M, Fritz M, Vie D, Hansma HG, Prater CB et al. (1994). "Tapping Mode Atomic-Force Microscopy in Liquids." *Appl. Phys. Lett.*, **64**(13):1738–40.

Harada Y, Asakura T. (1996). "Radiation forces on a dielectric sphere in the Rayleigh scattering regime." *Opt. Commun.*, **124**(5–6):529–41.

Heidemann SR, Buxbaum RE. (1994). "Mechanical tension as a regulator of axonal development." *Neurotoxicology*, **15**(1):95–107.

Heidemann SR, Kaech S, Buxbaum RE, Matus A. (1999). "Direct observations of the mechanical behaviors of the cytoskeleton in living fibroblasts." *J. Cell Biol.*, **145**(1):109–22.

Heilbronn A. (1922). "Eine Neue Methode zur Bestimmung der Viskosität lebender Protoplasten" *Jahrb. Wiss. Botan*, **61**:284–338.

Heilbrunn L. 1952. *An outline of general physiology*. Philadelphia: WB Saunders. 818 p.

Heilbrunn L. 1956. *The dynamics of living protoplasm*. New York: Academic Press. 327 p.

Helmke BP, Rosen AB, Davies PF. (2003). "Mapping mechanical strain of an endogenous cytoskeletal network in living endothelial cells." *Biophys. J.*, **84**(4):2691–9.

Helmke BP, Thakker DB, Goldman RD, Davies PF. (2001). "Spatiotemporal analysis of flow-induced intermediate filament displacement in living endothelial cells." *Biophys. J.*, **80**(1):184–94.

Henon S, Lenormand G, Richert A, Gallet F. (1999). "A new determination of the shear modulus of the human erythrocyte membrane using optical tweezers." *Biophys. J.*, **76**(2):1145–51.

Hertz H. (1882). "Ueber den Kontakt elastischer Koerper." *J. Reine. Angew. Mathematik.* **92**:156–71.

Hess ST, Huang SH, Heikal AA, Webb WW. (2002). "Biological and chemical applications of fluorescence correlation spectroscopy: A review." *Biochem.*, **41**(3):697–705.

Hochmuth RM. (2000). "Micropipette aspiration of living cells." *J. Biomech.*, **33**(1):15–22.

Howard J. 2001. *Mechanics of motor proteins and the cytoskeleton*. Sunderland: Sinauer Associates.

Hu SH, Chen JX, Fabry B, Numaguchi Y, Gouldstone A, Ingber DE, Fredberg JJ, Butler JP, Wang N. (2003). "Intracellular stress tomography reveals stress focusing and structural anisotropy in cytoskeleton of living cells." *A. J. Physiol. -Cell Physiol.*, **285**(5):C1082–90.

Isambert H, Venier P, Maggs AC, Fattoum A, Kassab R, Pantaloni D, Carlier MF. (1995). "Flexibility of actin filaments derived from thermal fluctuations. Effect of bound nucleotide, phalloidin, and muscle regulatory proteins." *J. Biol. Chem.*, **270**(19):11437–44.

Jackson JD. 1975. *Class electrodyn*. New York: Wiley.

Janmey PA, Weitz DA. (2004). "Dealing with mechanics: mechanisms of force transduction in cells." *Trends Biochem. Sci.*, **29**(7):364–70.

Jones JD, LubyPhelps K. (1996). "Tracer diffusion through F-actin: Effect of filament length and cross-linking." *Biophys. J.*, **71**(5):2742–50.

Jones JD, Ragsdale GK, Rozelle A, Yin HL, Luby-Phelps K. (1997). "Diffusion of vesicle-sized particles in living cells is restricted by intermediate filaments." *Mol. Biol. Cell*, **8**:1006.

Keller M, Schilling J, Sackmann E. (2001). "Oscillatory magnetic bead rheometer for complex fluid microrheometry" *Rev. Sci. Instr.*, **72**(9):3626–34.

Kellermayer MS, Smith SB, Granzier HL, Bustamante C. (1997). "Folding-unfolding transitions in single titin molecules characterized with laser tweezers." *Science*, **276**(5315):1112–16.

Klein-Nulend J, Bacabac RG, Veldhuijzen JP, Van Loon JJWA. (2003). "Microgravity and bone cell mechanosensitivity" *Adv. Space Res.*, 32:1551–9.

Klemenz A, Schwinger C, Brandt J, Kressler J. (2003). "Investigation of elasto-mechanical properties of alginate microcapsules by scanning acoustic microscopy." *J. Biomed. Mater. Res.*, **65A**(2):237–43.

Landau LD, Lifshitz EM. 1970. *Theory of elasticity*. New York: Pergamon Press.

Landau LD, Lifshitz EM, Pitaevskii LP. 1980. *Statist. phys*. Oxford, New York: Pergamon Press.

Lau AWC, Hoffman BD, Davies A, Crocker JC, Lubensky TC. (2003). "Microrheology, Stress Fluctuations and Active Behavior of Living Cells." *Phys. Rev. Lett.*, **91**:198101.

Lenormand G, Henon S, Richert A, Simeon J, Gallet F. (2001). "Direct measurement of the area expansion and shear moduli of the human red blood cell membrane skeleton." *Biophys. J.*, **81**(1):43–56.

Levine AJ, Lubensky TC. (2000). "One- and two-particle microrheology." *Phys. Rev. Lett.*, **85**(8):1774–7.

Levine AJ, Lubensky TC. (2001). "The Response Function of a Sphere in a Viscoelastic Two-Fluid Medium." *Phys. Rev. E.*, **63**:0415101–12.

Levine AJ, Lubensky TC. (2002). "Two-point microrheology and the electrostatic analogy." *Phys. Rev. E. Stat. Nonlin. Soft. Matter. Phys.*, **65**:011501.

Lim CT, Dao M, Suresh S, Sow CH, Chew KT. (2004). "Large deformation of living cells using laser traps." *Acta. Mater.*, **52**(7):1837–45.

Lincoln B, Erickson HM, Schinkinger S, Wottawah F, Mitchell D, Ulvick S, Bilby C, Guck J. (2004). "Deformability-based flow cytometry." *Cytometry*, **59A**(2):203–9.

Liverpool TB, Maggs AC. (2001). "Dynamic scattering from semiflexible polymers." *Macromol*, **34**(17):6064–73.

Luby-Phelps K. (1994). "Physical-Properties of Cytoplasm." *Curr. Opin. Cell Biol.*, **6**(1):3–9.

MacKintosh FC, Schmidt CF. (1999). "Microrheology." *Curr. Opin. Colloid. & Inter. Sci.*, **4**(4): 300–7.

Mahaffy RE, Park S, Gerde E, Kas J, Shih CK. (2004). "Quantitative analysis of the viscoelastic properties of thin regions of fibroblasts using atomic force microscopy." *Biophys. J.*, **86**(3): 1777–93.

Mahaffy RE, Shih CK, MacKintosh FC, Kas J. (2000). "Scanning probe-based frequency-dependent microrheology of polymer gels and biological cells." *Phys. Rev. Lett.*, **85**(4):880–3.

Maksym GN, Fabry B, Butler JP, Navajas D, Tschumperlin DJ, Laporte JD, Fredberg JJ. (2000). "Mechanical properties of cultured human airway smooth muscle cells from 0.05 to 0.4 Hz." *J. Appl. Phys.*, **89**(4):1619–32.

Mason TG, Gang H, Weitz DA. (1997). "Diffusing-wave-spectroscopy measurements of viscoelasticity of complex fluids." *J. Opt. Soc. Am. A. Opt. & Image Sci.*, **14**(1):139–49.

Mason TG, Gisler T, Kroy K, Frey E, Weitz DA. (2000). "Rheology of F-actin solutions determined from thermally driven tracer motion." *J. Rheol.*, **44**(4):917–28.

Mijailovich SM, Kojic M, Zivkovic M, Fabry B, Fredberg JJ. (2002). "A finite element model of cell deformation during magnetic bead twisting." *J. Appl. Phys.*, **93**(4):1429–36.

Mitchison J, Swann M. (1954). "The mechanical properties of the cell surface. I: The cell elastimeter." *J. Exp. Biol.*, **31**:445–59.

Ohura N, Yamamoto K, Ichioka S, Sokabe T, Nakatsuka H, Baba A, Shibata M, Nakatsuka T, Harii K, Wada, et al. (2003). "Global analysis of shear stress-responsive genes in vascular endothelial cells." *J. Atheroscler. Thromb.*, **10**(5):304–13.

Peetermans JA, Matthews EK, Nishio I, Tanaka T. (1987a). "Particle Motion in Single Acinar-Cells Observed by Microscope Laser-Light Scattering Spectroscopy." *Eur. Biophys. J.*, **15**(2): 65–9.

Peetermans JA, Nishio I, Ohnishi ST, Tanaka T. (1987b). "Single Cell Laser-Light Scattering Spectroscopy in a Flow Cell – Repeated Sickling of Sickle Red-Blood-Cells." *Biochim. Biophys. Acta.*, **931**(3):320–5.

Peterman EJ, Gittes F, Schmidt CF. (2003a). "Laser-induced heating in optical traps." *Biophys. J.*, **84**(2 Pt 1):1308–16.

Peterman EJG, Dijk MAv, Kapitein LC, Schmidt CF. (2003b). "Extending the Bandwidth of Optical-Tweezers Interferometry." *Rev. Sci. Instrum.*, **74**(7):3246–9.

Petersen NO, McConnaughey WB, Elson EL. (1982). "Dependence of locally measured cellular deformability on position on the cell, temperature, and cytochalasin B." *Proc. Natl. Acad. Sci. USA*, **79**(17):5327–31.

Pine DJ, Weitz DA, Chaikin PM, Herbolzheimer E. (1988). "Diffusing Wave Spectroscopy." *Phys. Rev. Lett.*, **60**:1134.

Pralle A, Prummer M, Florin EL, Stelzer EHK, Horber JKH. (1999). "Three-dimensional high-resolution particle tracking for optical tweezers by forward scattered light." *Micros. Res. Tech.*, **44**(5):378–86.

Puig-de-Morales M, Grabulosa M, Alcaraz J, Mullol J, Maksym GN, Fredberg JJ, Navajas D. (2001). "Measurement of cell microrheology by magnetic twisting cytometry with frequency domain demodulation." *J. Appl. Phys.*, **91**(3):1152–59.

Putman CAJ, Vanderwerf KO, Degrooth BG, Vanhulst NF, Greve J. (1994). "Viscoelasticity of Living Cells Allows High-Resolution Imaging by Tapping Mode Atomic-Force Microscopy." *Biophys. J.*, **67**(4):1749–53.

Reiser P. (1949). "The protoplasmic viscosity of muscle." *Protoplasma*, **39**:95–8.

Richelme F, Benoliel AM, Bongrand P. (2000). "Dynamic study of cell mechanical and structural responses to rapid changes of calcium level." *Cell Motil. Cytoskel.*, **45**(2):93–105.

Rockwell MA, Fechheimer M, Taylor DL. (1984). "A comparison of methods used to characterize gelation of actin in vitro." *Cell Motil.*, **4**(3):197–213.

Rubin CT, Lanyon LE. (1984). "Regulation of Bone-Formation by Applied Dynamic Loads." *J. Bone. Joint Surgery-Am. Vol.*, **66A**(3):397–402.

Sato M, Wong TZ, Brown DT, Allen RD. (1984). "Rheological properties of living cytoplasm: A preliminary investigation of squid axoplasm (Loligo pealei)." *Cell Motil.*, **4**(1):7–23.

Schmidt CF, Barmann M, Isenberg G, Sackmann E. (1989). "Chain Dynamics, Mesh Size, and Diffusive Transport in Networks of Polymerized Actin – A Quasielastic Light Scattering and Microfluorescence Study." *Macromolecules.*, **22**(9):3638–49.

Schmidt FG, Hinner B, Sackmann E, Tang JX. (2000). "Viscoelastic Properties of Semiflexible Filamentous Bacteriophage fd." *Phys. Rev. E.*, **62**(4):5509–17.

Schnurr B, Gittes F, MacKintosh FC, Schmidt CF. (1997). "Determining microscopic viscoelasticity in flexible and semiflexible polymer networks from thermal fluctuations." *Macromol.*, **30**:7781–92.

Sleep J, Wilson D, Simmons R, Gratzer W. (1999). "Elasticity of the red cell membrane and its relation to hemolytic disorders: An optical tweezers study." *Biophys. J.*, **77**(6):3085–95.

Storm C, Pastore JJ, MacKintosh FC, Lubensky TC, Janmey PA. (2005). "Nonlinear Elasticity in Biological Gels." *Nature*:in press.

Stossel T. (1990). "How Cells Crawl." *Am. Scientist*, **78**:408–23.

Svoboda K, Block SM. (1994). "Biological applications of optical forces." *Annu. Rev. Biophys. Biomol. Struct.*, **23**:247–85.

Thoumine O, Ott A. (1997). "Time scale dependent viscoelastic and contractile regimes in fibroblasts probed by microplate manipulation." *J. Cell Sci.*, **110** (Pt 17):2109–16.

Thoumine O, Ott A, Cardoso O, Meister JJ. (1999). "Microplates: A new tool for manipulation and mechanical perturbation of individual cells." *J. Biochem. Biophys. Methods*, **39**(1–2):47–62.

Trepat X, Grabulosa M, Buscemi L, Rico F, Fabry B, Fredberg JJ, Farre R. (2003). "Oscillatory magnetic tweezers based on ferromagnetic beads and simple coaxial coils." *Rev. Sci. Instr.*, **74**(9):4012–20.

Turner CH, Owan I, Takano Y. (1995). "Mechanotransduction in Bone – Role of Strain-Rate." *Am. J. Physiology-Endocrinol and Metab*, **32**(3):E438–42.

Valberg PA, Butler JP. (1987). "Magnetic Particle Motions within Living Cells – Physical Theory and Techniques." *Biophys. J.*, **52**(4):537–50.

Valberg PA, Feldman HA. (1987). "Magnetic Particle Motions within Living Cells – Measurement of Cytoplasmic Viscosity and Motile Activity." *Biophys. J.*, **52**(4):551–61.

Valentine MT, Kaplan PD, Thota D, Crocker JC, Gisler T, Prud'homme RK, Beck M, Weitz DA. (2001). "Investigating the microenvironments of inhomogeneous soft materials with multiple particle tracking." *Phys. Rev. E.*, **6406**(6):Art. No. 061506.

Viola F, Walker WF. (2003). "Radiation force imaging of viscoelastic properties with reduced artifacts." *IEEE Trans. Ultrason. Ferroelectr. Freq. Control*, **50**(6):736–42.

Vonna L, Wiedemann A, Aepfelbacher M, Sackmann E. (2003). "Local force induced conical protrusions of phagocytic cells." *J. Cell Sci.*, **116**(5):785–90.

Wagner O, Schuler H, Hofmann P, Langer D, Dancker P, Bereiter-Hahn J. (2001). "Sound attenuation of polymerizing actin reflects supramolecular structures: Viscoelastic properties of actin gels modified by cytochalasin D, profilin and alpha-actinin." *Biochem. J.*, **355**(Pt 3):771–8.

Wagner O, Zinke J, Dancker P, Grill W, Bereiter-Hahn J. (1999). "Viscoelastic properties of F-actin, microtubules, F-actin/alpha-actinin, and F-actin/hexokinase determined in microliter volumes with a novel nondestructive method." *Biophys. J.*, **76**(5):2784–96.

Wang N, Butler JP, Ingber DE. (1993). "Mechanotransduction across the Cell-Surface and through the Cytoskeleton." *Science*, **260**(5111):1124–27.

Waters CM, Sporn PH, Liu M, Fredberg JJ. (2002). "Cellular biomechanics in the lung." *Am. J. Physiol. Lung. Cell Mol. Physiol.*, **283**(3):L503–9.

Webb WW. (2001). "Fluorescence correlation spectroscopy: Inception, biophysical experimentations, and prospectus." *Appl. Opt.*, **40**(24):3969–83.

Weinbaum S, Zhang XB, Han YF, Vink H, Cowin SC. (2003). "Mechanotransduction and flow across the endothelial glycocalyx." *Proc. Natl. Acad. Sci. USA*, **100**(13):7988–95.

Weitz DA, Pine DJ. (1993). Diffusing-wave spectroscopy. In: Brown W, editor. *Dynamic light scattering*. Oxford: Oxford University Press.

Weitz DA, Zhu JX, Durian DJ, Gang H, Pine DJ. (1993). "Diffusing-Wave Spectroscopy – the Technique and Some Applications." *Physica. Scripta*, **T49B**:610–21.

Wolff J. 1986. *The law of bone remodelling*. Berlin: Springer.

Yagi K. (1961). "Mechanical and Colloidal Properties of Amoeba Protoplasm and Their Relations to Mechanism of Amoeboid Movement." *Comp. Biochem. Physiol.*, **3**(2):73–&.

3 The cytoskeleton as a soft glassy material

Jeffrey Fredberg and Ben Fabry

ABSTRACT: Using a novel method that was both quantitative and reproducible, Francis Crick and Arthur Hughes (Crick and Hughes, 1950) were the first to measure the mechanical properties inside single, living cells. They concluded their groundbreaking work with the words: "If we were compelled to suggest a model (of cell mechanics) we would propose Mother's Work Basket – a jumble of beads and buttons of all shapes and sizes, with pins and threads for good measure, all jostling about and held together by colloidal forces."

Thanks to advances in biochemistry and biophysics, we can now name and to a large degree characterize many of the beads and buttons, pins, and threads. These are the scores of cytoskeletal proteins, motor proteins, and their regulatory molecules. But the traditional reductionist approach – to study one molecule at a time in isolation – has so far not led to a comprehensive understanding of how cells are able to perform such exceptionally complex mechanical feats as division, locomotion, contraction, spreading, or remodeling. The question then arises, even if all of the cytoskeletal and signaling molecules were known and fully characterized, would this information be sufficient to understand how the cell orchestrates complex and highly specific mechanical functions? Or put another way, do molecular events playing out at the nanometer scale necessarily add up in a straightforward manner to account for mechanical events at the micrometer scale?

We argue here that the answer to these questions may be 'No.' We present a point of view that does not rely on a detailed knowledge of specific molecular functions and interactions, but instead focuses attention on dynamics of the microstructural arrangements between cytoskeletal proteins. Our thinking has been guided by recent advances in the physics of soft glassy materials. One of the more surprising findings that has come out of this approach is the discovery that, independent of molecular details, a single, measurable quantity (called the 'noise temperature') seems to account for transitions between fluid-like and solid-like states of the cytoskeleton. Although the interpretation and precise meaning of this noise temperature is still emerging, it appears to give a measure of the 'jostling' and the 'colloidal forces' that act within the cytoskeleton.

Introduction

Measurements of mechanical properties afford a unique window into the dynamics of protein-protein interactions within the cell, with elastic energy storage reflecting numbers of molecular interactions, energy dissipation reflecting their rate of turnover, and

The cytoskeleton as a soft glassy material

remodeling events reflecting their spatio-temporal reorganization (Fredberg, Jones et al., 1996). As discussed in Chapter 2, probes are now available that can measure each of these features with temporal resolution in the range of milliseconds and spatial resolution in the range of nanometers. Using such probes, this chapter demonstrates that a variety of phenomena that have been taken as the signature of condensed systems in the glassy state are prominently expressed by the cytoskeletal lattice of the living adherent cell. While highly specific interactions play out on the molecular scale, and homogeneous behavior results on the integrative scale, evidence points to metastability of interactions and nonequilibrium cooperative transitions on the mesoscale as being central factors linking integrative cellular function to underlying molecular events. Insofar as such fundamental functions of the cell – including embryonic development, contraction, wound healing, crawling, metastasis, and invasion – all stem from underlying cytoskeletal dynamics, identification of those dynamics as being glassy would appear to set these functions into an interesting context.

The chapter begins with a brief summary of experimental findings in living cells. These findings are described in terms of a remarkably simple empirical relationship that appears to capture the essence of the data with very few parameters. Finally, we show that this empirical relationship is predicted from the theory of soft glassy rheology (SGR). As such, SGR offers an intriguing perspective on mechanical behavior of the cytoskeleton and its relationship to the dynamics of protein-protein interactions.

Experimental findings in living cells

To study the rheology of cytoskeletal polymers requires a probe whose operative frequency range spans, insofar as possible, the internal molecular time scales of the rate processes in question. The expectation from such measurements is that the rheological behavior changes at characteristic relaxation frequencies, which in turn can be interpreted as the signature of underlying molecular interactions that dominate the response (Hill, 1965; Kawai and Brandt, 1980). Much of what follows in this chapter is an attempt to explain the failure to find such characteristic relaxation times in most cell types. The experimental findings of our laboratory, summarized below, are derived from single cell measurements using magnetic twisting cytometry (MTC) with optical detection of bead motion. Using this method, we were able to apply probing frequencies ranging from 0.01 Hz to 1 kHz. As shown by supporting evidence, these findings are not peculiar to the method; rather they are consistent with those obtained using different methods such as atomic force microscopy (Alcaraz, Buscemi et al., 2003).

Magnetic Twisting Cytometry (MTC)

The MTC device is a microrheometer in which the cell is sheared between a plate at the cell base (the cell culture dish upon which the cell is adherent) and a magnetic microsphere partially embedded into the cell surface, as shown in Fig. 3-1. We use ferrimagnetic microbeads (4.5 μm diameter) that are coated with a panel of antibody and nonantibody ligands that allow them to bind to specific receptors on the cell surface

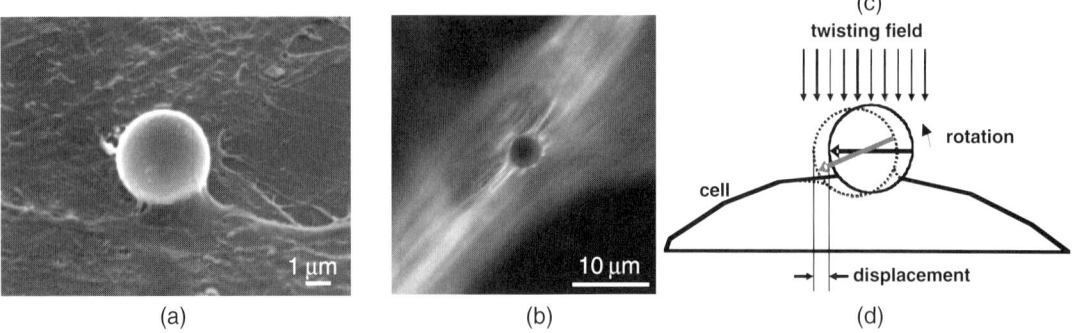

Fig. 3-1. (a) Scanning EM of a bead bound to the surface of a human airway smooth muscle cell. (b) Ferrimagnetic beads coated with an RGD-containing peptide bind avidly to the actin cytoskeleton (stained with fluorescently labeled phalloidin) of HASM cells via cell adhesion molecules (integrins). (c) A magnetic twisting field introduces a torque that causes the bead to rotate and to displace. Large arrows indicate the direction of the bead's magnetic moment before (black) and after (gray) twisting. If the twisting field is varied sinusoidally in time, then the microbead wobbles to and fro, resulting in a lateral displacement, (d), that can be measured. From Fabry, Maksym et al., 2003.

(including various integrin subtypes, scavenger receptors, urokinase receptors, and immune receptors). The beads are magnetized horizontally by a brief and strong magnetic pulse, and then twisted vertically by an external homogeneous magnetic field that varies sinusoidally in time. This applied field creates a torque that causes the beads to rotate toward alignment with the field, like a compass needle aligning with the earth's magnetic field. This rotation is impeded, however, by mechanical forces that develop within the cell as the bead rotates. Lateral bead displacements during bead rotation in response to the resulting oscillatory torque are detected by a CCD camera mounted on an inverted microscope.

Cell elasticity (g') and friction (g'') can then be deduced from the magnitude and phase of the lateral bead displacements relative to the torque (Fig. 3-2). Image acquisition with short exposure times of 0.1 ms is phase-locked to the twisting field so that 16 images are acquired during each twisting cycle. Heterodyning (a stroboscopic technique) is used at twisting frequencies >1 Hz up to frequencies of 1000 Hz. The images are analyzed using an intensity-weighted center-of-mass algorithm in which sub-pixel arithmetic allows the determination of bead position with an accuracy of 5 nm (rms).

Measurements of cell mechanics

The mechanical torque of the bead is proportional to the external magnetic field (which was generated using an electromagnet), the bead's magnetic moment (which was calibrated by measuring the speed of bead rotation in a viscous medium), and the cosine between the bead's magnetization direction with the direction of the twisting field. Consider the specific torque of a bead, T, which is the mechanical torque per bead volume, and has dimensions of stress (Pa). The ratio of the complex-specific torque $\tilde{T}$ to the resulting complex bead displacement $\tilde{d}$ (evaluated at the twisting frequency) then defines a complex modulus of the cell $\tilde{g} = \tilde{T}/\tilde{d}$, and has dimensions of

The cytoskeleton as a soft glassy material

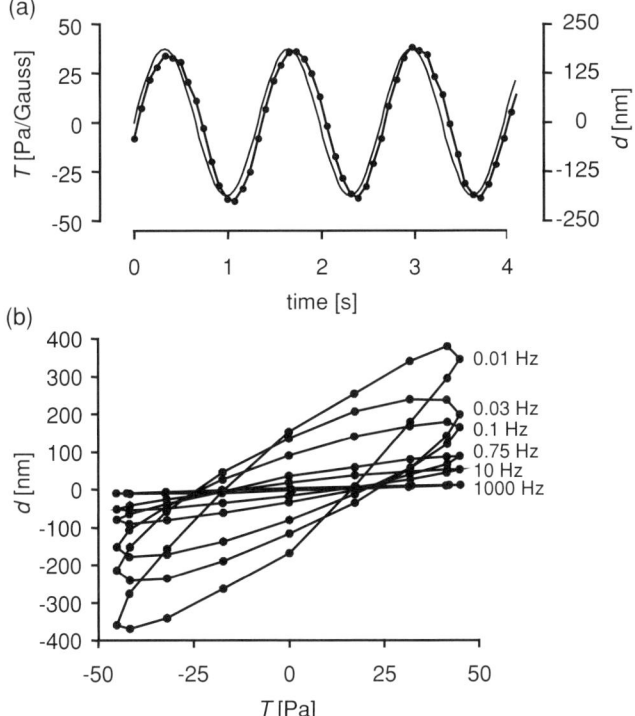

Fig. 3-2. (a) Specific torque T (solid line) and lateral displacement d (filled circles connected by a solid line) vs. time in a representative bead measured at a twisting frequency of 0.75 Hz. Bead displacement followed the sinusoidal torque with a small phase lag. The filled circles indicate when the image and data acquisition was triggered, which was 16 times per twisting cycle. (b) Loops of maximum lateral bead displacement vs. specific torque of a representative bead at different frequencies. With increasing frequency, displacement amplitude decreased. From Fabry, Maksym et al., 2003.

Pa/nm. These measurements can be transformed into traditional elastic shear (G') and loss (G'') moduli by multiplication of g' and g'' with a geometric factor that depends on the shape and thickness of the cell and the degree of bead embedding. Finite element analysis of cell deformation for a representative bead-cell geometry (assuming homogeneous and isotropic elastic properties with 10 percent of the bead diameter embedded in a cell 5 μm high) sets this geometric factor to 6.8 μm (Mijailovich, Kojic et al., 2002). This geometric factor need serve only as a rough approximation, however, because it cancels out in the scaling procedure described below, which is model independent. For each bead we compute the elastic modulus g' (the real part of $\tilde{g}$), the loss modulus g'' (the imaginary part of $\tilde{g}$), and the loss tangent η (the ratio g''/g') at a given twisting frequency. These measurements are then repeated over a range of frequencies.

Because only synchronous bead movements that occur at the twisting frequency are considered, nonsynchronous noise is suppressed by this analysis. Also suppressed are higher harmonics of the bead motion that may result from nonlinear material properties and that – if not properly accounted for – could distort the frequency dependence of the measured responses. However, we found no evidence of nonlinear

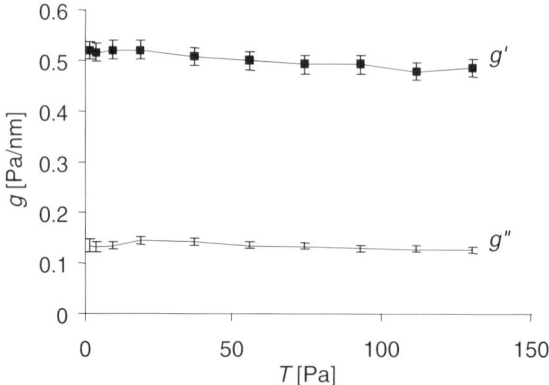

Fig. 3-3. g' and g'' vs. specific torque amplitude T. g' and g'' were measured in 537 HASM cells at $f = 0.75$ Hz. Specific torque amplitudes T varied from 1.8 to 130 Pa. g' and g'' were nearly constant, implying linear mechanical behavior of the cells in this range. Error bars indicate one standard error. From Fabry, Maksym et al., 2003.

cell behavior (such as strain hardening or shear thinning) at the level of stresses we apply with this technique, which ranges from about 1 Pa to about 130 Pa (Fig. 3-3). Throughout that range, which represents the physiological range, responses were linear.

Frequency dependence of g' and g''

The relationship of G' and G'' vs. frequency for human airway smooth muscle (HASM) cells under control conditions is shown in Fig. 3-4, where each data point represents the median value of 256 cells. Throughout the frequency range studied, G' increased with increasing frequency, f, according to a power law, f^{x-1} (as explained below, the formula is written in this way because the parameter x takes on a special meaning, namely, that of an effective temperature). Because the axes in Fig. 3-4 are logarithmic, a power-law dependency appears as a straight line with slope $x - 1$. The power-law exponent of G' was 0.20 ($x = 1.20$), indicating only a weak dependency of G' on frequency. G'' was smaller than G' at all frequencies except at 1 kHz. Like G', G'' also followed a weak power law with nearly the same exponent at low frequencies. At frequencies larger than 10 Hz, however, G'' exhibited a progressively stronger frequency dependence, approaching but never quite attaining a power-law exponent of 1, which would be characteristic of a Newtonian viscosity.

This behavior was at first disappointing because no characteristic time scale was evident; we were unable to identify a dominating relaxation process. The only characteristic time scale that falls out of the data is that associated with curvilinearity of the G'' data that becomes apparent in the neighborhood of 100 Hz (Fig. 3-4). As shown below, this curvilinearity is attributable to a small additive Newtonian viscosity that is entirely uncoupled from cytoskeletal dynamics. This additive viscosity is on the order of 1 Pa s, or about 1000-fold higher than that of water, and contributes to the energy dissipation (or friction) only above 100 Hz. Below 100 Hz, friction (G'') remained a

The cytoskeleton as a soft glassy material

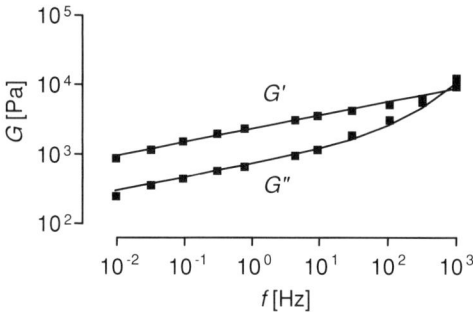

Fig. 3-4. G' and G'' (median of 256 human airway smooth muscle cells) under control conditions measured at frequencies between 0.01 Hz and 1000 Hz. The solid lines were obtained by fitting Eq. 3.2 to the data. G' and G'' (in units of Pa) were computed from the measured values of g' and g'' (in units of Pa/nm) times a geometric factor α of 6.8 µm. G' increased with increasing frequency, f, according to a power law, f^{x-1}, with $x = 1.20$. G'' was smaller than G' at all frequencies except at 1 kHz. At frequencies below 10 Hz, G'' also followed a weak power law with nearly the same exponent as did G'; above 10 Hz the power law exponent increased and approached unity. From Fabry, Maksym et al., 2003.

constant fraction (about 25 percent) of elasticity (G'). Such frictional behavior cannot be explained by a viscous dissipation process.

It is intriguing to note the combination of an elastic process (or processes) that increases with frequency according to a weak power-law over such a wide range of time scales, and a frictional modulus that, except at very high frequencies, is a constant, frequency-independent fraction of the elastic modulus. Similar behavior has been reported for a wide range of materials, biological tissue, and complex man-made structures such as airplane wings and bridges. Engineers use an empirical description – the structural damping equation (sometimes referred to as hysteretic damping law, or constant phase model) – to describe the mechanical behavior of such materials (Weber, 1841; Kohlrausch, 1866; Kimball and Lovell, 1927; Hildebrandt, 1969; Crandall, 1970; Fredberg and Stamenovic, 1989), but as regards mechanism, structural damping remains unexplained.

The structural damping equation

The mechanical properties of such a material can be mathematically expressed either in the time domain or in the frequency domain. In the time domain, the mechanical stress response to a unit step change in strain imposed at $t = 0$ is an instantaneous component attributable to a pure viscous response together with a component that rises instantaneously and then decays over time as a power law,

$$g(t) = \mu \delta(t) + g_0 (t/t_0)^{1-x}. \qquad (3.1)$$

g_0 is the ratio of stress to the unit strain measured at an arbitrarily chosen time t_0, μ is a Newtonian viscous term, and $\delta(t)$ is the Dirac delta function. The stress response to unit amplitude sinusoidal deformations can be obtained by taking the Fourier transform of the step response (Eq. 3.1) and multiplying by $j\omega$, which gives the complex modulus $\tilde{g}(\omega)$ as:

$$\tilde{g}(\omega) = g_0 \left(\frac{\omega}{\phi_0}\right)^{t-1} (1 - i\bar{\eta})\Gamma(2-x)\cos\frac{\pi}{2}(t-1) + i\omega\mu \qquad (3.2)$$

where $\bar{\eta} = \tan(x-1)\pi/2$ and ω is the radian frequency $2\pi f$ (Hildebrandt, 1969). g_0 and Φ_0 are scale factors for stiffness and frequency, respectively, Γ denotes the Gamma function, and i^2 is -1. g_0 and μ depend on bead-cell geometry. $\bar{\eta}$ has been called the structural damping coefficient (Fredberg and Stamenovic, 1989). The elastic modulus g' corresponds to the real part of Eq. 3.2, which increases for all ω according to the power-law exponent, $x - 1$. The loss modulus g'' corresponds to the imaginary part of Eq. 3.2 and includes a component that also increases as a power law with the same exponent. Therefore, the loss modulus is a frequency-independent fraction ($\bar{\eta}$) of the elastic modulus; such a direct coupling of the loss modulus to the elastic modulus is the characteristic feature of structural damping behavior (Fredberg and Stamenovic, 1989).

As mentioned already, the loss modulus includes a Newtonian viscous term, $j\omega\mu$, which turns out to be small except at very high frequencies. At low frequencies, the loss tangent η approximates $\bar{\eta}$. In the limit that x approaches unity, the power-law slope approaches zero, g' approaches g_0 and $\bar{\eta}$ approaches zero. In the limit that x approaches 2, the power-law slope approaches unity, G'' approaches μ and $\bar{\eta}$ approaches infinity. Thus, Eq. 3.2 describes a relationship between changes of the exponent of the power law and the transition from solid-like ($x = 1$, $\bar{\eta} = 0$) to fluid-like ($x = 2$, $\bar{\eta} = \infty$) behavior.

The structural damping equation describes the data in Fig. 3-4 exceedingly well and with only four free parameters: the scale factors g_0 and Φ_0, the Newtonian viscosity μ, and the power-law exponent $x - 1$. The structural damping coefficient $\bar{\eta}$ is not an independent parameter but depends on x only.

We now go on to show that three of the four parameters of the structural damping equation (g_0, Φ_0 and μ) can be considered constant, and that changes in the cell's mechanical behavior during contraction, relaxation, or other drug-induced challenges can be accounted for by changes of the parameter x alone.

Reduction of variables

When smooth muscle cells are activated with a contractile agonist such as histamine, they generate tension and their stiffness (G') increases, as has been shown in many studies (Warshaw, Rees et al., 1988; Fredberg, Jones et al., 1996; Hubmayr, Shore et al., 1996; Fabry, Maksym et al., 2001; Butler, Tolic-Norrelykke et al., 2002; Wang, Tolic-Norrelykke et al., 2002). Interestingly, in HASM cells this increase in G' after histamine activation (10^{-4} M) was more pronounced at lower frequencies. While G' still exhibited a weak power-law dependence on frequency, x fell slightly (Fig. 3-5). When cells were relaxed with DBcAMP (1 mM), the opposite happened: G' decreased, and x increased. When the actin cytoskeleton of the cells was disrupted with cytochalasin D (2 μM), G' decreased even more, while x increased further.

The cytoskeleton as a soft glassy material

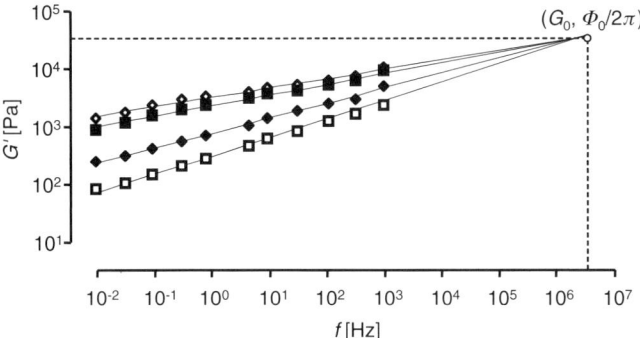

Fig. 3-5. G' vs. frequency in HASM cells under control conditions (■, $n = 256$), and after 10 min. treatment with the contractile agonist histamine [10^{-4} M] (◊, $n = 195$), the relaxing agonist DBcAMP [10^{-3} M] (◆, $n = 239$) and the actin-disrupting drug cytochalasin D [2×10^{-6} M] (□, $n = 171$). At all frequencies, treatment with histamine caused G' to increase, while treatment with DBcAMP and cytoD caused G' to decrease. Under all treatment conditions, G' increased with increasing frequency, f, according to a power law, f^{x-1}. x varied between 1.17 (histamine) and 1.33 (cytoD). A decreasing G' was accompanied by an increasing x, and vice versa. Solid lines are the fit of Eq. 3.2 to the data. Surprisingly, these lines appeared to cross at a coordinate close to $[G_0, \Phi_0/2\pi]$, well above the experimental frequency range. According to Eq. 3.2, an approximate crossover implies that in the HASM cell the values of G_0 and Φ_0 were invariant with differing treatment conditions. From Fabry, Maksym et al., 2003.

Remarkably, the G' data defined a family of curves that, when extrapolated, appeared to intersect at a single value (G_0) at a very high frequency (Φ_0) (Fig. 3-5). Such a common intersection, or fixed point, of the G' vs. frequency curves at a very high frequency means that G_0 and Φ_0 were invariant with different drug treatments.

With all drug treatments, G' and G'' tended to change in concert. The relationship between G'' and frequency remained a weak power law at lower frequencies, and the power-law exponent of G'' changed in concert with that of G'. At the highest frequencies, the curves of G'' vs. frequency for all treatments appeared to merge onto a single line with a power-law exponent approaching unity (Fig. 3-6).

The finding of a common intersection of the G' vs. f relationship stands up to rigorous statistical analysis, meaning that a three-parameter fit of the structural damping equation (G_0, Φ_0 and x) to the full set G' data (measured over five frequency decades and with different pharmacological interventions) is not statistically different from a fit with G_0 and Φ_0 being fixed, and with x being the only free parameter (Fabry, Maksym et al., 2003). The very same set of parameters – a fixed value for G_0 and Φ_0, respectively, and a drug-treatment-dependent x – also predicts the G'' vs. f relationship at frequencies below 100 Hz. Because the G'' data appears to merge onto a single line at higher frequencies, a single Newtonian viscosity μ that is common for all drug treatments can account for the data, although a rigorous statistical analysis indicates that a negligible but significant improvement of the fit can be achieved with different μ-values for different drug treatments (Fabry, Maksym et al., 2003). For all practical purposes, therefore, the mechanical behavior of HASM cells is restricted to vary only in a very particular way such that a single parameter, x, is sufficient to characterize the changes of both cell elasticity and friction.

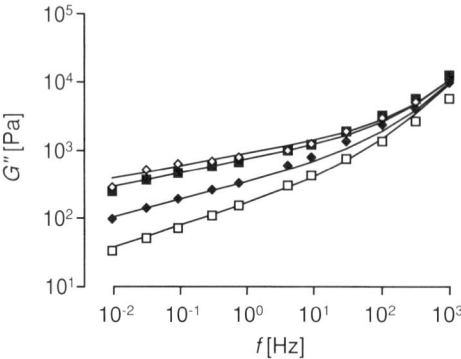

Fig. 3-6. G'' vs. frequency in HASM cells under control conditions (■, $n = 256$), and after 10 min. treatment with histamine [10^{-4} M] (◊, $n = 195$), DBcAMP [10^{-3} M] (♦, $n = 239$) and cytochalasin D [2×10^{-6} M] (□, $n = 171$). At all frequencies, treatment with histamine caused G'' to increase, while treatment with DBcAMP and cytoD caused G'' to decrease. Under all treatment conditions, G'' increased at frequencies below 10 Hz according to a power law, f^{x-1}, with exponents that were similar to that of the corresponding G'-data (Fig. 3-5). Above 10 Hz the power-law exponents increased and approached unity for all treatments; the G'' curves merged onto a single relationship. From Fabry, Maksym et al., 2003.

Universality

This surprising and particular behavior is not restricted to HASM cells. We found the very same behavior – a power-law relationship of G' and G'' vs. frequency, common intersection of the G' vs. f data for different drug treatments at a very high frequency, and merging of the G'' vs. f data onto a single line – in all other animal and human cell types we have investigated so far, including macrophages, neutrophils, various endothelial and epithelial cell types, fibroblasts, and various cancer cell lines (Fabry, Maksym et al., 2001; Fabry, Maksym et al., 2003; Puig-de-Morales, Millet et al., 2004).

Power-law behavior and common intersection were also found with an almost exhaustive panel of drugs that target the actin cytoskeleton and the activity of myosin light chain kinase (including BDM, ML-7, ML-9, W-7, various rho-kinase inhibitors, latrunculin, jasplakinolide) (Laudadio, Millet et al., 2005). Moreover, power-law behavior and common intersection are not peculiar to the details of the coupling between the bead and the cell, and can be observed in beads coated with different ligands (including RGD-peptide, collagen, vitronectin, fibronectin, urokinase, and acetylated low-density lipoprotein), and antibodies that specifically bind to various receptors (activating and nonactivating domains of various integrins and other cell adhesion molecules) (Puig-de-Morales, Millet et al., 2004). Neither is this behavior peculiar to the magnetic twisting technique. The power-law dependence of G' and G'' on frequency is consistent with data reported for atrial myocytes, fibroblasts, and bronchial endothelial cells measured with atomic force microscopy (AFM), for pellets of mouse embryonic carcinoma cells measured with a disk rheometer, for airway smooth muscle cells measured with oscillatory magneto-cytometry, and for kidney epithelial cells measured by laser tracking of Brownian motion of intracellular granules (Shroff, Saner et al., 1995; Goldmann and Ezzell, 1996; Mahaffy, Shih et al., 2000; Maksym,

The cytoskeleton as a soft glassy material

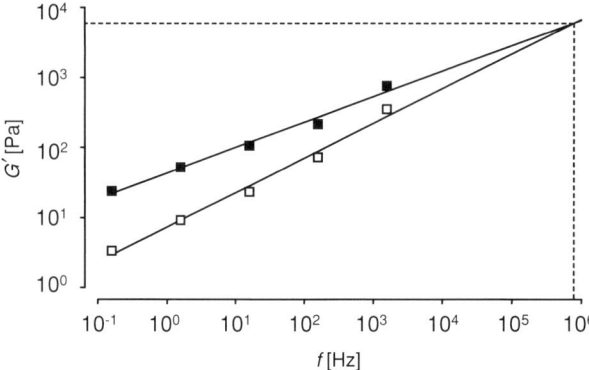

Fig. 3-7. G' vs. frequency in kidney epithelial cells measured with laser tracking microrheology under control conditions (■), and after 15 min. treatment with Latrunculin A [1×10^{-6} M] (□). Solid lines are the fit of Eq. 3.2 to the data, with $x = 1.36$ under control conditions, and $x = 1.5$ after Latrunculin A treatment. These lines crossed at $G_0 = 5.48$ kPa and $\Phi_0 = 6.64*10^5$ Hz. Adapted from Yamada, Wirtz et al., 2000.

Fabry et al., 2000; Yamada, Wirtz et al., 2000; Alcaraz, Buscemi et al., 2003). Using laser tracking microrheology, Yamada, Wirtz et al. also measured cell mechanics before and after microfilaments were disrupted with Latrunculin A, and obtained two curves of G' vs. f that intersected at a frequency comparable to the value we measured with our magnetic twisting technique, as shown in Fig. 3-7 (Yamada, Wirtz et al., 2000).

Finally, power-law behavior in the frequency domain (Eq. 3.2) corresponds to power-law behavior in the time domain (Eq. 3.1): when we measured the creep modulus of the cells by applying a step change of the twisting field, we did indeed find power-law behavior of the creep modulus vs. time, and a common intersection of the data at a very short time (Lenormand, Millet et al., 2004).

Scaling the data

Not surprisingly, although structural damping behavior always prevailed, substantial differences were observed in the absolute values of our G' and G'' measurements between different cell lines, and even larger (up to two orders of magnitude) differences when different bead coatings were used (Fabry, Maksym et al., 2001; Puig-de-Morales, Millet et al., 2004). Still larger (up to three orders of magnitude) differences were observed between individual cells of the same type even when they were grown within the same cell well (Fabry, Maksym et al., 2001). It is difficult to specify to what extent such differences reflect true differences in the "material" properties between different cells or between different cellular structures to which the beads are attached, vs. differences in the geometry (cell height, contact area between cell and bead, and so forth). Theoretically, these geometric details could be measured and then modeled, but in practice such measurements and models are inevitably quite rough (Mijailovich, Kojic et al., 2002).

Rather than analyzing "absolute" cell mechanics, it is far more practical (and insightful, as shown below) to focus attention instead upon relative changes in cell mechanics. To first order, such relative changes are independent of bead-cell geometry. Thus, the measurements need to be normalized or scaled appropriately such that each cell serves as its own control.

Two such scaled parameters have already been introduced. The first one is the slope of the power-law relationship $(x-1)$, which is the log change in G' or G'' per frequency decade (for example, the ratio of $G'_{10\ Hz}/G'_{1\ Hz}$). The second scaled parameter is the hysteresivity η, which is the ratio between G'' and G' at a single frequency.

Both scaling procedures cause factors (such as bead-cell geometry) that equally affect the numerator and denominator of those ratios to cancel out. Moreover, both scaling procedures can be performed on a bead-by-bead basis. The bead-by-bead variability of both x and η are negligible when compared with the huge variability in G' (Fabry, Maksym et al., 2001; Fabry, Maksym et al., 2001).

A third scaling parameter is the normalized stiffness G_n, which we define as the ratio between G' (measured at a given frequency, say 0.75 Hz) and G_0 (the intersection of the G' vs. f curves from different drug treatments, Fig. 3-5). Here, G_0 serves as an internal stiffness scale that is characteristic for each cell type and for each bead coating (receptor-ligand interaction), but that is unaffected by drug treatments. Thus, the normalized stiffness G_n allows us to compare the drug-induced responses of cells under vastly different settings, such as different bead coating, and so on.

Collapse onto master curves

The data were normalized as follows. We estimated x from the fit of Eq. 3.2 to the pooled (median over many cells) G' and G'' data. The hysteresivity η was estimated from ratio of G''/G' measured at 0.75 Hz (an arbitrary choice). The normalized cell stiffness G_n was estimated as G' measured at 0.75 Hz divided by g_0. log G_n vs. x and η vs. x graphs were then plotted (Fig. 3-8).

In human airway smooth muscle (HASM) cells, we found that drugs that increased x caused the normalized stiffness G_n to decrease (Fig. 3-8a, black symbols). The relationship between log G_n and x appears as nearly linear: $\ln G_n \sim -x$. The solid line in Fig. 3-8a is the prediction from the structural damping equation under the condition of a common intersection of all G' data at radian frequency Φ_0:

$$\ln G_n = (x-1)\ln(\omega/\Phi_0). \tag{3.3}$$

How close the normalized data fell to this prediction thus indicates how well a common intersection can account for those data.

Conversely, drugs that increased x caused the normalized frictional parameter η to increase (Fig. 3-8b, black symbols). The solid line in Fig. 3-8b is the prediction from (and not a fit of) the structural damping equation:

$$\eta = \tan(x-1)\pi/2. \tag{3.4}$$

How close the normalized data fell to this prediction thus indicates how well the structural damping equation can account for those data.

Surprisingly, the normalized data for the other cell types collapsed onto the very same relationships that were found for HASM cells (Fig. 3-8). In all cases, drugs that increased x caused the normalized stiffness G_n to decrease and hysteresivity η to

The cytoskeleton as a soft glassy material

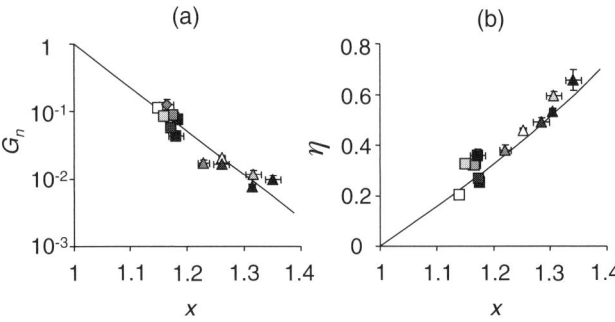

Fig. 3-8. Normalized stiffness G_n vs. x (left) and hysteresivity η (right) vs. x, of HASM cells (black, $n = 256$), human bronchial epithelial cells (light gray, $n = 142$), mouse embryonic carcinoma cells (F9) cells (dark gray, $n = 50$), mouse macrophages (J774A.1) (hatched, $n = 46$) and human neutrophils (gray, $n = 42$) under control conditions (■), treatment with histamine (□), FMLP (♦), DBcAMP (▲) and cytochalasin D (▲). x was obtained from the fit of Eq. 3.2 to the pooled (median) data. Drugs that increased x caused the normalized stiffness G_n to decrease and hysteresivity η to increase, and vice versa. The normalized data for all types collapsed onto the same relationships. The structural damping equation by Eq. 3.2 is depicted by the black solid curves: $\ln G_n = (x - 1) \ln(\omega/\Phi_0)$ with $\Phi_0 = 2.14 \times 10^7$ rad/s, and $\bar{\eta} = \tan((x - 1)\pi/2)$. Error bars indicate ± one standard error. If x is taken to be the noise temperature, then these data suggest that the living cell exists close to a glass transition and modulates its mechanical properties by moving between glassy states that are "hot," melted and liquid-like, and states that are "cold," frozen and solid-like. In the limit that x approaches 1 the system behaves as an ideal Hookean elastic solid, and in the limit that x approaches 2 the system behaves as an ideal Newtonian fluid (Eq. 3.2). From Fabry, Maksym et al., 2003.

increase. These relationships thus represent universal master curves in that a single parameter, x, defined the constitutive elastic and frictional behaviors for a variety of cytoskeletal manipulations, for five frequency decades, and for diverse cell types.

The normalized data from all cell types and drug treatments fell close to the predictions of the structural damping equation (Fig. 3-8). In the case of the η vs. x data (Fig. 3-8b), this collapse of the data indicates that the coupling between elasticity and friction, and their power-law frequency dependence, is well described by the structural damping equation. In the case of the log G_n vs. x data (Fig. 3-8a), the collapse of the data indicates that a common intersection of the G vs. f curves exists for all cell types, and that this intersection occurs approximately at the same frequency Φ_0.

Rigorous statistical analysis of the data from many different cell types has thus far supported the existence of a common intersection of the G' vs. f relationships measured after treatment with a large panel of cytoskeletally active drugs (Laudadio, Millet et al., 2005). This holds true regardless of the receptor-ligand pathway that was used to probe cell rheology (Puig-de-Morales, Millet et al., 2004). The same statistical analysis, however, hints that the crossover frequency Φ_0 may not be the same for all cell types and receptor-ligand pathways. Unfortunately, so far Φ_0 cannot be measured with high-enough accuracy to resolve such differences. Thus for all practical purposes we can regard Φ_0 as being the same for all cell types and for all receptor-ligand combinations (Fabry, Maksym et al., 2003; Puig-de-Morales, Millet et al., 2004).

The structural damping relationship has long been applied to describe the rheological data for a variety of biological tissues (Weber, 1835; Kohlrausch, 1847; Fung,

1967; Hildebrandt, 1969; Fredberg and Stamenovic, 1989; Suki, Peslin et al., 1989; Hantos, Daroczy et al., 1990; Navajas, Mijailovich et al., 1992; Fredberg, Bunk et al., 1993). Thus, it may seem natural (but still intriguing) that living cells, too, exhibit structural damping behavior, and in this regard the collapse of the η vs. x data from different cell types and drug treatments onto the same relationship (Fig. 3-8b) is a necessary consequence of such behavior. But it is utterly mystifying why the G_n vs. x data from different drug treatments should form any relationship at all and why, moreover, the data from different cell types and different receptor-ligand pathways should collapse onto the very same relationship.

Theory of soft glassy rheology

What are soft glassy materials

The master relationships shown in Figs. 3-8 and 3-9 demonstrate that when the mechanical properties of the cell change, they do so along a special trajectory. This trajectory is found to be identical in a large variety of cell types that are probed via different receptor-ligand pathways and over many frequency decades. In all those cases, changes of stiffness and friction induced by pharmacological interventions could be accounted for solely by changes in x. This parameter x appears to play a central organizing role leading to the collapse of all data onto master curves. But what is x?

A possible answer may come – surprisingly – from a theory of soft glassy materials that was developed by Sollich and colleagues (Sollich, Lequeux et al., 1997). In the remainder of this chapter a brief introduction is given to some fundamental principles and ideas about soft glassy rheology. Parallels between living cells and soft glassy materials are shown, and a discussion ensues on what insights may be gained from this into the mechanisms involved.

The class of soft glassy materials (SGM) comprises what would at first glance seem to be a remarkably diverse group of substances that includes foams, pastes, colloids, emulsions, and slurries. Yet the mechanical behavior of each of these substances is surprisingly alike. The common empirical criteria that define this class of materials are that they are very soft (in the range of Pa to kPa), that both G' and G'' increase with the same weak power-law dependencies on frequency, and that the loss tangent η is frequency insensitive and of the order 0.1 (Sollich, Lequeux et al., 1997; Sollich, 1998). The data presented so far establish that the cytoskeleton of living cells satisfies all of these criteria. Accordingly, we propose the working hypothesis that the cytoskeleton of the living cell can be added to the list of soft glassy materials.

Sollich reasoned that because the materials comprising this class are so diverse, the common rheological features must be not so much a reflection of specific molecules or molecular mechanisms as they are a reflection of generic system properties that play out at some higher level of structural organization (Sollich, 1998). The generic features that all soft glassy materials share are that each is composed of elements that are discrete, numerous, and aggregated with one another via weak interactions. In addition, these materials exist far away from thermodynamic equilibrium and are arrayed in a microstructural geometry that is inherently disordered and metastable. Note that the cytoskeleton of living cells shares all of these features.

Sollich's theory of SGMs

To describe the interaction between the elements within the matrix, Sollich developed a theory of soft glassy rheology (SGR) using earlier work by Bouchaud as a point of departure (Bouchaud, 1992). SGR theory considers that each individual element of the matrix exists within an energy landscape containing many wells, or traps, of differing-depth E. These traps are formed by interactions of the element with neighboring elements. In the case of living cells those traps might be plausibly thought to be formed by binding energies between neighboring cytoskeletal elements including but not limited to cross-links between actin filaments, cross-bridges between actin and myosin, hydrophilic interactions between various proteins, charge effects, or simple steric constraints.

In Bouchaud's theory of glasses, an element can escape its energy well and fall into another nearby well; such hopping events are activated by thermally driven random fluctuations. As distinct from Bouchaud's theory, in Sollich's theory of soft glassy materials each energy well is regarded as being so deep that the elements are unlikely to escape the well by thermal fluctuations alone. Instead, elements are imagined to be agitated, or jostled, by their mutual interactions with neighboring elements (Sollich, 1998). A clear notion of the source of the nonthermal agitation remains to be identified, but this agitation can be represented nonetheless by an effective temperature, or noise level, x.

Sollich's SGR theory follows from a conservation law for probability of an element being trapped in an energy well of depth E and local displacement (strain) l, at time t, denoted $P(E, l, t)$. Dynamics is then governed by a conservation equation for this probability, given by

$$\partial P(E, l, t)/\partial t + \gamma \partial P/\partial l = -g(E, l)P(E, l, t) + f(E)\Phi(t)\delta(l) \qquad (3.5)$$

where $\Phi(t) = \int dE\,dl\, g(E, l)P(E, l, t)$ (required for conservation of probability), and $\delta(l)$ is the Dirac delta function. Here $\gamma = dl/dt$ and $f(E)$ is the distribution of energy-well depths. Eq. 3.5 states that the material rate of change of P is given by the sum of two terms. The first term is depletion, equal to the probability of resident elements hopping out, given by the product of the probability of occupancy P and a transition rate $g(E, l)$. The second term is the accumulation rate, equal to the product of the total number of available transitions $\Phi(t)$ and the delta function constraint forcing elements to hop into wells at zero local strain.

Sollich takes $g(E, l) = \Phi_0 \exp(-(E + kl^2/2)/x)$ and $f(E) = \Phi_0 \exp(-E)$, respectively. Note that the transition rate $g(E, l)$ for hopping out of wells is distributed over E, which, in the nonlinear regime, is also a function of strain. Note further that the transition rate into wells of depth E only depends on strain through the constraint that $l = 0$ following a hop.

When $x > 1$, there is sufficient agitation in the matrix that the element can hop randomly between wells and, as a result, the system as a whole can flow and become disordered. When x approaches 1, however, the elements become trapped in deeper and deeper wells from which they are unable to escape: the system exhibits a glass transition and becomes a simple elastic solid with stiffness G_0.

Soft glassy rheology and structural damping

Remarkably, in the limit that the frequency is small (compared to Φ_0) and the imposed deformations are small (such that the rate for hopping out of wells is dominated by x), Sollich's theory leads directly to the structural damping equation (Eq. 3.2).

The data reported above establish firmly that the mechanical behavior of cells conforms well to Eq. 3.2. If Sollich's theory and underlying ideas are assumed to apply to the data reported here, then the parameters in Eq. 3.2 (x, G_0, Φ_0) can be identified as follows.

The parameter x is identified as being the noise temperature of the cytoskeletal matrix. The measured values of x in cells lie between 1.15 and 1.35, indicating that cells exist close to a glass transition.

G_0 is identified as being the stiffness of the cytoskeleton at the glass transition ($x = 1$). In this connection, Satcher and Dewey (1996) developed a static model of cell stiffness based on consideration of cell actin content and matrix geometry. All dynamic interactions were neglected in their model, as would be the case in SGR theory in the limit that x approaches 1, when all hopping ceases. As such, it might be expected that their model would predict this limiting value of the stiffness, G_0, as defined in Eq. 3.2. Indeed, we have found a remarkably good correspondence between their prediction (order of 10 kPa) and our estimate for G_0 (41 kPa in HASM cells).

Finally, Φ_0 is identified in Sollich's theory as being the maximum rate at which cytoskeletal elements can escape their traps. However, for soft glassy materials in general, and the case of living cells in particular, the factors that determine Φ_0 remain unclear. Statistical analysis of our data suggests that Φ_0 did not vary with drug treatments and possibly not even across cell type (Fig. 3-7). But why Φ_0 is invariant is not at all clear, and is not explained by SGR theory.

Open questions

Crucial aspects of soft glassy rheology (SGR) theory remain incomplete, however. First, the effective noise temperature x is a temperature to the extent that the rate at which elements can hop out of a trap assumes the form $\exp(-E/x)$, where x takes the usual position of a thermal energy $k_B T$ in the familiar Boltzmann exponential. By analogy, x has been interpreted by Sollich as reflecting jostling of elements by an unidentified but nonthermal origin. It appears an interesting question whether the ambiguity surrounding x might be resolved in the case of living cells (as opposed to the inert materials for which SGR theory was originally devised) by an obvious and ready source of nonthermal energy injection, namely those proteins that go through cyclic conformational changes and thus agitate the matrix by mechanisms that are ATP dependent.

Second, Sollich interprets E as an energy-well depth, but there are difficulties with this interpretation. In SGR theory the total energy is not a conserved quantity even in the zero-strain case, and the energy landscape has no spatial dimension, precluding explicit computation of microstructural rearrangements. Such microstructural rearrangements in the form of cytoskeletal reorganization during cell division, crawling,

The cytoskeleton as a soft glassy material

or intracellular transport processes are of fundamental interest in cell biology. The fluctuation-dissipation theorem implies a profound connection between dissipative phenomena as reflected in the measured values of G'' during oscillatory forcing, and the temporal evolution of the mean square displacement (MSD), or fluctuations, of free unforced particles in the medium. Experiments from our lab and others revealed a behavior of MSD that lies somewhere between that of simple diffusion in a homogeneous medium and ballistic behavior characteristic of short time displacements (An, Fabry et al., 2004). To what extent this is consistent with SGR theory also remains an open question (Lau, Hoffman et al., 2003).

Despite these questions, however, it is clear from Sollich's theory that for a soft glass to elastically deform, its elements must remain in energy wells; in order to flow, the elements must hop out of these wells. In the case of cells, these processes depend mainly on a putative energy level in the cytoskeletal lattice, where that energy is representative of the amount of molecular agitation, or jostling, present in the lattice relative to the depth of energy wells that constrain molecular motions. This energy level can be expressed as an effective lattice temperature (x) – as distinct from the familiar thermodynamic temperature. Even while the thermodynamic temperature is held fixed, this effective temperature can change, can be manipulated, and can be measured. The higher the effective temperature, the more frequently do elemental structures trapped in one energy well manage to hop out of that well only to fall into another. The hop, therefore, can be thought of as the fundamental molecular remodeling event.

It is interesting that, from a mechanistic point of view, the parameter x plays a central role in the theory of soft glassy materials. At the same time, from a purely empirical point of view, the parameter x is found to play a central organizing role leading to the collapse of all data onto master curves (Fig. 3-8). Whether or not the measured value of x might ultimately be shown to correspond to an effective lattice temperature, this empirical analysis would appear to provide a unifying framework for studying protein interactions within the complex integrative microenvironment of the cell body.

In the next section, the concepts developed so far are employed to tie together within such a unified framework diverse behaviors of cell physiology that were previously unexplained or regarded as unrelated.

Biological insights from SGR theory

Malleability of airway smooth muscle

The function of smooth muscle is to maintain shape and/or tone of hollow organs (Murphy, 1988). Typically, smooth muscle must do so over an extremely wide range of working lengths. Two unique features enable smooth muscle to do this. First, smooth muscle can develop its contractile forces almost independently of muscle length (Wang, Pare et al., 2001). To achieve this, the cytoskeletal lattice and associated contractile machinery of smooth muscle is disordered and highly malleable, quite unlike the ordered and fixed structure of striated muscle and the rather narrow range of lengths over which striated muscle can generate appreciable tension. Second, at the height of force development, smooth muscle can "latch" its contractile machinery, that is to say, down-regulate the rate of acto-myosin cycling, thereby leading to tone maintenance very economically in terms of energy metabolism (Hai and Murphy, 1989). To produce the same steady-state isometric force, for example, striated muscle hydrolyzes more ATP at a rate 300 times higher than does smooth muscle (Murphy, 1988).

Soft glassy rheology theory helps to piece together such information into an integrative context. Below, we present some earlier data from our laboratory on the mechanical and contractile properties of smooth muscle tissue; these data contained some previously unexplained loose ends. The glass hypothesis now offers a new and consistent explanation of these data.

Our earlier work focused on the contractile states of smooth muscle that were inferred from the responses to sinusoidal length or force perturbations. Fig. 3-9 summarizes a typical result obtained from sinusoidal length perturbations.

The stiffness E in those experiments is a measure of the number of force-generating acto-myosin bridges, while the hysteresivity η is a measure of internal mechanical friction and is closely coupled to the rate of cross-bridge cycling as reflected both in the unloaded shortening velocity and the rate of ATP utilization measured by NADH fluorimetry (Fredberg, Jones et al., 1996). The dramatic increase in force and stiffness after contractile stimulation (Fig. 3-10) therefore reflects an increase in the number of acto-myosin bridges. The progressive fall of η after contractile stimulus onset (Fig. 3-9) has been interpreted as reflecting rapidly cycling cross-bridges early in the contractile event converting to slowly cycling latch-bridges later in the contractile event (Fredberg, Jones et al., 1996). This molecular picture fits exceptionally well with computational analysis based on first principles of myosin-binding dynamics. According to this picture, imposed sinusoidal length oscillations (between 400 s and 900 s in Fig. 3-9) around a constant mean length lead to a disruption of acto-myosin cross-bridges and latch-bridges. This shifts the binding equilibrium of myosin toward a faster cycling rate such that with increasing oscillation amplitude hysteresivity increases, and muscle force and stiffness fall (Fredberg, 2000; Mijailovich, Butler et al., 2000). However, this picture is unable to explain why force and stiffness remained suppressed even after the length oscillations had stopped (Fig. 3-9).

Much the same behavior was found in a similar experiment in which force oscillations were imposed around a constant mean force (Fredberg, 2000): With increasing amplitude of the force oscillations, stiffness decreased and muscle length and hysteresivity increased (Fig. 3-10). Again, this behavior was exceptionally well explained by acto-myosin binding dynamics (Fredberg, 2000), but when the amplitude of the force oscillations was reduced, length and stiffness inexplicably did not return (Fig. 3-10).

A rather different perspective on these observations (Figs. 3-9 and 3-10) arises when they are viewed instead through the lens of glassy behavior. Accordingly, the relaxed smooth muscle cell is in a relatively "cold" state, with a noise temperature close to unity, but with the onset of contractile stimulation the cell very rapidly becomes "hot." After this hot initial transient, the cell then begins to gradually "cool" in the process of sustained contractile stimulation until, eventually, it approaches a steady-state that approximates a "frozen" state not only mechanically (high stiffness and low noise temperature) but also biochemically and metabolically (Gunst and Fredberg, 2003).

The cytoskeleton as a soft glassy material

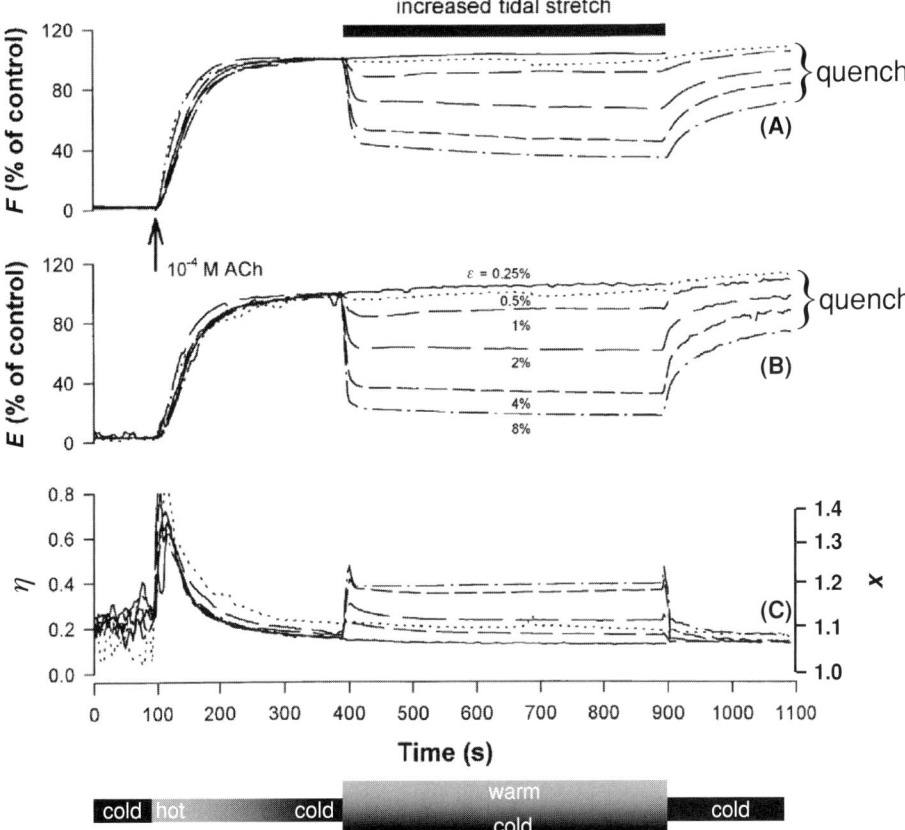

Fig. 3-9. Time courses of force (A), stiffness (B), and hysteresivity (C) in a bovine tracheal smooth muscle strip. To continuously track changes in stiffness and hysteresivity, sinusoidal length oscillations at a frequency of 0.2 Hz and with amplitude ε were superimposed throughout the measurement period. The tangent of the phase angle between the sinusoidal length oscillations and the resulting force oscillations (that is, the hysteresivity η) defines the noise temperature x (panel (C), right side) by a simple relationship (Eq. 3.5). The strip was stimulated at 100 s with acetylcholine (10^{-4} M). The solid trace in each panel corresponds to cyclic strain (ε) maintained throughout at 0.25%. Broken lines correspond to runs in the same muscle subjected to graded increments of ε from 0.25 to either 0.5, 1, 2, 4, or 8% over the time interval from 400 to 900 s. Adapted from Fredberg, Inouye et al., 1997.

The progressive decrease of stiffness and increase of hysteresivity with increasing amplitude of the imposed cyclic strain (Figs. 3-9 and 3-10) is consistent with a fluidization of the CSK matrix due to the application of a shear stress. SGR theory predicts that shear stress imposed at the macroscale adds to the agitation already present at the microscale, and thereby increases the noise temperature in the matrix (Sollich, 1998). This in turn allows elements to escape their cages more easily, such that friction and hysteresivity increases while stiffness decreases (Sollich, 1998; Fabry and Fredberg, 2003).

SGR theory now goes beyond the theory of perturbed myosin binding by predicting that the stretch-induced increase in noise temperature speeds up all internal molecular events, including accelerated plastic restructuring events within the CSK. When the

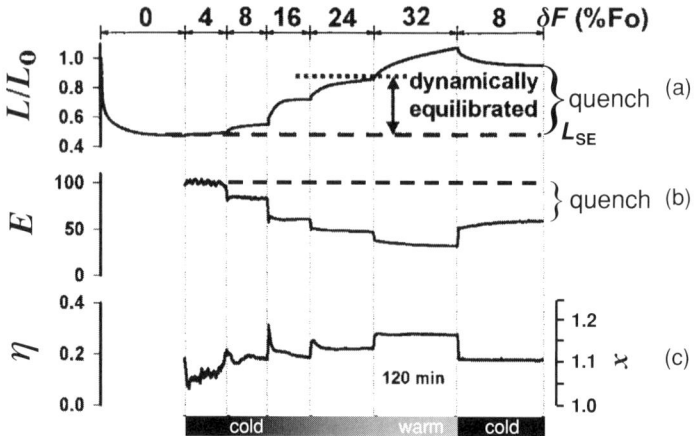

Fig. 3-10. Evolution of mechanical properties of bovine tracheal smooth muscle during contraction against a constant mean force on which force fluctuations (0.2 Hz) of graded amplitude δF were superimposed. (a) Mean muscle length L (relative to optimal length L_0). (b) Loop stiffness (percentage of maximum isometric value). (c) hysteresivity η and noise temperature x. L_{SE} is the statically equilibrated length of the muscle after 120 min. of unperturbed contraction against a constant load of 32% of maximum force (F_0). Adapted from Fredberg, 2000.

tidal stretches are terminated (Figs. 3-9 and 3-10), however, the noise temperature is suddenly lowered, and all plastic changes might become trapped, or quenched, so that the muscle is unable to return to maximum force and stiffness (Fig. 3.10), or maximum shortening (Fig. 3-10) (Fabry and Fredberg, 2003).

The glass hypothesis predicts, therefore, that the cell ought to be able to adapt faster to step-length changes imposed while the cell is transiently 'hot' (that is, early in activation), and far less so after it has cooled in the process of sustained activation (Fabry and Fredberg, 2003; Gunst and Fredberg, 2003). Indeed, Gunst and colleagues showed that a step change of muscle length alters the level of the subsequent force plateau to a degree that depends mostly on the timing of the length change with respect to stimulus onset (Gunst, Meiss et al., 1995; Gunst and Fredberg, 2003).

Conclusion

The behavior of soft glasses, and the underlying notion of the noise temperature, might provide a unifying explanation of the ability of the cytoskeletal lattice to deform, to flow, and to remodel. Such a view does not point to specific molecular processes that occur, but instead derives the mechanical properties from generic features: structural elements that are discrete, numerous, aggregated with one another via weak interactions, and arrayed in a geometry that is structurally disordered and metastable. We have proposed here that these features may comprise the basis of CSK rheology and remodeling.

References

Alcaraz, J., L. Buscemi, et al. (2003). Microrheology of human lung epithelial cells measured by atomic force microscopy. *Biophys. J.*, 84, 2071–9.

An, S. S., B. Fabry, et al. (2004). Role of heat shock protein 27 in cytoskeletal remodeling of the airway smooth muscle cell. *J. Appl. Physiol.*, 96, 1701–13.
Bouchaud, J. (1992). Weak ergodicity breaking and aging in disordered systems. *J. Phys. I.*, 2, 1705–13.
Butler, J. P., I. M. Tolic-Norrelykke, et al. (2002). Traction fields, moments, and strain energy that cells exert on their surroundings. *Am. J. Physiol. Cell Physiol.*, 282, C595–605.
Crandall, S. H. (1970). The role of damping in vibration theory. *J. Sound Vibr.*, 11, 3–18.
Crick, F. H. C. and A. F. W. Hughes (1950). The physical properties of cytoplasm. *Exp. Cell Res.*, 1, 37–80.
Fabry, B. and J. J. Fredberg (2003). Remodeling of the airway smooth muscle cell: are we built of glass? *Respir. Physiol. Neurobiol.*, 137, 109–24.
Fabry, B., G. N. Maksym, et al. (2001). Scaling the microrheology of living cells. *Phys. Rev. Lett.*, 87, 148102.
Fabry, B., G. N. Maksym, et al. (2003). Time scale and other invariants of integrative mechanical behavior in living cells. *Phys. Rev. E.*, 68, 041914.
Fabry, B., G. N. Maksym, et al. (2001). Time course and heterogeneity of contractile responses in cultured human airway smooth muscle cells. *J. Appl. Physiol.*, 91, 986–94.
Fredberg, J. J. (2000). Airway smooth muscle in asthma. Perturbed equilibria of myosin binding. *Am. J. Respir. Crit. Care Med.*, 161, S158–60.
Fredberg, J. J., D. Bunk, et al. (1993). Tissue resistance and the contractile state of lung parenchyma. *J. Appl. Physiol.*, 74, 1387–97.
Fredberg, J. J., D. Inouye, et al. (1997). Airway smooth muscle, tidal stretches, and dynamically determined contractile states. *Am. J. Respir. Crit. Care Med.*, 156, 1752–9.
Fredberg, J. J., K. A. Jones, et al. (1996). Friction in airway smooth muscle: mechanism, latch, and implications in asthma. *J. Appl. Physiol.*, 81, 2703–12.
Fredberg, J. J. and D. Stamenovic (1989). On the imperfect elasticity of lung tissue. *J. Appl. Physiol.*, 67, 2408–19.
Fung, Y. C. (1967). Elasticity of soft tissues in simple elongation. *Am. J. Physiol.*, 213, 1532–44.
Goldmann, W. H. and R. M. Ezzell (1996). Viscoelasticity in wild-type and vinculin-deficient (5.51) mouse F9 embryonic carcinoma cells examined by atomic force microscopy and rheology. *Exp. Cell Res.*, 226, 234–7.
Gunst, S. J. and J. J. Fredberg (2003). The first three minutes: smooth muscle contraction, cytoskeletal events, and soft glasses. *J. Appl. Physiol.*, 95, 413–25.
Gunst, S. J., R. A. Meiss, et al. (1995). Mechanisms for the mechanical plasticity of tracheal smooth muscle. *Am. J. Physiol.*, 268, C1267–76.
Hai, C. M. and R. A. Murphy (1989). Cross-bridge dephosphorylation and relaxation of vascular smooth muscle. *Am. J. Physiol.*, 256, C282–7.
Hantos, Z., B. Daroczy, et al. (1990). Modeling of low-frequency pulmonary impedance in dogs. *J. Appl. Physiol.*, 68, 849–60.
Hildebrandt, J. (1969). Comparison of mathematical models for cat lung and viscoelastic balloon derived by Laplace transform methods from pressure-volume data. *Bull. Math. Biophys.*, 31, 651–67.
Hill, A. V. (1965). *Trails and Trials in Physiology (pp. 14–15)*. London, E. Arnold.
Hubmayr, R. D., S. A. Shore, et al. (1996). Pharmacological activation changes stiffness of cultured human airway smooth muscle cells. *Am. J. Physiol.*, 271, C1660–8.
Kawai, M. and P. W. Brandt (1980). Sinusoidal analysis: a high resolution method for correlating biochemical reactions with physiological processes in activated skeletal muscles of rabbit, frog and crayfish. . *J Muscle Res. Cell Motil.*, 1, 279–303.
Kimball, A. L. and D. E. Lovell (1927). Internal friction in solids. *Phys. Rev.*, 30, 948–959.
Kohlrausch, F. (1866). Beiträge zur Kenntniss der elastischen Nachwirkung. *Ann. Phys. Chem.*, 128, 1–20, 207–227, 399–419.
Kohlrausch, R. (1847). Nachtrag ueber die elastische Nachwirkung beim Cocon- und Glasfaden, und die hygroskopische Eigenschaft des ersteren. *Ann. Phys. Chem.*, 72, 393–8.
Lau, A. W., B. D. Hoffman, et al. (2003). Microrheology, stress fluctuations, and active behavior of living cells. *Phys. Rev. Lett.*, 91, 198101.

Laudadio, R. E., E. J. Millet, et al. (2005). Rheology of the rat airway smooth muscle cell: scaling of responses to actin modulation. *Am. J. Physiol.*, in review.

Lenormand, G., E. Millet, et al. (2004). Linearity and time-scale invariance of the creep function in living cells. *J. Royal Soc. Interface*, 1, 91–7.

Mahaffy, R. E., C. K. Shih, et al. (2000). Scanning probe-based frequency-dependent microrheology of polymer gels and biological cells. *Phys. Rev. Lett.*, 85, 880–3.

Maksym, G. N., B. Fabry, et al. (2000). Mechanical properties of cultured human airway smooth muscle cells from 0.05 to 0.4 Hz. *J. Appl. Physiol.*, 89, 1619–32.

Mijailovich, S. M., J. P. Butler, et al. (2000). Perturbed equilibria of myosin binding in airway smooth muscle: bond-length distributions, mechanics, and ATP metabolism. *Biophys. J.*, 79, 2667–81.

Mijailovich, S. M., M. Kojic, et al. (2002). A finite element model of cell deformation during magnetic bead twisting. *J. Appl. Physiol.*, 93, 1429–36.

Murphy, R. A. (1988). Muscle cells of hollow organs. *News Physiol. Sci.*, 3, 124–8.

Navajas, D., S. Mijailovich, et al. (1992). Dynamic response of the isolated passive rat diaphragm strip. *J. Appl. Physiol.*, 73, 2681–92.

Puig-de-Morales, M., E. Millet, et al. (2004). Cytoskeletal mechanics in adherent human airway smooth muscle cells: probe specificity and scaling of protein-protein dynamics. *Am. J. Physiol. Cell Physiol.*, 287, C643–54.

Satcher, R. L., Jr. and C. F. Dewey, Jr. (1996). Theoretical estimates of mechanical properties of the endothelial cell cytoskeleton. *Biophys. J.*, 71, 109–18.

Shroff, S. G., D. R. Saner, et al. (1995). Dynamic micromechanical properties of cultured rat atrial myocytes measured by atomic force microscopy. *Am. J. Physiol.*, 269, C286–92.

Sollich, P. (1998). Rheological constitutive equation for a model of soft glassy materials. *Phys. Rev. E.*, 58, 738–59.

Sollich, P., F. Lequeux, et al. (1997). Rheology of soft glassy materials. *Phys. Rev. Lett.*, 78, 2020–2023.

Suki, B., R. Peslin, et al. (1989). Lung impedance in healthy humans measured by forced oscillations from 0.01 to 0.1 Hz. *J. Appl. Physiol.*, 67, 1623–9.

Wang, L., P. D. Pare, et al. (2001). Effect of chronic passive length change on airway smooth muscle length-tension relationship. *J. Appl. Physiol.*, 90, 734–40.

Wang, N., I. M. Tolic-Norrelykke, et al. (2002). Cell prestress. I. Stiffness and prestress are closely associated in adherent contractile cells. *Am. J. Physiol. Cell Physiol.*, 282, C606–16.

Warshaw, D. M., D. D. Rees, et al. (1988). Characterization of cross-bridge elasticity and kinetics of cross-bridge cycling during force development in single smooth muscle cells. *J. Gen. Physiol.*, 91, 761–79.

Weber, W. (1835). Ueber die Elasticitaet der Seidenfaeden. *Annalen der Physik und Chemie*, 34, 247–57.

Weber, W. (1841). Ueber die Elasticitaet fester Koerper. *Annalen der Physik und Chemie*, 54, 1–18.

Yamada, S., D. Wirtz, et al. (2000). Mechanics of living cells measured by laser tracking microrheology. *Biophys. J.*, 78, 1736–47.

4 Continuum elastic or viscoelastic models for the cell

Mohammad R. K. Mofrad, Helene Karcher, and Roger D. Kamm

ABSTRACT: Cells can be modeled as continuum media if the smallest operative length scale of interest is much larger than the distance over which cellular structure or properties may vary. Continuum description uses a coarse-graining approach that replaces the contributions of the cytoskeleton's discrete stress fibers to the local microscopic stress-strain relationship with averaged constitutive laws that apply at macroscopic scale. This in turn leads to continuous stress-strain relationships and deformation descriptions that are applicable to the whole cell or cellular compartments. Depending on the dynamic time scale of interest, such continuum description can be elastic or viscoelastic with appropriate complexity. This chapter presents the elastic and viscoelastic continuum multicompartment descriptions of the cell and shows a successful representation of such an approach by implementing finite element-based two- and three-dimensional models of the cell comprising separate compartments for cellular membrane and actin cortex, cytoskeleton, and nucleus. To the extent that such continuum models can capture stress and strain patterns within the cell, it can help relate biological influences of various types of force application and dynamics under different geometrical configurations of the cell.

Introduction

Cells can be modeled as continuum media if the smallest length scale of interest is significantly larger than the dimensions of the microstructure. For example when whole-cell deformations are considered, the length scale of interest is at least one or two orders of magnitude larger than the distance between the cell's microstructural elements (namely, the cytoskeletal filaments), and as such a continuum description may be appropriate. In the case of erythrocytes or neutrophils in micropipette aspiration, the macroscopic mechanical behavior has been successfully captured by continuum viscoelastic models. Another example is the cell deformation in magnetocytometry, the application of a controlled force or torque via magnetic microbeads tethered to a single cell. Because the bead size and the resulting deformation in such experiments are much larger than the mesh size of the cytoskeletal network, a continuum viscoelastic model has been successfully applied without the need to worry about the heterogeneous distribution of filamentous proteins in the cytoskeleton. It should be noted that in using a continuum model, there are no constraints in terms of isotropy or homogeneity of properties, as these can easily be incorporated to the extent they are known. Predictions of the continuum model, however, are only as good as the constitutive law – stress-strain relation – on which they are based. This could range from a simple linear elasticity model to a description that captures the viscoelastic behavior of a soft glassy material (see, for example Chapter 3). Accordingly, the continuum model tells us nothing about the microstructure, other than what might be indirectly inferred based on the ability of one constitutive law or another to capture the observed cellular strains. It is important that modelers recognize this limitation.

In essence, continuum mechanics is a coarse-graining approach that replaces the contributions of the cytoskeleton's discrete stress fibers to the local microscopic stress-strain relationship with averaged constitutive laws that apply at macroscopic scale. This in turn leads to continuous stress-strain relationships and deformation descriptions that are applicable to the whole cell or cellular compartments. Depending on the dynamic time scale of interest, such continuum descriptions can be elastic or viscoelastic with appropriate complexity.

This chapter presents elastic and viscoelastic continuum multicompartment descriptions of the cell and shows a successful representation of such approaches by implementing finite element-based two- and three-dimensional models of the cell comprising separate compartments for cellular membrane and actin cortex, cytoskeleton, and the nucleus. To the extent that such continuum models can capture stress and strain patterns within the cell, they can help us relate biological influences of various types of force application and dynamics under different geometrical configurations of the cell.

By contrasting the computational results against experimental data obtained using various techniques probing single cells – such as micropipette aspiration (Discher et al., 1998; Drury and Dembo, 2001), microindentation (Bathe et al., 2002), atomic force microscopy (AFM) (Charras et al., 2001), or magnetocytometry (Figs. 4-7, 4-8, Karcher et al., 2003; Mack et al., 2004) – the validity and limits of such continuum mechanics models will be assessed. In addition, different aspects of the model will be characterized by examining, for instance, the mechanical role of the membrane and actin cortex in the overall cell behavior. Lastly, the applicability of different elastic and viscoelastic models in the form of various constitutive laws to describe the cell under different loading conditions will be addressed.

Purpose of continuum models

Continuum models of the cell are developed toward two main purposes: analyzing experiments probing single cell mechanics, and evaluating the level of forces sensed by various parts of the cell *in vivo* or *in vitro*. In the latter case, a continuum model evaluates the stress and strain patterns induced in the cell by the experimental technique. Comparison of theoretical and computational predictions proposed by the continuum model against the experimental observations then allows for deduction of the cell's mechanical properties. In magnetocytometry, for example, the same torque or tangential force applied experimentally to a microbead attached atop a cell is imposed in continuum models of the cell. Material properties introduced in the model that reproduce the observed bead displacement yield possible mechanical properties of the probed cell (see Mijailovich et al., 2002, and Fig. 4-7 for torque application,

Continuum elastic or viscoelastic models for the cell

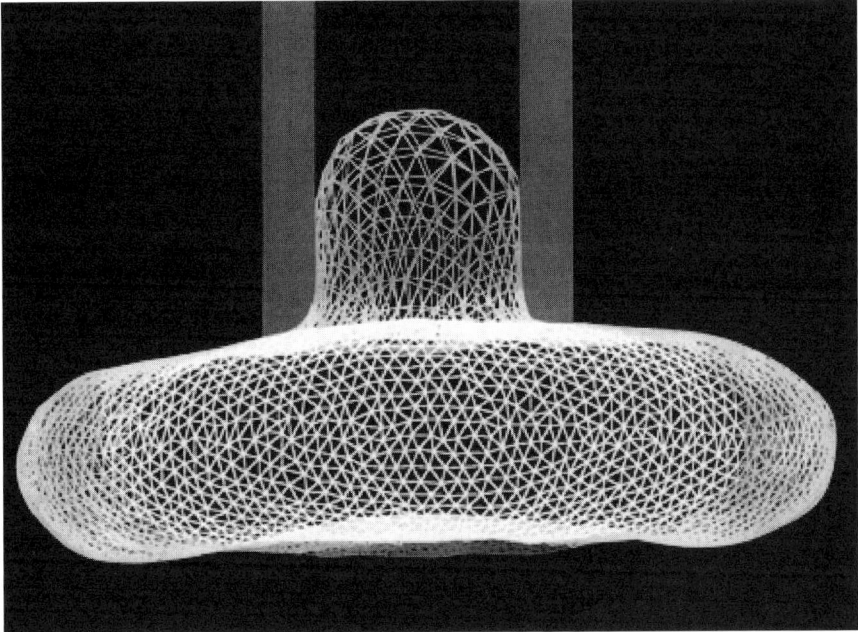

Fig. 4-1. Simulation of a small erythrocyte under aspiration. The micropipette, indicated by the solid gray shading, has an inside diameter of 0.9 μm. The surface of the cell is triangulated with 6110 vertex nodes that represent the spectrin-actin junction complexes of the erythrocyte cytoskeleton. The volume of the cell is 0.6 times the fully inflated volume, and the simulation is drawn from the stress-free model in the free shape ensemble. From Discher et al., 1998.

and Karcher et al., 2003, and Fig. 4-8 for tangential force application). Continuum models have also shed light on mechanical effects of other techniques probing single cells, such as micropipette aspiration (Figs. 4-1, 4-6, and for example, Theret et al., 1988; Yeung and Evans, 1989; Dong and Skalak, 1992; Sato et al., 1996; Guilak et al., 2000; Drury and Dembo, 2001), microindentation (for example, Bathe et al., 2002, probing neutrophils, Fig. 4-2 left), atomic force microscopy (AFM) (for example, Charras et al., 2001 and Charras and Horton, 2002, deducing mechanical

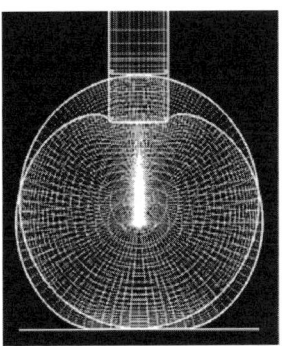 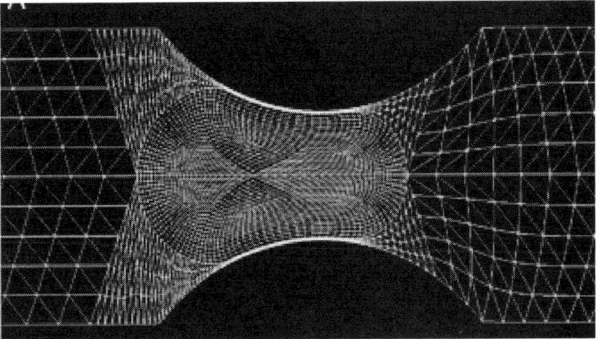

Fig. 4-2. Microindentation of a neutrophil (left) and passage through a capillary (right) (finite element model). From Bathe et al., 2002.

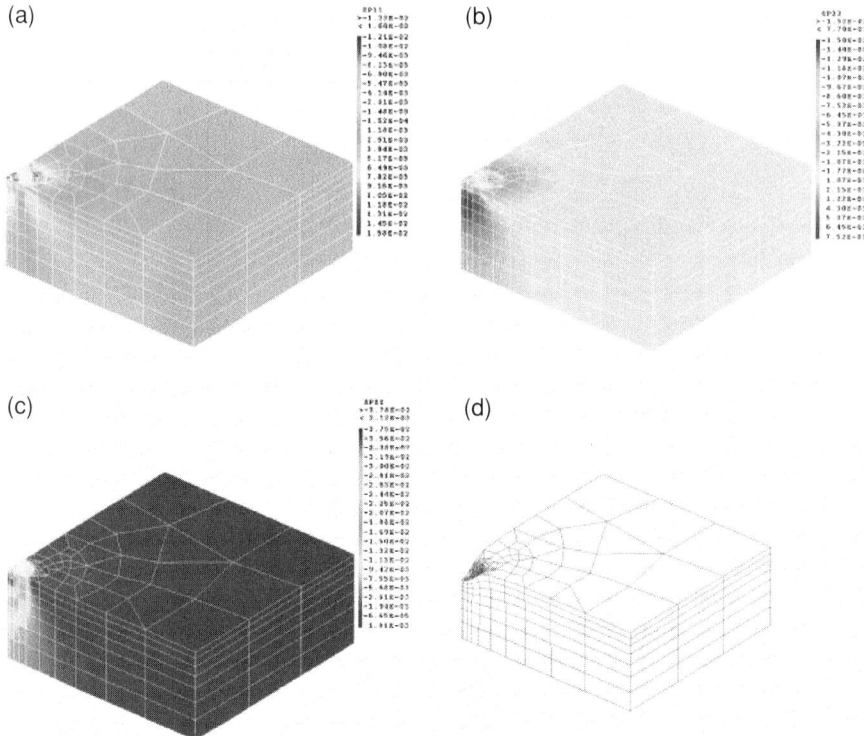

Fig. 4-3. Strain distributions elicited by AFM indentation. All of the scales are in strains. The numerical values chosen for this simulation were: $E = 10$ kPa, $v = 0.3$, $R = 15$ μm, $F = 1$ nN. (a) Radial strain distribution. The largest radial strains are found on the cell surface. A large strain gradient is present at the boundary between the region where the sphere is in contact with the cell surface and the region where it is not. (b) Tangential strain distribution. The largest tangential strains occurred at the cell surface in the area of indentation. (c) Vertical strain distribution. The largest vertical strains were located directly under the area of indentation within the cell thickness. (d) Deformations elicited by AFM indentation. The deformations have been amplified 15-fold in the z-direction. From Charras et al., 2001.

properties of osteoblasts, Figs. 4-3, 4-4), magnetocytometry (Figs. 4-7, 4-8, Karcher et al., 2003; Mack et al., 2004; Mijailovich et al., 2002), or optical tweezers (for example, Mills et al., 2004 stretching erythrocytes, Fig. 4-5). Finally, comparison of continuum models with corresponding experiments could help to distinguish active biological responses of the cell (such as remodeling and formation of pseudopods) from passive mechanical deformations, the only deformations captured by the model. This capability has not been exploited yet to the best of our knowledge.

In addition to helping interpret experiments, continuum models are also used to evaluate strains and stresses under biological conditions (for example, Fung and Liu, 1993, for endothelium of blood vessels). One example is found in the microcirculation where studies have examined the passage of blood cells through a narrow capillary (for example, Bathe et al., 2002, for neutrophils (Fig. 4-2 left), Barthes-Biesel, 1996, for erythrocytes) where finite element models have been used to predict the changes in cell shape and the cell's transit time through capillaries. In the case

Continuum elastic or viscoelastic models for the cell

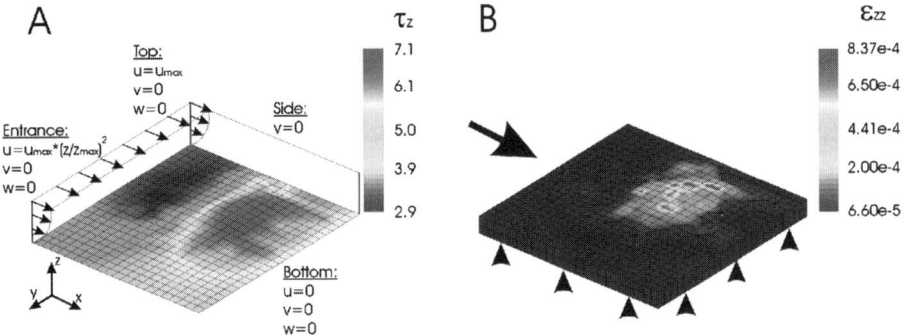

Fig. 4-4. The effect of fluid shear. (*a*) The shear stress resultant in the *z*-direction (τ_z) for a nominal 5 Pa shear stress on a flat substrate. The shear stresses are tensile and lower upstream and higher downstream. The imposed parabolic flow profile is shown at the entry and the boundary conditions are indicated on the graph. (*b*) The vertical strain distribution (ε_{zz}) for a cell submitted to fluid shear stresses. Black triangles indicate where the substrate was fully constrained. The cellular strains are maximal downstream from the cell apex and in the cellular region. In (*a*) and (*b*), the arrow indicates the direction of flow. From Charras and Horton, 2002.

of neutrophils, these inputs are crucial in understanding their high concentration in capillaries, neutrophil margination, and in understanding individual neutrophil activation preceding their leaving the blood circulation to reach infection sites. Neutrophil concentration depends indeed on transit time, and activation has recently been shown experimentally to depend on the time scale of shape changes (Yap and Kamm, 2005). Similarly, continuum models can shed light on blood cells' dysfunctional microrheology arising from changes in cell shape or mechanical properties (for example, time-dependent stiffening of erythrocytes infected by malaria parasites in Mills et al., 2004 (Fig. 4-5)).

Other examples include the prediction of forces exerted on a migrating cell in a three-dimensional scaffold gel (Zaman et al., 2005), prediction of single cell attachment and motility on a substrate, for example the model for fibroblasts or the unicellular organism Ameboid (Gracheva and Othmer, 2004), or individual protopod dynamics based on actin polymerization (Schmid-Schönbein, 1984).

Principles of continuum models

A continuum cell model provides the displacement, strain, and stress fields induced in the cell, given its initial geometry and material properties, and the boundary conditions it is subjected to (such as displacements or forces applied on the cell surface). Laws of continuum mechanics are used to solve for the distribution of mechanical stress and deformation in the cell. Continuum cell models of interest lead to equations that are generally not tractable analytically. In practice, the solution is often obtained numerically via discretization of the cell volume into smaller computational cells using (for example) finite element techniques.

A typical continuum model relies on linear momentum conservation (applicable to the whole cell volume). Because body forces within the cell are typically small, and,

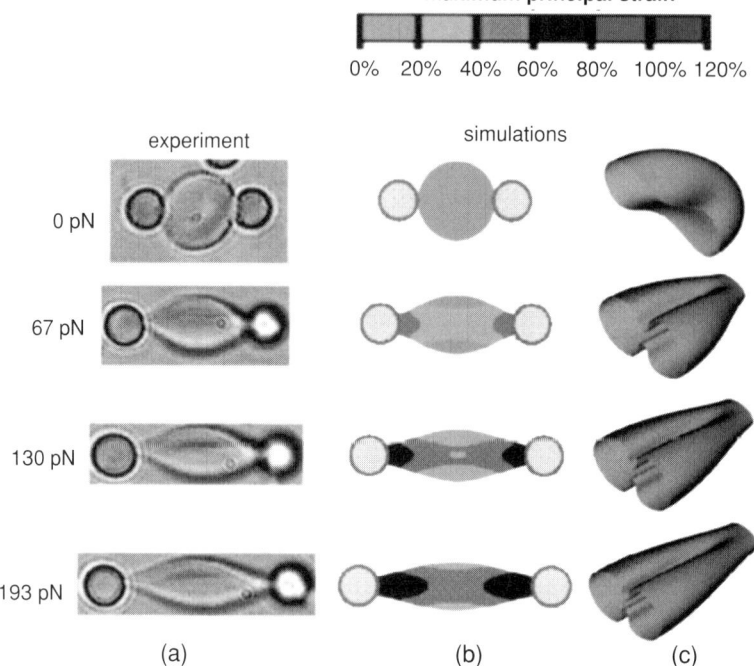

Fig. 4-5. Images of erythrocytes being stretched using optical tweezer at various pulling forces. The images in the left column are obtained from experimental video photography whereas the images in the center column (top view) and in the right column (half model 3D view) correspond to large deformation computational simulation of the biconcave red cell. The middle column shows a plan view of the stretched biconcave cell undergoing large deformation at the forces indicated on the left. The predicted shape changes are in reasonable agreement with observations. The contours in the middle column represent spatial variation of constant maximum principal strain. The right column shows one half of the full 3D shape of the cell at different imposed forces. Here, the membrane is assumed to contain a fluid with preserved the internal volume. From Mills et al., 2004.

at the scale of a cell, inertial effects are negligible in comparison to stress magnitudes the conservation equation simply reads:

$$\nabla \cdot \underline{\underline{\sigma}} = \underline{0}$$

with $\underline{\underline{\sigma}}$ = Cauchy's stress tensor.

Boundary conditions

For the solution to uniquely exist, either a surface force or a displacement (possibly equal to zero) should be imposed on each point of the cell boundary. Continuity of normal surface forces and of displacement imposes necessary conditions to ensure uniqueness of the solution.

Mechanical and material characteristics

Mechanical properties of the cell must be introduced in the model to link strain and stress fields. Because a cell is composed of various parts with vastly different mechanical properties, the model ideally should distinguish between the main parts

Continuum elastic or viscoelastic models for the cell

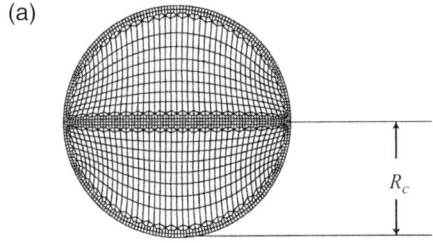

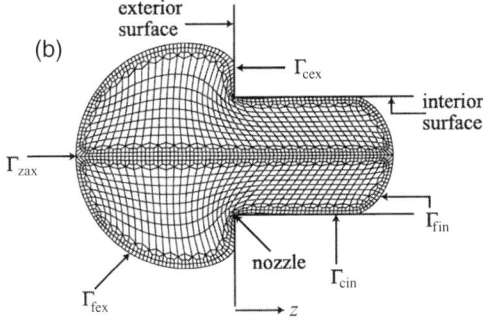

Fig. 4-6. Geometry of a typical computational domain at two stages. (*a*) The domain in its initial, round state. (*b*) The domain has been partially aspirated into the pipet. Here, the interior, exterior, and nozzle of the pipet are indicated. Γ_{fin}, free-interior; Γ_{fex}, free-exterior; Γ_{cin}, constrained-interior; and Γ_{cex}, constrained-exterior boundaries. There is a fifth, purely logical boundary, Γ_{zax}, which is the axis of symmetry. From Drury and Dembo, 2001.

of the cell, namely the plasma membrane, the nucleus, the cytoplasm, and organelles, which are all assigned different mechanical properties. This often leads to the introduction of many poorly known parameters. A compromise must then be found between the number of cellular compartments modeled and the number of parameters introduced.

The *cytoskeleton* is difficult to model, both because of its intricate structure and because it typically exhibits both solid- and fluid-like characteristics, both active and passive. Indeed, a purely solid passive model would not capture functions like crawling, spreading, extravasion, invasion, or division. Similarly, a purely fluid model would fail in describing the ability to maintain the structural integrity of cells, unless the membrane is sufficiently stiff.

The *nucleus* has generally been found to be stiffer and more viscous than the cytoskeleton. Probing isolated chondrocyte nuclei with micropipette aspiration Guilak et al. (2000) found nuclei to be three to four times stiffer and nearly twice as viscous as the cytoplasm. Its higher viscosity results in a slower time scale of response, so that the nucleus can often be considered as elastic, even when the rest of the cell requires viscoelastic modeling. Nonetheless, the available data on nuclear stiffness seem to be rather divergent, with values ranging from 18 Pa to nearly 10 kPa (Tseng et al., 2004; Dahl et al., 2005), due perhaps to factors such as differences in cell type, measurement technique, length scale of measurement, and also method of interpretation.

The *cellular membrane* has very different mechanical properties from the rest of the cell, and hence, despite its thinness, often requires separate modeling. It is more

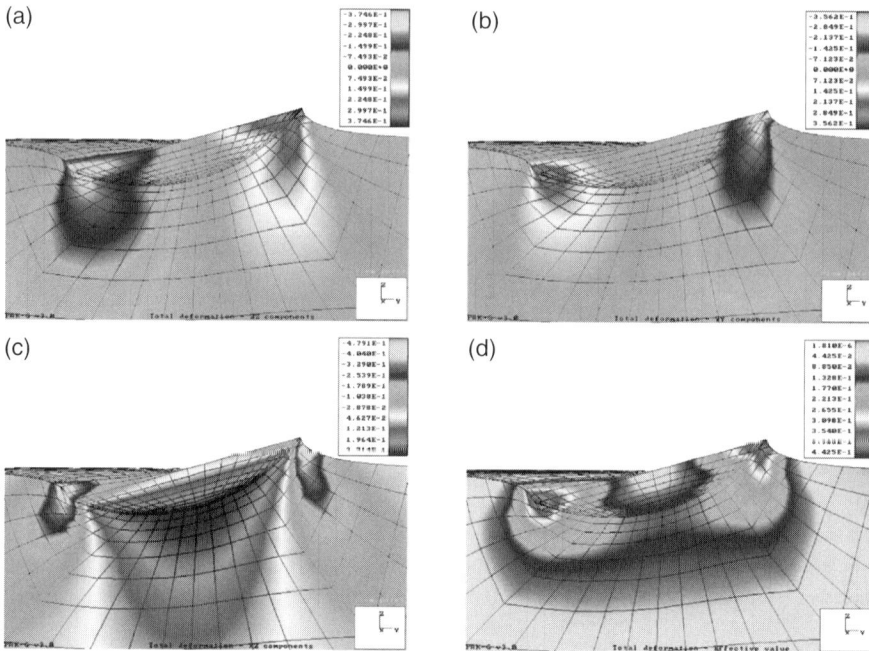

Fig. 4-7. Deformed shapes and strain fields in a cell 5 μm in height for bead embedded 10% of its diameter. Shown are strain fields of the components of strain: ε_{zz} (a), ε_{yy} (b), ε_{yz} (c), and the effective strain ε_{eff} (d). The effective strain is defined as: $\varepsilon_{eff} = \sqrt{\frac{2}{3}\varepsilon_{ij} - \varepsilon_{ij}}$, where ε_{ij} are strain components in Cartesian system x_i (x, y, z). From Mijailovitch et al., 2002.

fluid-like (Evans, 1989; Evans and Yeung, 1989) and should be modeled as a viscoelastic material with time constants of the order of tens of μs.

The *cortex*, that is, the shell of cytoskeleton that is just beneath the membrane, is in most cell types stiffer than the rest of the cytoskeleton. Bending stiffness of the membrane and cortex has been measured in red blood cells (Hwang and Waugh, 1997; Zhelev et al., 1994). A cortical tension when the cell is at its (unstimulated) resting state has also been observed in endothelial cells and leukocytes (Schmid-Schönbein et al., 1995).

Example of studied cell types

Blood cells: leukocytes and erythrocytes

Blood cells are subjected to intense mechanical stimulation from both blood flow and vessel walls, and their rheological properties are important to their effectiveness in performing their biological functions in the microcirculation. Modeling of neutrophils' viscoelastic large deformations in narrow capillaries or in micropipette experiments has shed light on their deformation and their passage time through a capillary or entrance time in a pipette. Examples of such studies are Dong et al. (1988), Dong and Skalak (1992), Bathe et al. (2002), and Drury and Dembo (2001) (see Fig. 4.6), who used finite element techniques and/or analytical methods to model the large deformations in neutrophils. Shape recovery after micropipette aspiration – a measure

Continuum elastic or viscoelastic models for the cell

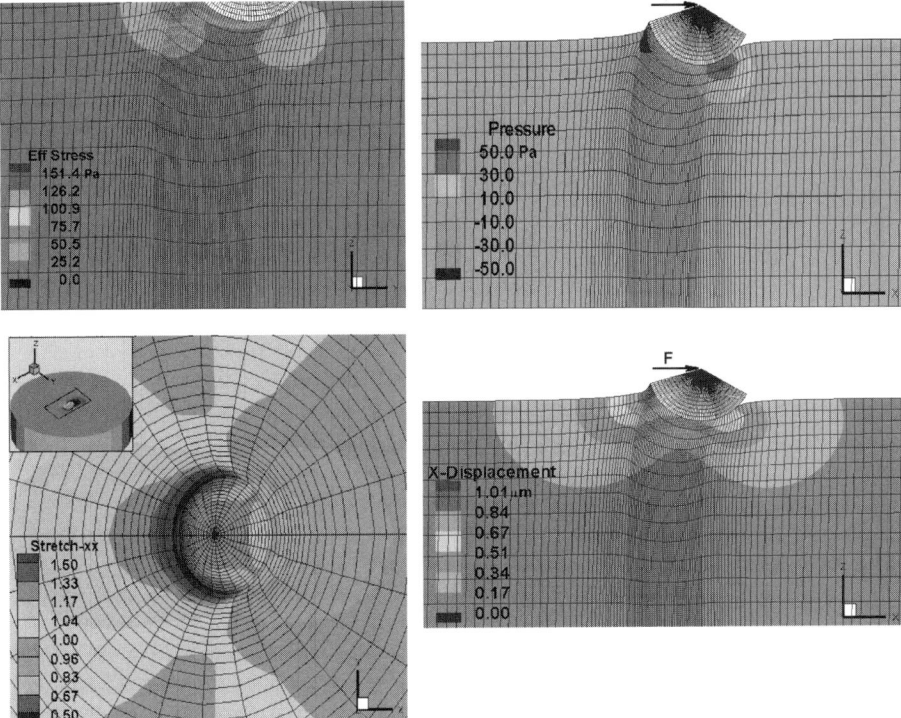

Fig. 4-8. Computational finite element models of a cell monolayer being pulled at 500 pN using magnetic cytometry experiment. Top panels show the pressure and effective stress fields induced in the cell after 2 s. (effective stress is a scalar invariant of the stress tensor excluding the compressive part). Lower left panel shows the membrane xx-stretch (in the direction of the applied force), while the lower right panel shows the induced deformation in the cytoskeleton in the direction of the applied force. From Karcher et al., 2003.

of the neutrophil's viscoelastic properties and its active remodeling – was for example investigated with a theoretical continuum model consisting of two compartments: a cytoplasm modeled as a Newtonian liquid, and a membrane modeled with a Maxwell viscoelastic fluid in the first time of recovery and a constant surface tension for the later times (Tran-Son-Tay et al., 1991). Erythrocytes have typically been modeled as viscoelastic membranes filled with viscous fluids, mostly to understand microcirculation phenomena, but also to explain the formation of "spikes" or crenations on their surface (Landman, 1984).

Adherent cells: fibrobasts, epithelial cells, and endothelial cells

Many types of cell, anchored to a basal substrate and sensitive to mechanical stimuli – like fibrobasts and epithelial and endothelial cells – have been probed by magnetocytometry, the forcing of a µm-sized bead attached atop a single cell through a certain type of membrane receptor (such as integrins).

Continuum modeling of this experiment was successfully developed to analyze the detailed strain/stress fields induced in the cell by various types of bead forcing (oscillatory or ramp forces of various magnitudes) (Mijailovitch et al., 2002; Fig. 4-7).

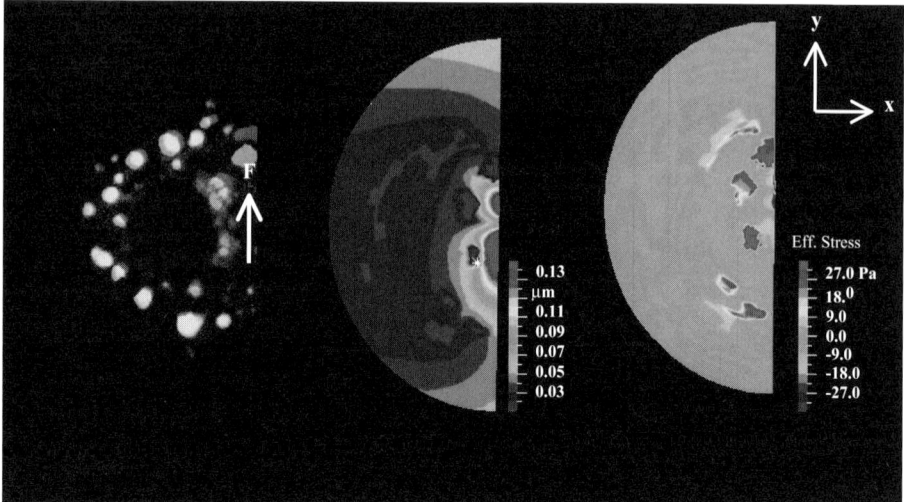

Fig. 4-9. A continuum, viscoelastic finite element simulation representing experimental cell contact sites on the basal cell surface estimated the focal adhesion shear stress distribution during magneto-cytometry. Left panel shows merged experimental fluorescent images depicting the focal adhesion sites. Middle and right panels show displacement and shear stress in the basal membrane of the cell. Zero displacement and elevated shear stresses are evident in the focal adhesion regions. From Mack et al., 2004.

(Karcher et al., 2003; Fig. 4-8). Modeling the cell with two Maxwell viscoelastic compartments representing, respectively, the cytoskeleton and the membrane/cortex, the authors found that the membrane/cortex contributed a negligible mechanical effect on the bead displacement at the time scales corresponding to magnetocytometry.

Comparison with experiments on NIH 3T3 fibroblasts led to a predicted viscoelastic time scale of ∼1 s and a shear modulus of ∼1000 Pa for these cells. In addition, the model showed that the degree to which the bead is embedded in the cell, a parameter difficult to control and measure in experiments (Laurent et al., 2002; Ohayon et al., 2004), dramatically changes the magnitude of stress and strain, although it influences their pattern very little. Continuum modeling also allowed for modulation of cell height and material properties to investigate the behavior of different adherent cell types. It also demonstrated that the response of the cell when forced with the microbead was consistent with that of a linear elastic model, quite surprising in view of the locally large strains.

The cell attachment to its substrate by the basal membrane was later modified to investigate force transmission from the bead to the basal membrane (Mack et al., 2004) (Fig. 4-9). Only experimentally observed points of attachments, that is, focal adhesion sites, were fixed in the model, allowing for the rest of the cell substrate to move freely. Forcing of the bead on the apical surface of NIH 3T3 fibroblasts preferentially displaced focal adhesion sites closer to the bead and induced a larger shear on the corresponding fixed locations in the model, implying that focal adhesion translation correlates with the local level of force they sense.

An alternative experiment to probe cell deformation and adhesion consists of plating them on a compliant substrate. Finite element modeling of cells probed by this

Continuum elastic or viscoelastic models for the cell

technique was recently used to evaluate the stress and strain experienced at the nuclear envelope, thereby investigating the mechanical interplay between the cytoskeleton and the nucleus. The ultimate goal of this study was to identify potential sources of mechanical dysfunction in fibroblasts deficient in specific structural nuclear membrane proteins (Hsiao, 2004). The model showed that the effect of nuclear shape, relative material properties of the nucleus and cytoskeleton, and focal adhesion size were important parameters in determining the magnitude of stress and strain at the nucleus/cytoskeleton interface.

Limitations of continuum model

Continuum models of the cell aim at capturing its passive dynamics. In addition to the limitations mentioned above, current models do not yet typically account for active biology: deformations and stresses experienced as a direct consequence of biochemical responses of the cell to mechanical load cannot be predicted by current continuum models. However, by contrasting the predicted purely mechanical cell response to experimental observations, one could isolate phenomena involving active biology, such as cell contraction or migration, from the passive mechanical response of the cell. Alternatively, continuum models might be envisioned that account for active processes through time-dependent properties or residual strains that are linked to biological processes. (See also Chapter 10.)

Another limitation of continuum models stems from lack of description of cytoskeletal fibers. As such, they are not applicable for micromanipulations of the cell with a probe of the same size or smaller than the cytoskeletal mesh ($\sim$0.1–1.0 μm). This includes most AFM experiments. In addition, the continuum models exclude small Brownian motions due to thermal fluctuations of the cytoskeleton, which would correspond to fluctuations of the network nodes in a continuum model and have been shown to play a key role in cell motility (Mogilner and Oster, 1996).

Finally, continuum models have so far employed a limited number of time constants to characterize the cell's behavior. However, cells have recently been shown to exhibit behaviors with power-law rheology implying a continuous spectrum of time scales (Fabry et al., 2001; Desprat et al., 2004, and Chapter 3). Modeling the cell with no intrinsic time constant has successfully captured this behavior (for example, Djordjevič et al., 2003), though this type of model cannot and does not aim at predicting or describing force or strain distribution within the cell. One of the challenges, therefore, to the use of continuum models for the prediction of intracellular stress and strain patterns is to develop cell material models that capture this complex behavior. In the meantime, models involving a finite number of time constants consistent with the time scale of the experimental technique can be used, recognizing their limitations.

Conclusion

Continuum mechanical models have proven useful in exploiting and interpreting results of a number of experimental techniques probing single cells or cell monolayers. They can help identify the stress and strain patterns induced within the cell by experimental perturpations, or the material properties of various cell compartments. In addition, continuum models enable us to predict the forces experienced within cells *in vivo*, and to then form hypotheses on how cells might sense and transduce forces into behavior such as changes in shape or gene expression.

The time scale of cell stimulation in experiments *in vivo* often requires that we take into account the time-dependent response of the cell, that is, to model it or some of its components as viscous or viscoelastic. Likewise, it is often necessary to model cell compartments with different materials, as their composition gives them very distinct mechanical properties.

Such continuum models have proven useful in the past, and will continue to play a role in cell modeling. As we gain more accurate experimental data on cellular rheology, these results can be incorporated into continuum models of improved accuracy of representation. As such, they are useful "receptacles" of experimental data with the capability to then predict the cellular response to mechanical stimulus, provided one accepts the limitations, and recognizes that they provide little by way of insight into the microstructural basis for macroscopic rheology.

References

Barthes-Biesel, D. 1996. Rheological models of red blood cell mechanics. In Advances in Hemodynamics. How HT, editor. Ed. JAI Press. 31–65.

Bathe, M., A. Shirai, C. M. Doerschuk, and R. D. Kamm. 2002. Neutrophil transit times through pulmonary capillaries: The effects of capillary geometry and fMLP-stimulation. *Biophys. J.*, 83(4):1917–1933.

Charras, G. T., and M. A. Horton. 2002. Determination of cellular strains by combined atomic force microscopy and finite element modeling. *Biophys. J.*, 83(2):858–879.

Charras, G. T., P. P. Lehenkari, and M. A. Horton. 2001. Atomic force microscopy can be used to mechanically stimulate osteoblasts and evaluate cellular strain distributions. *Ultramicroscopy*, 86(1–2):85–95.

Dahl, K. N., Engler, A. J., Pajerowski, J. D. & Discher, D. E. 2005. "Power-law rheology of isolated nuclei with deformation mapping of nuclear substitution." *Biophys. J.*, 89(4):2855–2864.

Desprat, N., A. Richert, J. Simeon, and A. Asnacios. 2004. Creep function of a single living cell. *Biophys. J.*, :biophysj.104.050278.

Discher, D. E., D. H. Boal, and S. K. Boey. 1998. Simulations of the erythrocyte cytoskeleton at large deformation. II. Micropipette aspiration. *Biophys. J.*, 75(3):1584–1597.

Djordjević, V. D., J. Jarić, B. Fabry, J. J. Fredberg, and D. Stamenović. 2003. Fractional derivatives embody essential features of cell rheological behavior. *Ann. Biomed. Eng.*, 31:692–699.

Dong, C., and R. Skalak. 1992. Leukocyte deformability: Finite element modeling of large viscoelastic deformation. *J. Theoretical, Biol.*, 158(2):173–193.

Dong, C., R. Skalak, K. L. Sung, G. W. Schmid-Schönbein, and S. Chien. 1988. Passive deformation analysis of human leukocytes. *J. Biomech. Eng.*, 110(1):27–36.

Drury, J. L., and M. Dembo. 2001. Aspiration of human neutrophils: Effects of shear thinning and cortical dissipation. *Biophys. J.*, 81(6):3166–3177.

Evans, E. 1989. Structure and deformation properties of red blood cells: Concepts and quantitative methods. *Methods Enzymol.*, 173:3–35.

Evans, E., and A. Yeung. 1989. Apparent viscosity and cortical tension of blood granulocytes determined by micropipette aspiration. *Biophys. J.*, 56(1):151–160.

Fabry, B., G. Maksym, J. Butler, M. Glogauer, D. Navajas, and J. Fredberg. 2001. Scaling the microrheology of living cells. *Phys. Rev. Lett.*, 87:148102.

Fung, Y. C., S. Q. Liu. 1993. Elementary mechanics of the endothelium of blood vessels. *ASME J. Biomech. Eng.*, 115: 1–12.

Gracheva, M. E., and H. G. Othmer. 2004. A continuum model of motility in ameboid cells. *Bull. of Math. Bio.*, 66(1):167–193.

Guilak, F., J. R. Tedrow, and R. Burgkart. 2000. Viscoelastic properties of the cell nucleus. *Biochem. and Biophy. Res. Comm.*, 269(3):781–786.

Hsiao, J. 2004. Emerin and inherited disease. Masters thesis. Division of Health Science and Technology, Harvard-MIT, Cambridge.

Hwang, W., and R. Waugh. 1997. Energy of dissociation of lipid bilayer from the membrane skeleton of red blood cells. *Biophys. J.*, 72(6):2669–2678.

Karcher, H., J. Lammerding, H. Huang, R. Lee, R. Kamm, and M. Kaazempur-Mofrad. 2003. A three-dimensional viscoelastic model for cell deformation with experimental verification. *Biophys. J.*, 85(5):3336–3349.

Landman, K. 1984. A continuum model for a red blood cell transformation: Sphere to crenated sphere. *J. Theor. Biol.*, 106(3):329–351.

Laurent, V., S. Henon, E. Planus, R. Fodil, M. Balland, D. Isabey, and F. Gallet. 2002. Assessment of mechanical properties of adherent living cells by bead micromanipulation: Comparison of magnetic twisting cytometry vs optical tweezers. *J. Biomech. Eng.*, 124(August):408–421.

Mack, P. J., M. R. Kaazempur-Mofrad, H. Karcher, R. T. Lee, and R. D. Kamm. 2004. Force-induced focal adhesion translocation: Effects of force amplitude and frequency. *Am. J. Physiol. Cell Physiol*:00567.02003.

Mijailovich, S. M., M. Kojic, M. Zivkovic, B. Fabry, and J. J. Fredberg. 2002. A finite element model of cell deformation during magnetic bead twisting. *J. Appl. Physiol.*, 93(4):1429–1436.

Mills, J. P., L. Qie, M. Dao, C. T. Lim, and S. Suresh. 2004. Nonlinear elastic and viscoelastic deformation of the human red blood cell with optical tweezers. *Mech. and Chem. of Biosys.*, 1(3):169–180.

Mogilner, A., and G. Oster. 1996. Cell motility driven by actin polymerization. *Biophys. J.*, 71(6):3030–3045.

Ohayon, J., P. Tracqui, R. Fodil, S. Féréol, V. M. Laurent, E. Planus, and D. Isabey. 2004. Anaylses of nonlinear responses of adherent epithelial cells probed by magnetic bead twisting: A finite-element model based on a homogenization approach. *J. Biomech. Eng.*, 126(6):685–698.

Sato, M., N. Ohshima, and R. M. Nerem. 1996. Viscoelastic properties of cultured porcine aortic endothelial cells exposed to shear stress. *J. Biomech.*, 29(4):461.

Schmid-Schönbein, G. W., and R. Skalak. 1984. Continuum mechanical model of leukocytes during protopod formation. *J. Biomech. Eng.*, 106(1):10–18.

Schmid-Schönbein, G. W., T. Kosawada, R. Skalak, S. Chien. 1995. Membrane model of endothelial cells and leukocytes. A proposal for the origin of cortical stress. *ASME J. Biomech. Eng.*, 117:171–178.

Theret, D. P., M. J. Levesque, M. Sato, R. M. Nerem, and L. T. Wheeler. 1988. The application of a homogeneous half-space model in the analysis of endothelial cell micropipette measurements. *ASME J. Biomech. Eng.*, 110:190–199.

Tran-Son-Tay, R., D. Needham, A. Yeung, and R. M. Hochmuth. 1991. Time-dependent recovery of passive neutrophils after large deformation. *Biophys. J.*, 60(4):856–866.

Tseng, Y., Lee, J. S., Kole, T. P., Jiang, I. & Wirtz, D. 2004. Micro-organization and viscoelasticity of interphase nucleus revealed by particle nanotracking. *J. Cell Sci.*, 117, 2159–2167.

Yap, B., and R. D. Kamm. 2005. Mechanical deformation of neutrophils into narrow channels induces pseudopod projection and changes in biomechanical properties. *J. Appl. Physiol.*, 98: 1930–1939.

Yeung, A., and E. Evans. 1989. Cortical cell – liquid core model for passive flow of liquid-like spherical cells into micropipettes. *Biophys. J.*, 56:139–149.

Zaman, M. H., R. D. Kamm, P. T., Matsudaira, and D. A. Lauffenburger. 2005. Computational model for cell migration in 3-dimensional matrices. *Biophys. J.*, 89(2):1389–1397.

Zhelev, D., D. Needham, and R. Hochmuth. 1994. Role of the membrane cortex in neutrophil deformation in small pipettes. *Biophys. J.*, 67(2):696–705.

5 Multiphasic models of cell mechanics

Farshid Guilak, Mansoor A. Haider, Lori A. Setton,
Tod A. Laursen, and Frank P.T. Baaijens

ABSTRACT: Cells are highly complex structures whose physiology and biomechanical properties depend on the interactions among the varying concentrations of water, charged or uncharged macromolecules, ions, and other molecular components contained within the cytoplasm. To further investigate the mechanistic basis of the mechanical behaviors of cells, recent studies have developed models of single cells and cell–matrix interactions that use multiphasic constitutive laws to represent the interactions among solid, fluid, and in some cases, ionic phases of cells. The goals of such studies have been to characterize the relative contributions of different physical mechanisms responsible for empirically observed phenomena such as cell viscoelasticity or volume change under mechanical or osmotic loading, and to account for the coupling of mechanical, chemical, and electrical events within living cells. This chapter describes several two-phase (fluid-solid) or three-phase (fluid-solid-ion) models, originally developed for studying soft hydrated tissues, that have been extended to describe the biomechanical behavior of individual cells or cell–matrix interactions in various tissue systems. The application of such "biphasic" or "triphasic" continuum-based approaches can be combined with other structurally based models to study the interactions of the different constitutive phases in governing cell mechanical behavior.

Introduction

Cells of the human body are regularly subjected to a complex mechanical environment, consisting of temporally and spatially varying stresses, strains, fluid flow, osmotic pressure, and other biophysical factors. In many cases, the mechanical properties and the rheology of cells play a critical role in their ability to withstand mechanical loading while performing their physiologic functions. In other cases, mechanical factors serve as important signals that influence, and potentially regulate, cell phenotype in both health and disease. An important goal in the field of cell mechanics thus has been the study of the mechanical properties of the cell and its biomechanical interactions with the extracellular matrix. Accordingly, such approaches have required the development of constitutive models based on realistic cellular structure and composition to better describe cell behavior.

Based on empirical studies of cell mechanical behavior, continuum models of cell mechanics generally have assumed either fluid or solid composition and cell properties, potentially including cortical tension at the membrane (Evans and Yeung,

Multiphasic models of cell mechanics

1989; Dong et al., 1991; Needham and Hochmuth, 1992; Karcher et al., 2003). In other approaches, the elastic behavior of the cell has been described using structural models such as the "tensegrity" approach (Ingber, 2003). Most such models have employed constitutive models that assume cells consist of a single-phase material (that is, fluid or solid), as detailed in Chapters 4, and 6. However, a number of recent studies have developed models of single cells and cell–matrix interactions that use multiphasic constitutive laws to account for interactions among solid, fluid, and in some cases, ionic phases of cells. The goals of such studies have been to characterize the relative importance of the mechanisms accounting for empirically observed phenomena such as cell volume change under mechanical or osmotic loading, the mechanistic basis responsible for cell viscoelasticity, and the coupling of various mechanical, chemical, and electrical events within living cells. The presence of these behaviors, which arise from interactions among different phases, often cannot be described by single-phase models.

The cell cytoplasm may consist of varying concentrations of water, charged or uncharged macromolecules, ions, and other molecular components. Furthermore, due to the highly charged and hydrated nature of its various components (Maughan and Godt, 1989; Cantiello et al., 1991), the cytoplasm's gel-like properties have been described under several different contexts see for example Chapter 7 and Pollack (2001). Much of the supporting data for the application of multiphasic models of cells has come from the study of volumetric and morphologic changes of cells in response to mechanical or osmotic loading. The majority of work in this area has been performed on cells of articular cartilage (chondrocytes), likely due to the fact that these cells are embedded within a highly charged and hydrated extracellular matrix that has been modeled extensively using multiphasic descriptions. For example (see Fig. 5-1), chondrocytes in articular cartilage exhibit significant changes in shape and volume that occur in coordination with the deformation and dilatation of the extracellular matrix (Guilak, 1995; Guilak et al., 1995; Buschmann et al., 1996). By using generalized continuum models of cells and tissue, the essential characteristics of cell and tissue mechanics and their mechanical interactions can be better understood. In this chapter, we describe several experimental and theoretical approaches for studying the multiphasic behavior of living cells.

Biphasic (solid–fluid) models of cell mechanics

Viscoelastic behavior in cells can arise from both flow-dependent (fluid–solid interactions and fluid viscosity) and flow-independent mechanisms (for example, intrinsic viscoelasticity of the cytoskeleton). Previous studies have described the cytoplasm of "solid-like" cells as a gel or as a porous-permeable, fluid-saturated meshwork (Oster, 1984; Oster, 1989; Pollack, 2001) such that the forces within the cell exhibit a balance of stresses arising from hydrostatic and osmotic pressures and the elastic properties of the cytoskeleton. This representation of cell mechanical behavior is consistent with the fundamental concepts of the biphasic theory, which has been used to represent the mechanical behavior of soft hydrated tissues as being that of a two-phase material. This continuum mixture theory approach has been adopted in several studies to model volumetric and viscoelastic cell behaviors and to investigate potential mechanisms

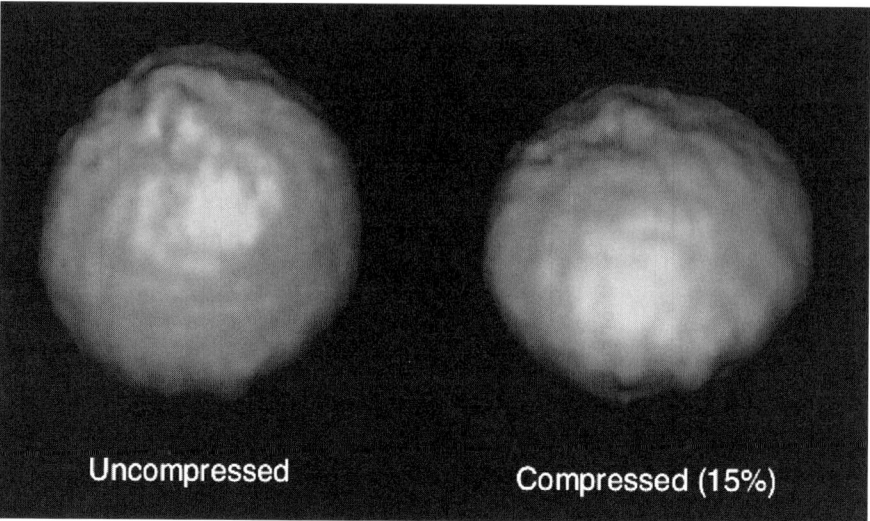

Fig. 5-1. Three-dimensional reconstructions of viable chondrocytes within the extracellular matrix before (left) and after (right) compression of the tissue to 15% surface-to-surface tissue strain. Significant changes in chondrocyte height and volume were observed, showing that cellular deformation was coordinated with deformation of the tissue extracellular matrix.

responsible for cell mechanical behavior (Bachrach et al., 1995; Shin and Athanasiou, 1999; Guilak and Mow, 2000; Baaijens et al., 2005; Trickey et al., 2006).

Modern mixture theories (Truesdell and Toupin, 1960; Bowen, 1980) provide a foundation for multiphasic modeling of cell mechanics as well as of soft hydrated tissues. The biphasic model (Mow et al., 1980; Mow et al., 1984), based on Bowen's theory of incompressible mixtures (Bowen, 1980), has been widely employed in modeling the mechanics of articular cartilage and other musculoskeletal tissues, such as intervertebral disc (Iatridis et al., 1998), bone (Mak et al., 1997), or meniscus. (Spilker et al., 1992). In such models, the cell or tissue is idealized as a porous and permeable solid material that is saturated by a second phase consisting of interstitial fluid (water with dissolved ions). Viscoelastic behavior can arise from intrinsic viscoelasticity of the solid phase, or from diffusive drag between the solid and fluid phases.

In the biphasic theory, originaly developed to describe the mechanical behavior of soft, hydrated tissues (Mow et al., 1980), the momentum balance laws for the solid and fluid phases, respectively, are written as:

$$\nabla \cdot \boldsymbol{\sigma}^s + \boldsymbol{\Pi} = 0, \quad \nabla \cdot \boldsymbol{\sigma}^f - \boldsymbol{\Pi} = 0 \qquad (5.1)$$

where $\boldsymbol{\sigma}^s$ and $\boldsymbol{\sigma}^f$ are partial Cauchy stress tensors that measure the force per unit mixture area on each phase. The symbol $\boldsymbol{\Pi}$ denotes a momentum exchange vector that accounts for the interphase drag force as fluid flows past solid in the mixture. Note that, in biphasic models of cells or cartilage, the contribution of inertial terms to the momentum balance equations is negligible, as the motion is dominated by elastic deformation and diffusive drag and occurs at relatively low frequencies. The mixture is assumed to be intrinsically incompressible and saturated, so that:

$$\nabla \cdot (\phi^s \dot{\mathbf{u}}^s + \phi^f \mathbf{v}^f) = 0, \quad \text{where } \phi^s + \phi^f = 1, \qquad (5.2)$$

$\mathbf{u}^s$ is the solid displacement, $\mathbf{v}^f$ is the fluid velocity, and ϕ^s is the solid volume fraction. For example, under the assumption of infinitesimal strain, with isotropic solid phase and inviscid fluid phase, while the momentum exchange is described by Darcy's Law, the resulting constitutive laws are:

$$\boldsymbol{\sigma}^s = -\phi^s p\mathbf{I} + \lambda^s tr(\mathbf{e})\mathbf{I} + 2\mu^s \mathbf{e}, \quad \boldsymbol{\sigma}^f = -\phi^f p\mathbf{I}, \quad \Pi = K(\mathbf{v}^f - \dot{\mathbf{u}}^s) \quad (5.3)$$

where $\mathbf{I}$ is the identity tensor, p is a pore pressure used to enforce the incompressibility constraint, $\mathbf{e} = 1/2[\nabla \mathbf{u}^s + (\nabla \mathbf{u}^s)^T]$ is the infinitesimal strain tensor, λ^s, μ^s are Lamé coefficients for the solid phase, and K is a diffusive drag coefficient. The Lamé coefficients λ^s, μ^s are associated with "drained" elastic equilibrium states that occur under static loading when all fluid flow has ceased in the mixture. An alternate set of elastic moduli are the Young's modulus E^s and Poisson ratio ν^s ($0 \leq \nu^s < 0.5$) where $\mu^s = \frac{E^s}{2(1+\nu^s)}$ is the solid phase shear modulus and $\lambda^s = \frac{E^s \nu^s}{(1+\nu^s)(1-2\nu^s)}$.

For this linear biphasic model, by substituting Eq. 5.3 into Eq. 5.1, the governing equations Eq. 5.1 and Eq. 5.2 constitute a system of seven equations in the seven unknowns $\mathbf{u}^s, \mathbf{v}^f, p$. The fluid velocity $\mathbf{v}^f$ is commonly eliminated to yield a "u-p formulation" consisting of the four equations:

$$\partial_t (\nabla \cdot \mathbf{u}^s) = k\nabla^2 p, \quad \mu^s \left[\frac{1}{1-2\nu^s} \nabla(\nabla \cdot \mathbf{u}^s) + \nabla^2 \mathbf{u}^s \right] = \nabla p \quad (5.4)$$

where $k = (\phi^f)^2/K$ is the permeability. The fluid velocity is then given by:

$$\mathbf{v}^f = \partial_t \mathbf{u}^s - \frac{\phi^f}{K} \nabla p \quad (5.5)$$

The governing equations Eqs. 5.4–5.5 illustrate a common formulation of the linear isotropic biphasic model. Within the framework of Eqs. 5.1–5.2, this fundamental model can be extended to account for additional mechanisms via modification of the constitutive relations in Eq. 5.3. Such mechanisms have included transverse isotropy of the solid phase, large deformation, solid matrix viscoelasticity, nonlinear strain-dependent permeability, intrinsic fluid viscosity, and tension-compression nonlinearity (Lai et al., 1981; Holmes et al., 1985; Cohen et al., 1998).

Biphasic poroviscoelastic models of cell mechanics

In other approaches, both the flow-dependent and flow-independent viscoelastic behaviors have been taken into account to describe transient cell response to loading. This type of approach has been used previously to separate the influence of "intrinsic" viscoelastic behavior of the solid extracellular matrix of tissues such as articular cartilage from the time- and rate-dependent effects due to fluid–solid interactions in the tissue (Mak, 1986a; Mak, 1986b; Setton et al., 1993; DiSilvestro and Suh, 2002).

For example, in modeling the creep response of chondrocytes during both full and partial micropipette aspiration (Baaijens et al., 2005; Trickey et al., in press), it was found that an elastic biphasic model cannot capture the time-dependent response of chondrocytes accurately (Baaijens et al., 2005). To examine the relative contributions of intrinsic solid viscoelasticity (solid–solid interactions) as compared to biphasic viscoelastic behavior (fluid–solid interactions), a large strain, finite element simulation of the micropipette aspiration experiment was developed to model the cell using finite strain incompressible and compressible elastic models, a two-mode compressible viscoelastic model, a biphasic elastic, or a biphasic viscoelastic model.

Assuming isotropic and constant permeability, the governing equations Eq. 5.1 and 5.2 may be rewritten (Mow et al., 1980; Sengers et al., 2004) as:

$$\nabla \cdot \sigma - \nabla p = 0 \quad (5.6)$$
$$\nabla \cdot v - \nabla \cdot k \nabla p = 0$$

where v denotes the solid velocity. If a viscoelastic model is used to investigate the time-dependent behavior, a two-mode model may be used. The stress tensor is split into an elastic part and a viscoelastic part:

$$\sigma = \sigma_e + \tau. \quad (5.7)$$

If a finite strain formulation is used, a suitable constitutive model for the compressible elastic contribution can be given by

$$\sigma_e = \kappa(J-1)\mathbf{I} + \frac{G}{J}(\mathbf{B} - J^{2/3}\mathbf{I}), \quad (5.8)$$

where, using the deformation tensor $\mathbf{F} = (\nabla_0 x)^T$ with $\vec{\nabla}_0$ the gradient operator with respect to the reference configuration, the volume ratio is given by $J = \det(\mathbf{F})$, and the right Cauchy-Green tensor by $\mathbf{B} = \mathbf{F} \cdot \mathbf{F}^T$. The material parameters κ and G denote the compressibility modulus and the shear modulus, respectively. The viscoelastic response is modeled using a compressible Upper Convected Maxwell model:

$$\overset{\nabla}{\tau} + \frac{1}{\lambda}\tau = 2G_v \mathbf{D}^d \quad (5.9)$$

where the operator ∇ denotes the upper-convected time derivative (Baaijens, 1998), and $\mathbf{D}^d$ is the deviatoric part of the rate of deformation tensor, defined by:

$$\mathbf{D}^d = \mathbf{D} - \frac{1}{3}tr(\mathbf{D})\mathbf{I}, \quad \text{where } \mathbf{D} = \frac{1}{2}(\dot{\mathbf{F}} \cdot \mathbf{F}^{-1} + \mathbf{F}^{-T} \cdot \dot{\mathbf{F}}^T) \quad (5.10)$$

G_v is the modulus and λ is the relaxation time of the viscoelastic mode.

Multiphasic and triphasic models (solid–fluid–ion)

In response to alterations in their osmotic environment, cells passively swell or shrink. The capability of the biphasic model to describe this osmotic response is limited to the determination of effective biphasic material parameters that vary with extracellular osmolality. The triphasic continuum mixture model (Lai et al., 1991) provides a framework that has the capability to more completely describe mechanochemical coupling via both mechanical and chemical material parameters in the governing equations. This model has been successfully employed in quantitative descriptions of mechanochemical coupling in articular cartilage, where the aggrecan of the extracellular matrix gives rise to a net negative fixed-charge density within the tissue. Similar approaches have been used to describe other charged hydrated soft tissues (Huyghe et al., 2003).

Multiphasic models of cell mechanics

To date, there has been only limited application of the triphasic model to cell mechanics (Gu et al., 1997). At mechanochemical equilibrium *in vitro*, many cells are known to exhibit a passive volumetric response corresponding to an ideal osmometer. For an ideal osmometer, cell volume and inverse osmolality (normalized to their values in the iso-osmotic state) are linearly related via the Boyle van't Hoff relation. The resulting states of mechanochemical equilibrium of the cell exhibit an internal balance between mechanical and chemical stresses. The triphasic model, in the absence of electrical fields, gives rise to the mixture momentum equation:

$$\nabla \cdot \boldsymbol{\sigma} = 0, \quad \text{where: } \boldsymbol{\sigma} = -p\mathbf{I} + \boldsymbol{\sigma}_E \tag{5.11}$$

where σ is the mixture stress and σ_E is the extra stress in the solid phase. The balance of electrochemical potentials for intracellular and extracellular water gives rise to the Donnan osmotic pressure relation:

$$p = RT(\varphi c - \varphi^* c^*) \tag{5.12}$$

where T is the temperature, φ and φ^* are the intracellular and extracellular osmotic activity coefficients, and c and c^* are the intracellular and extracellular ion concentrations, respectively (R is the universal gas constant). To close the system of governing equations (Eqs. 5.11–5.12), the intracellular extra stress and the intracellular ion concentration need to be characterized. While the former may be postulated via a constitutive description for the subcellular components (such as nucleus, cytoskeleton, membrane), the latter necessitates a detailed analysis of electrochemical ion potentials inside a cell that accounts for intracellular ionic composition and biophysical mechanisms such as the selective permeability of the bilayer lipid membrane, and the nontransient activity of ion pumps and ion channels.

The strength of this type of mixture theory approach was recently illustrated in a study modeling the transient swelling and recovery behavior of a single cell subjected to an osmotic stress with neutrally charged solutes (Ateshian et al., 2006). A generalized "triphasic" formulation and notation (Gu et al., 1998) were used to account for multiple solute species and incorporated partition coefficients for the solutes in the cytoplasm relative to the external solution. Numerical simulations demonstrate that the volume response of the cell to osmotic loading is very sensitive to the partition coefficient of the solute in the cytoplasm, which controls the magnitude of cell volume recovery. Furthermore, incorporation of tension in the cell membrane significantly affected the mechanical response of the cell to an osmotic stress. Of particular interest was the fact that the resulting equations could be reduced to the classical equations of Kedem and Katchalsky (1958) in the limit when the membrane tension is equal to zero and the solute partition coefficient in the cytoplasm is equal to unity. These findings emphasize the strength of using more generalized mixture approaches that can be selectively simplified in their representation of various aspects of cell mechanical behavior.

Analysis of cell mechanical tests

Similar to other experiments of cell mechanics, the analysis of cell multiphasic properties has involved the comparison and matching of different experimental configurations to the theoretical response as predicted by analytical or numerical models of each individual experimental configuration. Given the complexity of the governing equations for multiphasic or poroelastic models, analytical solutions have only been possible for simplified geometries that approximate various testing configurations. In most cases, numerical methods such as finite element techniques are used in combination with optimization methods to best fit the predicted behavior to the actual cellular response in order to determine the intrinsic material properties of the cell.

Micropipette aspiration

Cells

The micropipette aspiration technique has been used extensively to study the mechanical properties of both fluid-like and solid-like cells (Hochmuth, 2000), including circulating cells such as red blood cells (Evans, 1989) and neutrophils (Sung et al., 1982; Dong et al., 1991; Ting-Beall et al., 1993), or adhesion-dependent cells such as fibroblasts (Thoumine and Ott, 1997), endothelial cells (Theret et al., 1988; Sato et al., 1990), or chondrocytes (Jones et al., 1999; Trickey et al., 2000; Guilak et al., 2002). This technique involves the use of a small glass pipette to apply controlled suction pressures to the cell surface while measuring the ensuing transient deformation via video microscopy. The analysis of such experiments has required the development of a variety of theoretical models that assume cells behave as viscous liquid drops (Yeung and Evans, 1989), potentially possessing cortical tension (Evans and Yeung, 1989), or as elastic or viscoelastic solids (Theret et al., 1988; Sato et al., 1990; Haider and Guilak, 2000; Haider and Guilak, 2002) and specifically, to model the solid-like response of cells to micropipette aspiration, Theret et al. (1988) developed an elegant analytical solution of an associated contact problem to calculate the Young's modulus (E) of an incompressible cell. This elastic model was subsequently extended to a standard linear solid (Kelvin) model, thus incorporating viscoelastic cell properties (Sato et al., 1990). These models idealized the cell as an elastic or viscoelastic incompressible and homogeneous half-space. Experimentally, the length of cell aspiration was measured at several pressure increments, and the Young's modulus (E) was determined as a function of the applied pressure (Δp), the length of aspiration of the cell into the micropipette (L), and the radius of the micropipette (a) as $E = 3a\Phi\Delta p/(2\pi L)$, where Φ is a function of the ratio of the micropipette thickness to its inner radius (Fig. 5-2).

In recent studies, the micropipette aspiration test has been modeled assuming that cells exhibit biphasic behavior. The cell was modeled using finite strain incompressible and compressible elastic models, a two-mode compressible viscoelastic model, or a biphasic elastic or viscoelastic model. Comparison of the model to the experimentally measured response of chondrocytes to a step increase in aspiration pressure showed that a two-mode compressible viscoelastic formulation could predict the creep response of chondrocytes during micropipette aspiration (Fig. 5-3). Similarly, a biphasic two-mode viscoelastic analysis could predict all aspects of the cell's creep response to a step aspiration. In contrast, a purely biphasic elastic formulation was not capable

Multiphasic models of cell mechanics

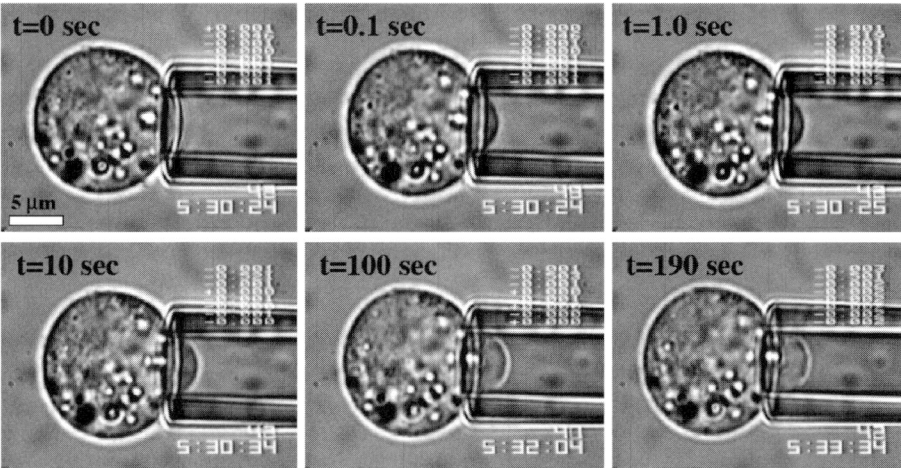

Fig. 5-2. Viscoelastic creep response of a chondrocyte to a step increase in pressure in the micropipette aspiration test. Images show the chondrocyte in the resting state under a tare pressure at time zero, followed by increasing cell displacement over time after step application of the test pressure. From Haider and Guilak, 2000.

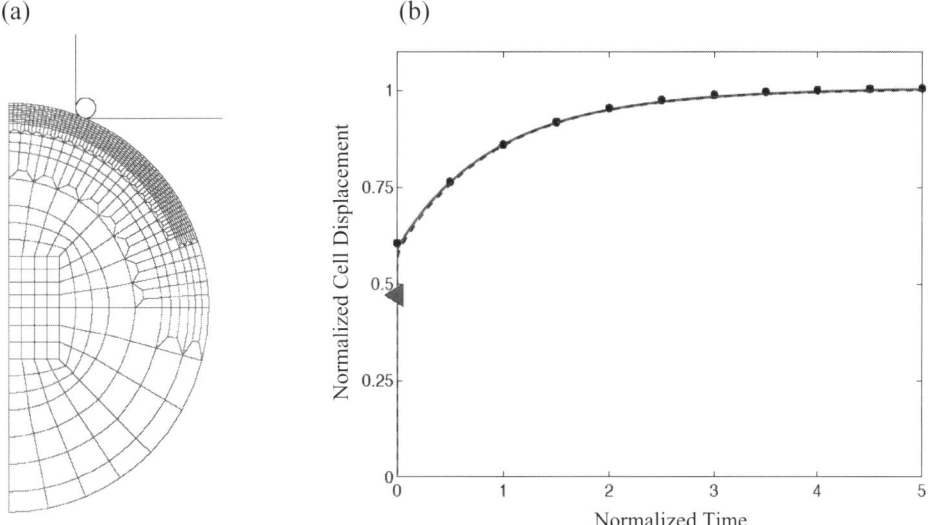

Fig. 5-3. (a) Biphasic, viscoelastic finite element mesh used to model the micropipette aspiration test. The mesh contains 906 elements and has 3243 nodes, with biquadratic interpolation for the displacement and a bilinear, continuous interpolation for the pressure. Along the left boundary, symmetry conditions are applied, while along the portion of the cell boundary within the micropipette, suction pressure is applied. Sliding contact conditions along the interface with the micropipette are enforced through the use of a Lagrange multiplier formulation. (b) Normalized aspiration length of a chondrocyte as a function of the normalized time. The solid dots indicate the experimental behavior of an average chondrocyte. The solid line corresponds to the two-mode viscoelastic model. The dashed line corresponds to the biphasic, two-mode viscoelastic model. For the latter model, the triangle marks the end of the pressure application and the start of the creep response. The creep response of the cell was well described by both the two-mode viscoelastic model and the biphasic viscoelastic model. From Baaijens et al., 2005.

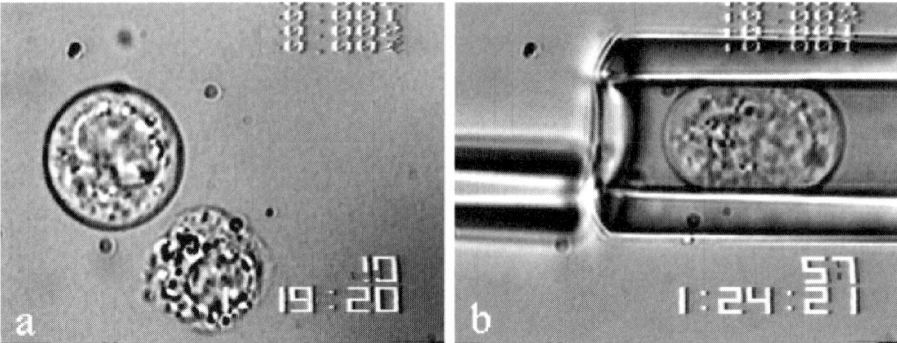

Fig. 5-4. Micropipette aspiration test to examine the volumetric response of cells to mechanical deformation. Video images of a chondrocyte and micropipette before (a) and after (b) complete aspiration of the cell. Cells show a significant decrease in volume, which when matched to a theoretical model can be used to determine Poisson's ratio as one measure of compressibility. From Trickey et al., 2006.

of predicting the complete creep response, suggesting that the viscoelastic response of the chondrocytes under micropipette aspiration is predominantly due to intrinsic viscoelastic phenomena and is not due to the biphasic behavior.

Other studies have also used the micropipette technique to determine the volume change of chondrocytes after complete aspiration into a micropipette (Jones et al., 1999). While many cells are assumed to be incompressible with a Poisson ratio of 0.5, these studies demonstrated that certain cells, such as chondrocytes, in fact exhibit a certain level of compressibility (Fig. 5-4), presumably due to the expulsion of intracellular fluid. Isolated cells were fully aspirated into a micropipette and allowed to reach mechanical equilibrium. Cells were then extruded from the micropipette and cell volume and morphology were measured over time. By simulating this experimental procedure with a finite element analysis modeling the cell as either a biphasic or viscoelastic material, the Poisson ratio and viscoelastic recovery properties of the cell were determined. The Poisson ratio of chondrocytes was found to be ~0.38, suggesting that cells may in fact show volumetric changes in response to mechanical compression. The finding of cell compressibility in response to mechanical loading is consistent with previous studies showing significant loss of cell volume in chondrocytes embedded within the extracellular matrix (Guilak et al., 1995). Taken together with micropipette studies, these studies suggest that cell volume changes are due to biphasic mechanical effects resulting in fluid exudation from the cell, while cellular viscoelasticity is more likely due to intrinsic behavior of the cytoplasm and not to flow-dependent effects.

Biphasic properties of the pericellular matrix

In vitro experimental analysis of the mechanics of isolated cells provides a simplified and controlled environment in which theoretical models, and associated numerical solutions, can be employed to measure and compare cell properties via material parameters. Ultimately, however, most biophysical analyses of cell mechanics are motivated by a need to extrapolate the *in vitro* findings to a characterization of the physiological

Multiphasic models of cell mechanics

(*in vivo*) environment of the cell. For cells such as articular chondrocytes, it is possible to isolate a functional cell–matrix unit and analyze its mechanical properties *in vitro*, thus providing a link to the physiological setting (Poole, 1992). This modeling approach is briefly described here in the context of the articular chondrocyte.

Chondrocytes in articular cartilage are completely surrounded by a narrow region of tissue, termed the pericellular matrix (PCM). The PCM is characterized by the presence of type VI collagen (Poole, 1992), which is not found elsewhere in cartilage under normal circumstances, and a higher concentration of aggrecan relative to the extracellular matrix (ECM), as well as smaller amounts of other collagen types and proteins. The chondrocyte together with the pericellular matrix and the surrounding capsule has been termed the chondron (Poole, 1992; Poole, 1997). While the function of the PCM is not known, there has been considerable speculation that the chondron plays a biomechanical role in articular cartilage (Szirmai, 1974). For example, it has been hypothesized that the chondron provides a protective effect for the chondrocyte during loading (Poole et al., 1987), and others have suggested that the chondron serves as a mechanical transducer (Greco et al., 1992; Guilak and Mow, 2000).

To determine mechanical properties of the PCM, the solution of a layered elastic contact problem that models micropipette aspiration of an isolated chondron was developed. This theoretical solution was applied to measure an elastic Young's modulus for the PCM in human chondrons isolated from normal and osteoarthritic sample groups (Alexopoulos et al., 2003). The mean PCM Young's modulus of chondrons isolated from the normal group (66.5 ± 23.3 kPa) was found to be a few orders of magnitude larger than the chondrocyte modulus (~ 1 kPa) and was found to drop significantly in the osteoarthritic group (41.3 ± 21.1 kPa, $p < 0.001$). These findings support the hypothesis that the PCM serves a protective mechanical role that may be significantly altered in the presence of disease. In a multiscale finite element analysis (Guilak and Mow, 2000), the macroscopic solution for transient deformation of a cartilage layer under a step load was computed and used to solve a separate microscale problem to detemine the mechanical environment of a single chondrocyte. In this study, the inclusion of a PCM layer in the microscale model significantly altered the mechanical environment of a single cell. A mathematical model for purely radial deformation in a chondron was developed and analyzed under dynamic loading in the range 0–3 Hz (Haider, 2004). This study found that the presence of a thin, highly stiff PCM that is less permeable than the chondrocyte enhances the transmission of compressive strain mechanical signals to the cell while, simultaneously, protecting it from excessive solid stress.

Using the micropipette aspiration test coupled with a linear biphasic finite element model, recent studies have reported the biphasic material properties of the PCM of articular chondrocytes (Alexopoulos et al., 2005) (Fig. 5-5). Chondrons were mechanically extracted from nondegenerate and osteoarthritic (OA) human cartilage. Micropipette aspiration was used to examine the creep behavior of the pericellular matrix, which was matched using optimization to a biphasic finite element model (Fig. 5-6). The transient mechanical behavior of the PCM was well-described by a biphasic model, suggesting that the viscoelastic response of the PCM is attributable to flow-dependent effects, similar to that of the ECM. With osteoarthritis, the mean

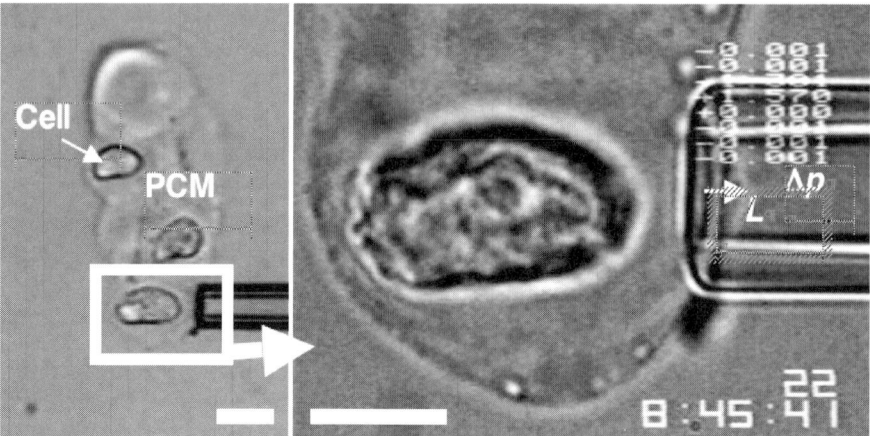

Fig. 5-5. Micropipette aspiration of the pericellular matrix (PCM) of chondrocytes. This region surrounds cells in articular cartilage, similar to a glycocalyx, but contains significant amounts of extracellular matrix collagens, proteoglycans, and other macromolecules. The mechanical properties of this region appear to have a significant influence on the stress-strain and fluid-flow environment of the cell. From Alexopoulos et al., 2005.

Young's modulus of the PCM was significantly decreased (38.7 ± 16.2 kPa vs. 23.5 ± 12.9 kPa, $p < 0.001$), and the permeability was significantly elevated ($4.19 \pm 3.78 \times 10^{-17}$ m^4/N·s vs. $10.2 \pm 9.38 \times 10^{-17}$ m^4/N·s, $p < 0.001$). The Poisson ratio was similar for both nondegenerate and osteoarthritic PCM (0.044 ± 0.063 vs. 0.030 ± 0.068, $p > 0.6$). These findings suggest that the PCM may undergo enzymatic and mechanical degradation with osteoarthritis, similar to that occurring in the ECM. In combination with previous theoretical models of cell–matrix interactions in cartilage, these findings suggest that changes in the properties of the PCM may have an important influence on the biomechanical environment of the cell.

Together, these studies support the utility of *in vitro* mechanical analyses of isolated functional cell–matrix units. Because cartilage, in particular, is avascular and aneural, characterization of PCM mechanical and chemical properties is a key step toward characterizing the *in vivo* state of the cell and its metabolic response to alterations in the local cellular environment. The triphasic model provides a framework for developing extended models of the PCM that can delineate effects of the distinct mechanochemical composition of the PCM, relative to the ECM, on the local environment of the cell.

Indentation studies of cell multiphasic properties

In addition to micropipette aspiration, various techniques for cellular indentation have been used to measure the modulus of adherent cells, including cell indentation (Daily et al., 1984; Duszyk et al., 1989; Zahalak et al., 1990), scanning probe microscopy (Radmacher et al., 1992; Shroff et al., 1995), or cytoindentation (Shin and Athanasiou, 1999; Koay et al., 2003). (See also the discussion of experimental approaches in Chapter 2.) The conceptual basis of these techniques is generally similar, in that a rigid probe is used to indent the cell and the ensuing creep or stress-relaxation behavior

(a) **CELL AND PERICELLULAR MATRIX MODEL**

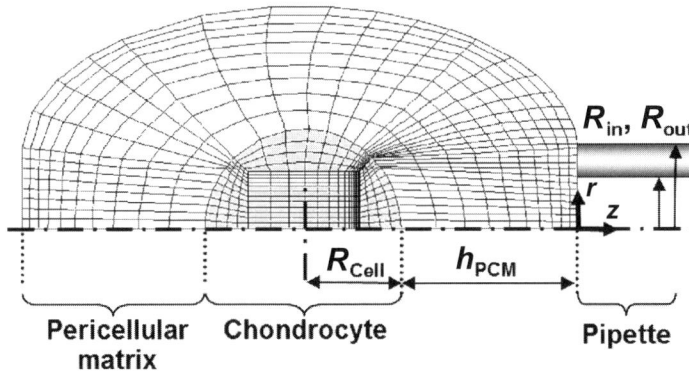

(b) **TRANSIENT RESPONSE**

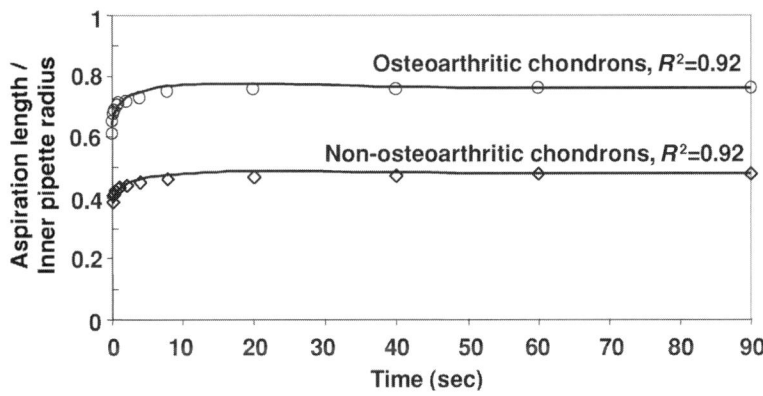

Fig. 5-6. (a) Biphasic finite element mesh of the micropipette aspiration experiment. The cell and pericellular matrix were modeled using an axisymmetric mesh with bilinear quadrilateral elements (342 nodes, 314 elements). (b) Transient response of normal and osteoarthritic chondrons (cell with the pericellular matrix), and the associated biphasic prediction of their mechanical behavior. The transient mechanical behavior of the PCM was well-described by a biphasic model, suggesting that the viscoelastic response of the pericellular matrix is attributable to flow-dependent effects, similar to that of the extracellular matrix. From Alexopoulos et al., 2005 with permission.

is recorded. These techniques have generally used elastic or viscoelastic models to calculate the equilibrium or dynamic moduli of cells over a range of frequencies. In one set of cell indentation experiments, MG63 osteoblast-like cells were modeled with either a linear elasticity solution of half-space indentation or the linear biphasic theory under the assumption that the viscoelastic behavior of each cell was due to the interaction between the solid cytoskeletal matrix and the cytoplasmic fluid (Shin and Athanasiou, 1999). The intrinsic biphasic material properties (aggregate modulus, Poisson's ratio, and permeability) were determined by curve-fitting the experimental surface reaction force and deformation with a linear biphasic finite element code in conjunction with optimization routines. These cells exhibited a compressive aggregate modulus of 2.05 ± 0.89 kPa with a Poisson ratio of 0.37 ± 0.03. These properties are on the same order of magnitude as the elastic properties determined using other techniques (Trickey et al., in press), although the permeability of $1.18 \pm 0.65 \times 10^{-10}$ m^4/N·s is several orders of magnitude higher than that estimated for chondrocytes using micropipette aspiration (Trickey et al., 2000).

Analysis of cell–matrix interactions using multiphasic models

Previous studies suggest that cells have the ability to respond to the local stress-strain state within the extracellular matrix, thus suggesting that cellular response reflects the history of the time-dependent and spatially varying changes in the mechanical environment of the cells. The use of multiphasic models for cells has been of particular value in theoretical models of cell–matrix interactions that seek to model the stress-strain and fluid-flow environment of single cells within a tissue matrix. However, the relationship between the stress-strain and fluid-flow fields at the macroscopic "tissue" level and at the microscopic "cellular" level are not fully understood. To directly test such hypotheses, it would be important to have accurate knowledge of the local stress and deformation environment of the cell. In this respect, theoretical models of cells and tissues are particularly valuable in that they may be used to provide information on biophysical parameters that cannot be measured experimentally *in situ* at the cellular level, for example, the stress-strain, physicochemical, and electrical states in the immediate vicinity of the cell.

Based on existing experimental data on the deformation behavior and biomechanical properties of articular cartilage and chondrocytes, a multiscale biphasic finite element model was developed of the chondrocyte as a spheroidal inclusion embedded within the extracellular matrix of a cartilage explant (Fig. 5-7). In these studies, the cell membrane was neglected, and it was assumed that the cell was freely permeable to water to allow for changes in volume via transport of interstitial water in an out of the cell. Finite element analysis of the stress, strain, fluid flow, and hydraulic fluid pressure were made of a configuration simulating a cylindrical cartilage specimen (5 mm × 1 mm) subjected to a step load in an unconfined compression experiment. A parametric analysis was performed by varying the mechanical properties of the cell over 5–7 orders of magnitude relative to the properties of the ECM. Using a range of chondrocyte biphasic properties reported in the literature ($E \sim 0.5 - 1$ kPa, $k \sim 10^{-10} - 10^{-15}$ m^4/N·s, $\nu \sim 0.1 - 0.4$) (Shin and Athanasiou, 1999; Trickey et al., 2000; Trickey et al., 2006), the distribution of stress at the cellular level was found to be time varying and inhomogeneous, and it differed significantly from that in the bulk extracellular matrix. At early time points (<100 s) following application of the load, the chondrocytes were exposed primarily to shear stress and strain and hydraulic fluid pressure, with little volume change. At longer time periods, changes in cell shape and volume were predicted coincident with exudation of the interstitial fluid (Fig. 5-7). The large difference ($\sim$3 orders of magnitude) in the elastic properties of the chondrocyte and of the extracellular matrix results in the presence of stress concentrations at the cell–matrix border and a nearly two-fold increase in strain and dilatation (volume change) at the cellular level, as compared to that at the macro-level. The presence of a narrow "pericellular matrix" with different properties than that of the chondrocyte or extracellular matrix significantly altered the principal stress and strain magnitudes

Multiphasic models of cell mechanics

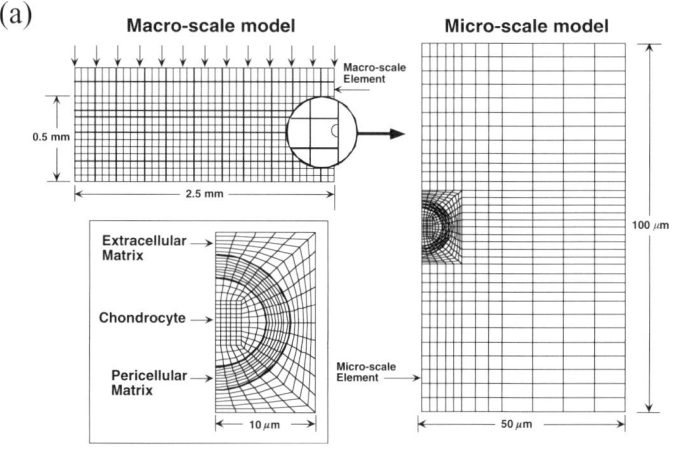

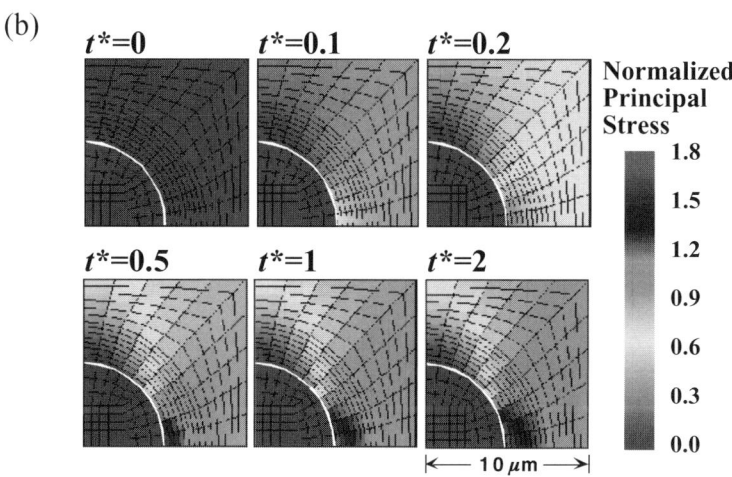

Fig. 5-7. (a) A biphasic multiscale finite element method was used to model the mechanical environment of a single cell within the cartilage extracellular matrix. The "macro-scale" response of a cartilage explant in a state of unconfined compression was the first model. From this solution, a linear interpolation of the time-history of the kinematic boundary conditions within a 50 × 100 μm region were then applied to a "micro-scale" finite element mesh that incorporated a chondrocyte (10 μm diameter) and its pericellular matrix (2.5 μm thick) embedded within and attached to the extracellular matrix. Using this technique, it is assumed that due to their low volume fraction (<10%), the cells do not contribute mechanically to the macroscopic properties and behavior of the extracellular matrix. (b) Predictions of the compressive stress in the cell and extracellular matrix versus time. A gray-scale image of one-quarter of the cell is shown within the matrix. Stress is normalized to the far-field extracellular matrix stress at equilibrium, and time is normalized to the biphasic gel time ($t^* = t/\tau_{gel}$). At early times following loading, low magnitudes of solid stress were observed, as the total stress in the tissue was borne primarily by pressurization of the interstitial fluid. With time, stresses were transferred to the solid phase and increased stress concentrations are observed at the cell–matrix boundary. From Guilak and Mow, 2000.

within the chondrocyte, suggesting a functional biomechanical role for this tissue region. These findings suggest that even under simple compressive loading conditions, chondrocytes are subjected to a complex local mechanical environment consisting of tension, compression, shear, and hydraulic pressure. Knowledge of the magnitudes and distribution of local stress/strain and fluid-flow fields in the extracellular matrix around the chondrocytes is an important step in the interpretation of studies of mechanical signal transduction in cartilage explant culture models.

In other tissues, anisotropic behavior may play an important role in defining the micromechanical environment of the cell. For example, cellular response to mechanical loading varies between the anatomic zones of the intervertebral disc, and this difference may be related to differences in the structure and mechanics of both cells and extracellular matrix, which are expected to cause differences in the physical stimuli (such as pressure, stress, and strain) in the cellular micromechanical environment (Guilak et al., 1999; Baer et al., 2003). In other studies, finite element analyses have been used to model flow-dependent viscoelasticity using the biphasic theory for soft tissues; finite deformation effects using a hyperelastic constitutive law for the solid phase; and material anisotropy by including a fiber-reinforced continuum law in the hyperelastic strain energy function. The model predicted that the cellular micromechanical environment varies dramatically depending on the local tissue stiffness and anisotropy. Furthermore, the model predicted that stress-strain and fluid-flow environment is strongly influenced by cell shape, suggesting that the geometry of cells *in situ* may be an adaptation to reduce cellular strains during tissue loading.

With similar multiscaling techniques, other studies have used triphasic constitutive models to predict the physicochemical environment of cells within charged, hydrated tissues (Likhitpanichkul et al., 2003). These studies also show that in addition to nonhomogeneous stress-strain and fluid-flow fields within the extracellular matrix, cells may also be exposed to time- and spatially varying osmotic pressure and electric fields due to the coupling between electrical, chemical, and mechanical events in the cell and in the surrounding tissues. Such methods may provide new insight into the physical regulatory mechanisms that influence cell behavior *in situ*.

Summary

Multiphasic approaches have important advantages and disadvantages relative to more classic single-phase models. The disadvantages are based primarily on the added complexity required for computational models. In most cases, analytical solutions are intractable and numerical methods such as finite element modeling are required. Furthermore, additional experimental tests are necessary to determine the intrinsic mechanical contributions of the different phases to the overall behavior of the cell. However, multiphasic models may provide a more realistic representation of the physical events that govern cell mechanical behavior. Furthermore, as most current multiphasic models are based on a continuum approach, the constitutive models describing each phase can be selected independently to best describe the empirically observed behavior of the cell. A multiphasic approach may also be combined with other structurally based models (such as, tensegrity models), and thus may provide a versatile modeling approach for examining the interactions of the different constitutive phases governing cell mechanical behavior.

References

Alexopoulos, L.G., Haider, M.A., Vail, T.P., Guilak, F. (2003). Alterations in the mechanical properties of the human chondrocyte pericellular matrix with osteoarthritis. *J. Biomech. Eng.*, 125, 323–333.

Alexopoulos, L.G., Williams, G.M., Upton, M.L., Setton, L.A., Guilak, F. (2005). Osteoarthritic changes in the biphasic mechanical properties of the chondrocyte pericellular matrix in articular cartilage. *J. Biomech.*, 38, 509–517.

Ateshian, G.A., Likhitpanichkul, M., Hung, C.T. (2006). A mixture theory analysis for passive transport in osmotic loading of cells. *J. Biomech.*, 39, 464–475.

Baaijens, F.P. (1998). Mixed finite element methods for viscoelastic flow analysis: A review. *J. of Non-Newtonian Fluid Mechanics* 79, 361–385.

Baaijens, F.P.T., Trickey, W.R., Laursen, T.A., Guilak, F. (2005). Large deformation finite element analysis of micropipette aspiration to determine the mechanical properties of the chondrocyte. *An. Biomed. Eng.*, 33, 494–501.

Bachrach, N.M., Valhmu, W.B., Stazzone, E., Ratcliffe, A., Lai, W.M., Mow, V.C. (1995). Changes in proteoglycan synthesis of chondrocytes in articular cartilage are associated with the time-dependent changes in their mechanical environment. *J. Biomech.*, 28, 1561–1570.

Baer, A.E., Laursen, T.A., Guilak, F., Setton, L.A. (2003). The micromechanical environment of intervertebral disc cells determined by a finite deformation, anisotropic, and biphasic finite element model. *J. Biomech. Eng.*, 125, 1–11.

Bowen, R.M. (1980). Incompressible porous media models by use of the theory of mixtures. *Internat. J. Eng. Sci.*, 18, 1129–1148.

Buschmann, M.D., Hunziker, E.B., Kim, Y.J., Grodzinsky, A.J. (1996). Altered aggrecan synthesis correlates with cell and nucleus structure in statically compressed cartilage. *J. Cell Sci.*, 109, 499–508.

Cantiello, H.F., Patenaude, C., Zaner, K. (1991). Osmotically induced electrical signals from actin filaments. *Biophys. J.*, 59, 1284–1289.

Cohen, B., Lai, W.M., Mow, V.C. (1998). A transversely isotropic biphasic model for unconfined compression of growth plate and chondroepiphysis. *J. Biomech. Eng.*, 120, 491–496.

Daily, B., Elson, E.L., Zahalak, G.I. (1984). Cell poking. Determination of the elastic area compressibility modulus of the erythrocyte membrane. *Biophys. J.*, 45, 671–682.

DiSilvestro, M.R., Suh, J.K. (2002). Biphasic poroviscoelastic characteristics of proteoglycan-depleted articular cartilage: simulation of degeneration. *An. Biomed. Eng.*, 30, 792–800.

Dong, C., Skalak, R., Sung, K.L. (1991). Cytoplasmic rheology of passive neutrophils. *Biorheology*, 28, 557–567.

Duszyk, M., Schwab, B.D., Zahalak, G.I., Qian, H., Elson, E.L. (1989). Cell poking: quantitative analysis of indentation of thick viscoelastic layers. *Biophys. J.*, 55, 683–690.

Evans, E., Yeung, A. (1989). Apparent viscosity and cortical tension of blood granulocytes determined by micropipet aspiration. *Biophys. J.*, 56, 151–160.

Evans, E.A., 1989. Structure and deformation properties of red blood cells: Concepts and quantitative methods. *Methods Enzymol.*, 173, 3–35.

Greco, F., Specchia, N., Falciglia, F., Toesca, A., Nori, S. (1992). Ultrastructural analysis of the adaptation of articular cartilage to mechanical stimulation. *Italian J. of Orthopaedics and Traumatology*, 18, 311–321.

Gu, W.Y., Lai, W.M., Hung, C.T., Liu, Z.P., Mow, V.C. (1997). Analysis of transient swelling and electrical responses of an isolated cell to sudden osmotic loading. *Adv. in Bioengin.*, BED36, 189–190.

Gu, W.Y., Lai, W.M., Mow, V.C. (1998). A mixture theory for charged-hydrated soft tissues containing multi-electrolytes: Passive transport and swelling behaviors. *J. Biomech. Eng.*, 120, 169–180.

Guilak, F. (1995). Compression-induced changes in the shape and volume of the chondrocyte nucleus. *J. Biomech.*, 28, 1529–1542.

Guilak, F., Erickson, G.R., Ping Ting-Beall, H. (2002). The effects of osmotic stress on the viscoelastic and physical properties of articular chondrocytes. *Biophys. J.*, 82, 720–727.

Guilak, F., Mow, V.C. (2000). The mechanical environment of the chondrocyte: A biphasic finite element model of cell–matrix interactions in articular cartilage. *J. Biomech.*, 33, 1663–1673.

Guilak, F., Ratcliffe, A., Mow, V.C. (1995). Chondrocyte deformation and local tissue strain in articular cartilage: A confocal microscopy study. *J. Orthop. Res.*, 13, 410–421.

Guilak, F., Ting-Beall, H.P., Baer, A.E., Trickey, W.R., Erickson, G.R., Setton, L.A. (1999). Viscoelastic properties of intervertebral disc cells – Identification of two biomechanically distinct cell populations. *Spine*, 24, 2475–2483.

Haider, M.A. (2004). A radial biphasic model for local cell–matrix mechanics in articular cartilage. *SIAM J. Appl. Math.*, 64, 1588–1608.

Haider, M.A., Guilak, F. (2000). An axisymmetric boundary integral model for incompressible linear viscoelasticity: Application to the micropipette aspiration contact problem. *J. Biomech. Eng.*, 122, 236–244.

Haider, M.A., Guilak, F. (2002). An axisymmetric boundary integral model for assessing elastic cell properties in the micropipette aspiration contact problem. *J. Biomech. Eng.*, 124, 586–595.

Hochmuth, R.M. (2000). Micropipette aspiration of living cells. *J. Biomech.*, 33, 15–22.

Holmes, M.H., Lai, W.M., Mow, V.C. (1985). Singular perturbation analysis of the nonlinear, flow-dependent compressive stress relaxation behavior of articular cartilage. *J Biomech. Eng.*, 107, 206–218.

Huyghe, J.M., Houben, G.B., Drost, M.R., van Donkelaar, C.C. (2003). An ionised/non-ionised dual porosity model of intervertebral disc tissue. *Biomech. and Model. in Mechanobiol.*, 2, 3–19.

Iatridis, J.C., Setton, L.A., Foster, R.J., Rawlins, B.A., Weidenbaum, M., Mow, V.C. (1998). Degeneration affects the anisotropic and nonlinear behaviors of human anulus fibrosus in compression. *J. Biomech.*, 31, 535–544.

Ingber, D.E. (2003). Tensegrity I. Cell structure and hierarchical systems biology. *J. Cell Sci.*, 116, 1157–1173.

Jones, W.R., Ting-Beall, H.P., Lee, G.M., Kelley, S.S., Hochmuth, R.M., Guilak, F. (1999). Alterations in the Young's modulus and volumetric properties of chondrocytes isolated from normal and osteoarthritic human cartilage. *J. Biomech.*, 32, 119–127.

Karcher, H., Lammerding, J., Huang, H., Lee, R.T., Kamm, R.D., Kaazempur-Mofrad, M.R. (2003). A three-dimensional viscoelastic model for cell deformation with experimental verification. *Biophys. J.*, 85, 3336–3349.

Kedem, O., Katchalsky, A. (1958). Thermodynamic analysis of the permeability of biological membranes to non-electrolytes. *Biochim. Biophys. Acta* 27, 229–246.

Koay, E.J., Shieh, A.C., Athanasiou, K.A. (2003). Creep indentation of single cells. *J. Biomech. Eng.*, 125, 334–341.

Lai, W.M., Hou, J.S., Mow, V.C. (1991). A triphasic theory for the swelling and deformation behaviors of articular cartilage. *J. Biomech. Eng.*, 113, 245–258.

Lai, W.M., Mow, V.C., Roth, V. (1981). Effects of nonlinear strain-dependent permeability and rate of compression on the stress behavior of articular cartilage. *J. Biomech. Eng.*, 103, 61–66.

Likhitpanichkul, M., Sun, D.D., Guo, X.E., Lai, W.M., Mow, V.C. (2003). The mechano-electrochemical environment of chondrocytes in articular cartilage explants under unconfined compression: Emphasis on the cell matrix interactions. *Proceedings of the 2003 Summer Bioengineering Conference, Key Biscayne, FL*.

Mak, A.F. (1986a). The apparent viscoelastic behavior of articular cartilage – The contributions from the intrinsic matrix viscoelasticity and interstitial fluid flows. *J. Biomech. Eng.*, 108, 123–130.

Mak, A.F. (1986b). Unconfined compression of hydrated viscoelastic tissues: A biphasic poroviscoelastic analysis. *Biorheology*, 23, 371–383.

Mak, A.F.T., Huang, D.T., Zhang, J.D., Tong, P. (1997). Deformation-induced hierarchial flows and drag forces in bone canaliculi and matrix microporosity. *J. Biomech.*, 20, 11–18.

Maughan, D.W., Godt, R.E. (1989). Equilibrium distribution of ions in a muscle fiber. *Biophys. J.*, 56, 717–722.

Mow, V.C., Holmes, M.H., Lai, W.M. (1984). Fluid transport and mechanical properties of articular cartilage: A review. *J. Biomech.*, 17, 377–394.

Mow, V.C., Kuei, S.C., Lai, W.M., Armstrong, C.G. (1980). Biphasic creep and stress relaxation of articular cartilage in compression: Theory and experiments. *J. Biomech. Eng.*, 102, 73–84.

Needham, D., Hochmuth, R.M. (1992). A sensitive measure of surface stress in the resting neutrophil. *Biophys. J.*, 61, 1664–1670.

Oster, G. (1984). The mechanochemistry of cytogels. *Physica*, 12D, 333–350.

Oster, G. (1989). Cell motility and tissue morphogenesis. In: *Cell Shape: Determinants, Regulation, and Regulatory Role.* Stein, W.D. and Bronner, F. (Eds.) San Diego, CA, Academic Press, pp. 33–61.

Pollack, G.H. (2001). *Cells, gels and the engines of life: A new, unifying approach to cell function.* Seattle, WA, Ebner & Sons.

Poole, C.A. (1992). Chondrons: the chondrocyte and its pericellular microenvironment. In: *Articular Cartilage and Osteoarthritis.* Kuettner, K.E., Schleyerbach, R., Peyron, J.G. and Hascall, V.C. (Eds.) New York, London: Academic Press, pp. 201–220.

Poole, C.A. (1997). Articular cartilage chondrons: Form, function and failure. *J. Anat.*, 191 (Pt 1), 1–13.

Poole, C.A., Flint, M.H., Beaumont, B.W. (1987). Chondrons in cartilage: Ultrastructural analysis of the pericellular microenvironment in adult human articular cartilages. *J. Orthop. Res.*, 5, 509–522.

Radmacher, M., Tillmann, R.W., Fritz, M., Gaub, H.E. (1992). From molecules to cells: Imaging soft samples with the atomic force microscope. *Science*, 257, 1900–1905.

Sato, M., Theret, D.P., Wheeler, L.T., Ohshima, N., Nerem, R.M. (1990). Application of the micropipette technique to the measurement of cultured porcine aortic endothelial cell viscoelastic properties. *J. Biomech. Eng.*, 112, 263–268.

Sengers, B.G., Oomens, C.W., Baaijens, F.P. (2004). An integrated finite-element approach to mechanics, transport and biosynthesis in tissue engineering. *J. Biomech. Eng.*, 126, 82–91.

Setton, L.A., Zhu, W., Mow, V.C. (1993). The biphasic poroviscoelastic behavior of articular cartilage: Role of the surface zone in governing the compressive behavior. *J. Biomech.*, 26, 581–592.

Shin, D., Athanasiou, K. (1999). Cytoindentation for obtaining cell biomechanical properties. *J. Orthop. Res.*, 17, 880–890.

Shroff, S.G., Saner, D.R., Lal, R. (1995). Dynamic micromechanical properties of cultured rat atrial myocytes measured by atomic force microscopy. *Am. J. Physiol.*, 269, C286–292.

Szirmai, J.A., 1974. The concept of the chondron as a biomechanical unit. In: *Biopolymer und Biomechanik von Bindegewebssystemen.* Hartmann, F. (Ed.) Berlin, Academic Press, pp. 87.

Spilker, R.L., Donzelli, P.S., Mow, V.C. (1992). A transversely isotropic biphasic finite element model of the meniscus. *J. Biomech.*, 112, 1027–1045.

Sung, K.L., Schmid-Schonbein, G.W., Skalak, R., Schuessler, G.B., Usami, S., Chien, S. (1982). Influence of physicochemical factors on rheology of human neutrophils. *Biophys. J.*, 39, 101–106.

Theret, D.P., Levesque, M.J., Sato, M., Nerem, R.M., Wheeler, L.T. (1988). The application of a homogeneous half-space model in the analysis of endothelial cell micropipette measurements. *J. Biomech. Eng.*, 110, 190–199.

Thoumine, O., Ott, A. (1997). Comparison of the mechanical properties of normal and transformed fibroblasts. *Biorheology*, 34, 309–326.

Ting-Beall, H.P., Needham, D., Hochmuth, R.M. (1993). Volume and osmotic properties of human neutrophils. *Blood*, 81, 2774–2780.

Trickey, W.R., Baaijens, F.P.T., Laursen, T.A., Alexopoulos, L.G., Guilak, F. (2006). Determination of the Poisson's ratio of the cell: Recovery properties of chondrocytes after release from complete micropipette aspiration. *J. Biomech.*, (in press) 39, 78–87.

Trickey, W.R., Lee, M., Guilak, F. (2000). Viscoelastic properties of chondrocytes from normal and osteoarthritic human cartilage. *J. Orthop. Res.*, 18, 891–898.

Truesdell, C., Toupin, R., 1960. The classical field theories. In: *Handbuch der Physik*. Flugge, S. (Ed.) Berlin, Springer-Verlag, pp. 226–793.

Yeung, A., Evans, E. (1989). Cortical shell-liquid core model for passive flow of liquid-like spherical cells into micropipets. *Biophys. J.*, 56, 139–149.

Zahalak, G.I., McConnaughey, W.B., Elson, E.L. (1990). Determination of cellular mechanical properties by cell poking, with an application to leukocytes. *J. Biomech. Eng.*, 112, 283–294.

6 Models of cytoskeletal mechanics based on tensegrity

Dimitrije Stamenović

ABSTRACT: Cell shape is an important determinant of cell function and it provides a regulatory mechanism to the cell. The idea that cell contractile stress may determine cell shape stability came with the model that depicts the cell as tensed membrane that surrounds viscous cytoplasm. Ingber has further advanced this idea of the stabilizing role of the contractile stress. However, he has argued that tensed intracellular cytoskeletal lattice, rather than the cortical membrane, confirms shape stability to adherent cells. Ingber introduced a special class of tensed reticulated structures, known as tensegrity architecture, as a model of the cytoskeleton. Tensegrity architecture belongs to a class of stress-supported structures, all of which require preexisting tensile stress ("prestress") in their cable-like structural members, even before application of external loading, in order to maintain their structural integrity. Ordinary elastic materials such as rubber, polymers, or metals, by contrast, require no such prestress. A hallmark property that stems from this feature is that structural rigidity (stiffness) of the matrix is nearly proportional to the level of the prestress that it supports. As distinct from other stress-supported structures falling within the class, in tensegrity architecture the prestress in the cable network is balanced by compression of internal elements that are called struts. According to Ingber's cellular tensegrity model, cytoskeletal prestress in generated by the cell contractile machinery and by mechanical distension of the cell. This prestress is carried mainly by the cytoskeletal actin network, and is balanced partly by compression of microtubules and partly by traction at the extracellular adhesions.

The idea that the cytoskeleton maintains its structural stability through the agency of contractile stress rests on the premise that the cytoskeleton is a static network. In reality, the cytoskeleton is a dynamic network, which is exposed to dynamic loads and in which the dynamics of various biopolymers contribute to its rheological properties. Thus, the static model of the cytoskeleton provides only a limited insight into its mechanical properties (for example, near-steady-state conditions). However, our recent measurements have shown that cell rheological (dynamic) behavior may also be affected by the contractile prestress, suggesting thereby that the tensegrity idea may also account for some features of cell rheology.

This chapter describes the basic idea of the cellular tensegrity hypothesis, how it applies to problems in cellular mechanics, and what its limitations are.

Introduction

A new model of cell structure to explain how the internal cytoskeleton of adherent cells mediates alterations in cell functions caused by changes in cell shape was proposed in the early 1980s by Donald Ingber and colleagues (Ingber et al., 1981; Ingber and Jamieson, 1985). This model is based on a building system known as *tensegrity architecture* (Fuller, 1961). The essential premise of what is known as the cellular tensegrity model is that the cytoskeletal lattice carries preexisting tensile stress, termed prestress, whose role is to confer shape stability to the cell. A second premise is that this cytoskeletal prestress is partly balanced by forces that arise at cell adhesions to the extracellular matrix and partly by internal, compression-supporting cytoskeletal structures (for example, microtubules). The cytoskeletal prestress is generated actively, by the cytoskeletal contractile apparatus. Additional prestress is generated passively by cell mechanical distension through adhesions to the substrate, by cytoplasmic swelling pressure (turgor), and by forces generated by filament polymerization. The prestress is primarily carried by the cytoskeletal actin network and to a lesser extent by the intermediate filament network (Ingber, 1993; 2003a).

There is a growing body of experimental data that is consistent with the cellular tensegrity model. The strongest piece of evidence in support of the tensegrity model is the observed proportional relationship between cell stiffness and the cytoskeletal contractile stress (Wang et al., 2001; 2002). Experimental data also show that microtubules carry compression that, in turn, balances a substantial portion of the prestress, which is another key feature of tensegrity architecture (Wang et al., 2001; Stamenović et al., 2002a). Together, these two findings have provided so far the most convincing evidence in support of the cellular tensegrity model.

In successive sections, this chapter describes basic concepts, definitions, and underlying mechanisms of the cellular tensegrity model; describes experimental data that are consistent and those that are not consistent with the tensegrity model; and describes results from mathematical modeling of typical tensegrity-based models of cell mechanics and compares predictions from those models to experimental data from living cells. Then the chapter briefly discusses the usefulness of the tensegrity idea in studying the dynamic behavior of cells and ends with a summary.

The cellular tensegrity model

It is well established that cell shape is critical for the control of many cell behaviors, including growth, motility, differentiation, and apoptosis and that the effects of cell shape are mediated through changes in the intracellular cytoskeleton (see Ingber, 2003a and 2003b). To explain how cells generate mechanical stresses in response to alterations in their shape and how those stresses affect cellular function, various models of cellular mechanics have been advanced, as other chapters here extensively discuss. All these models can be divided into two distinct classes: continuum models, and discrete models.

Continuum models (Theret et al., 1988; Evans and Yeung, 1989; Fung and Liu, 1993; Schmid-Schönbein et al., 1995; Bausch et al., 1998; Fabry et al., 2001a) assume that the stress-bearing elements within the cell are small compared to the length scales of interest and that they uniformly fill the space within the cell body. The microscale behavior of these elements is given by equations that describe local deformation and mass, and momentum and energy balance (see Chapter 3 of this book). This leads to descriptions of stress and strain patterns that are continuous in space within the cell.

Continuum models can run from simple to very complex and multicompartmental, from elastic to viscoelastic (Chapter 4) or even poroelastic (Chapter 5).

Discrete models (Porter, 1984; Ingber and Jamieson, 1985; Forgacs, 1995; Satcher and Dewey, 1996; Stamenović et al., 1996; Boey et al., 1998) consider discrete stress-bearing elements of the cell that are finite in size, sometimes spanning distances that are comparable to the cell size (for example, microtubules). The cell is depicted as being composed of a large number of these discrete elements that do not fill the space. The behavior of each discrete element is subject to conditions of mechanical equilibrium and geometrical compatibility at every node. At this point, a coarse-graining average can be applied and local stresses and strains can be obtained as continuous field variables. Within the class of discrete models there is a special subclass, known as stress-supported (or prestressed) structures. While ordinary elastic materials such as rubber, polymers, or metals, by contrast, require no such prestress, all stress-supported structures require tensile prestress in their structural members, even before the application of external loading, in order to maintain their structural integrity. A hallmark property that stems from this feature is that structural rigidity (stiffness) of the matrix is proportional to the level of the prestress that it supports (Volokh and Vilnay, 1997; Stamenović and Ingber, 2002). Tensegrity architecture falls within this class. As in the case of continuum models, discrete models, of which the tensegrity architecture is one, can range from very simple to very complex, multimodular, and multicompartmental.

It has long been known that many cell types exist under tension (prestress) (Harris et al., 1980; Albrecht-Buehler, 1987; Heidemann and Buxbaum, 1990; Kolodney and Wysolmerski, 1992; Evans et al., 1993). Theoretical models that depict the cell as a tensed (that is, prestressed) membrane that surrounds viscous cytoplasm have been proposed in the past (Evans and Yeung, 1989; Fung and Liu, 1993; Schmid-Schönbein et al., 1995). However, none of those studies show that this cell prestress may play a key role in regulating cell deformability. In the early 1980s, Donald Ingber (Ingber et al., 1981; Ingber and Jamieson, 1985) introduced a novel model of cytoskeletal mechanics based on architecture that secures structural stability through the agency of prestress. This model has become known as the cellular tensegrity model. Basic features and mechanisms of this model and how they apply to mechanics of cells are described in the coming sections.

Definitions, basic mechanisms, and properties of tensegrity structures

Tensegrity architecture is a building principle introduced by R. Buckminster Fuller (Fuller, 1961). He defined tensegrity as a system through which structures are stabilized by continuous tension carried by the structural members (like a camp tent or a spider web) rather than continuous compression (like a stone arch). Fuller referred to this architecture as "tensional integrity," or "tensegrity" (Fig. 6-1).

The central mechanism by which tensegrity and other prestressed structures develop restoring stress in the presence of external loading is by geometrical rearrangement (that is, by change in spacing and orientation and to a lesser degree by change in length) of their pre-tensed members. The greater the pre-tension carried by these members, the less geometrical rearrangement they undergo under an applied load, and thus the less deformable (more rigid) the structure will be. In the absence of prestress, these

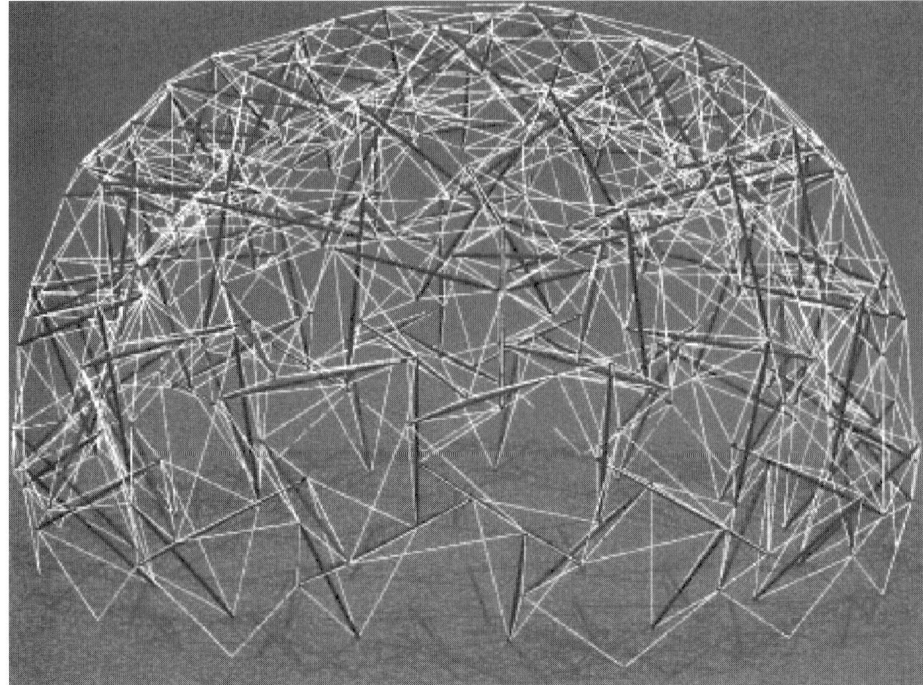

Fig. 6-1. A cable-and-strut tensegrity dome ("Dome Image ©1999 Bob Burkhardt"). In this structure, tension in the cables (white lines) is partly balanced by the compression of the struts (thick black lines) and partly by the attachments to the substrate. At each free node one strut meets several cables. Adapted with permission from Burkhardt, 2004.

structures become unstable and collapse. This explains why the structural stiffness increases in proportion with the level of the prestress.

An interesting (although not an intrinsic) property of tensegrity structures is a long-distance transfer of mechanical disturbances. Ingber referred to this phenomenon as the "action at a distance" effect (Ingber, 1993; 2003a). Because tensegrity structures resist externally applied loads by geometrical rearrangements of their structural members, any local disturbance should result in a global rearrangement of the structural lattice and should be manifested at points distal from the point of an applied load. This is quite different from continuum models where local disturbances produce only local responses, which dissipate inversely with the distance from the point of load application. In complex and multimodular tensegrity structures, this action at a distance may not be easily observable because the effect of an applied mechanical disturbance may be dissipated through the multi-connectedness of structural members and fade away at points distal from the point of load application.

The cellular tensegrity model

In the cellular tensegrity model, actin filaments and intermediate filaments of the cytoskeleton are envisioned as tensile elements (cables) that carry the prestress. Microtubules and thick cross-linked actin bundles, on the other hand, are viewed as compression elements (struts) that partly balance the prestress. The rest of the prestress is balanced by the extracellular matrix, which is physically connected to the cytoskeleton through the focal adhesion complex. In highly spread cells, however, intracellular compression-supporting elements may become redundant and the extracellular matrix may balance the entire prestress. In other words, the cytoskeleton and the extracellular matrix are viewed as a single, synergetic, mechanically stabilized system, or the "extended cytoskeleton" (Ingber, 1993). Thus, although the cellular tensegrity model allows for the presence of internal compression-supporting elements, they are neither necessary nor sufficient for the overall stress balance in the cell–extracellular matrix system.

Do living cells behave as predicted by the tensegrity model?

This section presents a survey of experimental data that are consistent with the cellular tensegrity model, as well as those that are not.

Circumstantial evidence

Data obtained from *in vitro* biophysical measurements on isolated actin filaments (Yanagida et al., 1984; Gittes et al., 1993; MacKintosh et al., 1995) and microtubules (Gittes et al., 1993; Kurachi et al., 1995) indicate that actin filaments are semiflexible, curved, of high tensile modulus (order of 1 GPa), and of the persistence length (a measure of stiffness of a polymer molecule that can be described as a mean radius of curvature of the molecule at some temperature due to thermal fluctuations) on the order of 10 μm. On the other hand, microtubules appeared straight, as rigid tubes, of nearly the same modulus as actin filaments but of much greater persistent length, order of 10^3 μm. Based on these persistence lengths, actin filaments should appear curved and microtubules should appear straight on the whole cell level if they were not mechanically loaded. However, immunofluorescent images of the cytoskeleton lattice of living cells (Fig. 6-2) show that actin filaments appear straight, whereas microtubules appeared curved (Kaech et al., 1996; Eckes et al., 1998, 2003a). It follows, therefore, that some type of mechanical force must act on these molecular filaments in living cells: conceivably, the tension in actin filaments straighten them while compression in microtubules result in their bending (caused by buckling). On the other hand, Satcher et al. (1997) found that in endothelial cells the average pore size of the actin cytoskeleton ranges from 50–100 nm, which is much smaller than the persistence length of actin filaments. This, in turn, suggests that the straight appearance of actin cytoskeletal filaments is the result of their very short length relative to their persistence length.

It is well established that the prestress borne by the cytoskeleton is transmitted to the substrate through transmembrane integrin receptors. Harris et al. (1980) showed that in response to the contraction of fibroblasts cultured on a flexible silicon rubber substrate, the substrate wrinkles. Similarly, contracting fibroblasts that adhere to a polyacrylamide gel substrate cause the substrate to deform (Pelham and Wang, 1997). Severing focal adhesion attachments of endothelial cells to the substrate by trypsin results in a quick retraction of these cells (Sims et al., 1992), suggesting that

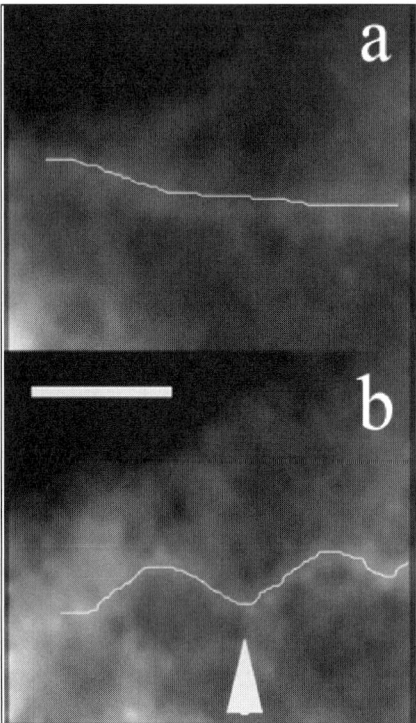

Fig. 6-2. Local buckling of a green fluorescent protein-labeled microtubule (arrowhead) in living endothelial cells following cell contraction induced by thrombin. The microtubule appears fairly straight prior to cell contraction (a) and assumes a typical sinusoidal buckled shape following contraction (b). The white lines are drawn to enhance the shape of microtubule; the scale bar is 2 μm. Adapted with permission from Wang et al., 2001.

the cytoskeleton carries prestress and that this prestress is transmitted to and balanced by traction forces that act at the cell-anchoring points to the substrate.

Experimental observations support the existence of a mechanical coupling between tension carried by the actin network and compression of microtubules, analogous to the tension-compression synergy in the cable-and-strut tensegrity model. For example, as migrating cultured epithelial cells contract, their microtubules in the lamellipodia region buckle as they resist the contractile force exerted on them by the actin network (Waterman-Storer and Salmon, 1997). Extension of a neurite, which is filled with microtubules, is opposed by pulling forces of the actin microfilaments that surround those microtubules (Heidemann and Buxbaum, 1990). Microtubules of endothelial cells, which appear straight in relaxed cells, appear buckled immediately following contraction of the actin network (Fig. 6-2) (Wang et al., 2001). In their mechanical measurements on fibroblasts, Heidemann and co-workers also observed the curved shape of microtubules. However, they associated these configurations with fluid-like behavior of microtubules because they observed slow recovery of microtubules following mechanical disturbances applied to the cell surface (Heidemann et al., 1999; Ingber et al., 2000). Contrary to these observations, Wang et al. (2001) observed relatively quick recovery of microtubules in endothelial cells following mechanical disturbances.

Cells of various types probed with different techniques exhibit a stiffening effect, such that cell stiffness increases progressively with increasingly applied mechanical load (Petersen et al., 1982; Sato et al., 1990; Alcaraz et al., 2003). This, in turn, implies that stress-strain behavior of cells is nonlinear, such that stress increases faster than strain. This stiffening is also referred to as a strain or stress hardening. In discrete structures, this nonlinearity is primarily a result of geometrical rearrangement and recruitment of structural members in the direction of applied load, and less due to nonlinearity of individual structural members (Stamenović et al., 1996). In their early works, Ingber and colleagues considered this stiffening to be a key piece of evidence in support of the cellular tensegrity idea, as various physical (Wang et al., 1993) and mathematical (Stamenović et al., 1996; Coughlin and Stamenović, 1998) tensegrity models exhibit this behavior under certain types of loading. It turns out that this is an inconclusive piece of evidence that neither supports nor refutes the tensegrity model for the following reasons. First, the stress/strain hardening behavior characterizes various types of solid materials, many of which are not at all related to tensegrity. Second, the stress/strain hardening behavior is not an intrinsic property of tensegrity structures because they can also exhibit softening – that is, under a given loading their stiffness may decrease with increasingly applied load (Coughlin and Stamenović, 1998; Volokh et al., 2000) – or they may, under certain conditions, have constant stiffness, independent of the applied load (Stamenović et al., 1996). Third, recent mechanical measurements in living airway smooth muscle cells showed that their stress-strain behavior is linear over a wide range of applied stress, and thus they exhibit neither stiffening nor softening (Fabry et al., 1999; 2001).

Based on the above circumstantial evidence and differing interpretations of the evidence, it is clear that rigorous experimental validation of the cellular tensegrity model was needed to demonstrate a close association between cell stiffness and cytoskeletal prestress, and to show that cells exhibit the action-at-a-distance behavior. Also essential is quantitative assessment of the contribution of the substrate vs. compression of microtubules in balancing the prestress, and also understanding the role of intermediate filaments in the context of the tensegrity model. New advances in cytometry techniques made it possible to provide direct, quantitative data for these behaviors. These data are described below.

Prestress-induced stiffening

An a priori prediction of all prestressed structures is that their stiffness increases in nearly direct proportion with prestress (Volokh and Vilnay, 1997). A number of experiments in various cell types have shown evidence of prestress-induced stiffening. For example, it has been shown that mechanical (Wang and Ingber, 1994; Pourati et al., 1998; Cai et al., 1998), pharmacological (Hubmayr et al., 1996; Fabry et al., 2001), and genetic (Cai et al., 1998) modulations of cytoskeletal prestress are paralleled by changes in cell stiffness. Advances in the traction cytometry technique (Fig. 6-3) made it possible to quantitatively measure various indices of cytoskeletal prestress (Pelham and Wang, 1997; Butler et al., 2002; Wang et al., 2002). These data are then correlated with data obtained from measurements of cell stiffness. It was found (see Fig. 6-4) that in cultured human airway smooth muscle cells whose contractility was altered by graded doses of contractile and relaxant agonists, cell stiffness (G) increases in direct proportion with the contractile stress (P); $G \approx 1.04P$ (Wang et al., 2001; 2002). Although this association between cell stiffness and contractile stress does not preclude other interpretations, it is the hallmark of structures that secure

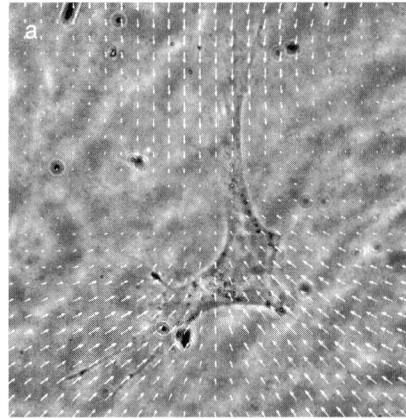
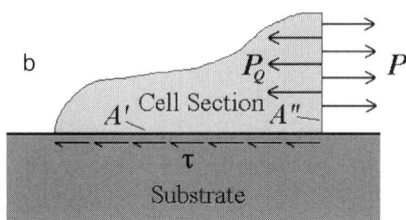

Fig. 6-3. (a) A human airway smooth muscle cell cultured on a flexible polyacrylamide gel substrate. As the cell contracts (histamine 10 μM), the substrate deforms, causing fluorescent microbead markers embedded in the gel to move (arrows). From measured displacement field of the markers and known elastic properties of the gel, one can calculate traction (τ) that arises at the cell-gel interface (Butler et al., 2002). Because the cytoskeletal prestress (P) is balanced partly by τ, one can asses P (Wang et al., 2002). (b) A free body diagram of a cell section depicting a three-way force balance between the cytoskeleton (P), substrate (P_S), and microtubules (P_Q): $P_S = P - P_Q$ where P_S indicates the part of P that is balanced by the substrate and P_Q indicates the part of P that is balanced by compression-supporting microtubules. At equilibrium, the force balance requires that $\tau A' = P_S A''$ where A' and A'' are interfacial and cross-sectional areas of the cell section, respectively. Because τ, A' and A'' can be directly measured, one can obtain P_S. A' and A'' were measured for many optical cross-sections of the cell. For each section, P_S was calculated and the average value was obtained (Wang et al., 2002). Note that in the absence of internal compression structures (for example, upon disruption of microtubules), $P_Q = 0$ and the entire prestress P is balanced by τ (i.e., $P_S = P$).

shape stability through the agency of the prestress. Other possible interpretations of this finding are discussed below.

In addition to generating contractile force, it has been shown that pharmacological agonists also induce polymerization of the actin network (Mehta and Gunst, 1999; Tang et al., 1999). Thus, the observed stiffening in response to contractile agonists could be nothing more than the result of actin polymerization. However, An et al. (2002) have shown that agonist-induced actin polymerization in smooth muscle cells accounts only for a portion of the observed stiffening, whereas the remaining portion of the stiffening is associated with contractile force generation. Another potential mechanism that could explain the data in Fig. 6-4 is the effect of cross-bridge recruitment. It is known from studies of isolated smooth muscle strips in uniaxial extension that both muscle stiffness and muscle force are directly proportional to the number of attached cross-bridges (Fredberg et al., 1996). Thus, the proportionality between the cell stiffness and the prestress could reflect nothing more than the effect of changes in the number of attached cross-bridges in response to pharmacological stimulation. A result that goes against this possibility is obtained from a theoretical model of the myosin cross-bridge kinetics (Mijailovich et al., 2000). This model predicts a qualitatively different oscillatory response from the one measured in airway smooth muscle cells (Fabry et al., 2001). Thus, the kinetics of cross-bridges cannot explain all aspects of cytoskeletal mechanics.

Action at a distance

To investigate whether the cytoskeleton exhibits the action-at-a-distance effect, Maniotis et al. (1997) performed experiments in which the tip of a glass micropipette

Models of cytoskeletal mechanics based on tensegrity

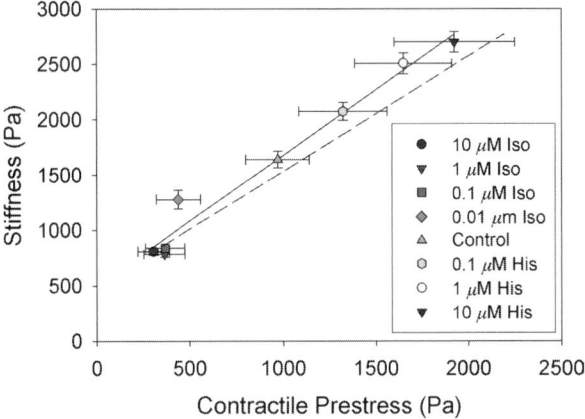

Fig. 6-4. Cell stiffness (G) increases linearly with increasing cytoskeletal contractile stress. Measurements were done in cultured human airway smooth muscle cells. Cell contractility was modulated by graded doses of histamine (constrictor) and graded doses of isoproterenol (relaxant). Stiffness was measured using the magnetic cytometry technique and the prestress was measured by the traction cytometry technique (Wang et al., 2002). The slope of the regression line is 1.18 (solid line). The measured prestress represents the portion of the cytoskeletal prestress that is balanced by the substrate (that is, P_S from Fig. 6-3). Because in those cells microtubules balance on average ~14% of P_S (Stamenović et al., 2002a), the slope of the stiffness vs. the total prestress (P) relationship should be reduced by 14% and thus equals 1.04 (dashed line). The stiffness vs. prestress relationship displays a nonzero intercept. This is due to a bias in the method used to calculate the prestress (in other words, the cell cross-sectional area A'' from Fig. 6-3 is an overestimate) (Wang et al., 2002). In the absence of this artifact, the stiffness vs. prestress relationship would display close-to-zero intercept, that is, $G \approx 1.04P$ (Wang et al., 2002). (Redrawn from Wang et al. (2002) and Stamenović et al. (2002b); G is rescaled to take into account the effect of bead internalization. From Mijailovich et al., 2002.

coated with fibronectin and bound to integrin receptors of living endothelial cells was pulled laterally. Because integrins are physically linked to the cytoskeleton, then if the cytoskeleton is organized as a discrete tensegrity structure, pulling on integrins should produce an observable deformation distal from the point of load application. The authors observed that the nuclear border moved along the line of applied pulling force, which is a manifestation of the action at a distance. A more convincing piece of evidence for this phenomenon was provided by Hu et al. (2003). These investigators designed the intracellular tomography technique that enabled them to observe displacement distribution within the cytoskeleton region in response to locally applied shear disturbance. Lumps of displacement concentrations were found at distances greater than 20 μm from the point of application of the shear loading, which is indicative of the action-at-a-distance effect (Fig. 6-5). Interestingly, when the actin lattice was disrupted (cytochalasin D), the action-at-a-distance effect disappeared (data not shown), suggesting that connectivity of the actin network is essential for transmission of mechanical signals throughout the cytoskeleton. The action-at-a-distance effect has also been observed in neurons (Ingber et al., 2000) and in endothelial cells (Helmke et al., 2003). On the other hand, Heidemann et al. (1999) failed to observe this phenomenon in living fibroblasts when they applied various mechanical disturbances by a glass micropipette to the cell surface through integrin receptors. They found that

112 D. Stamenović

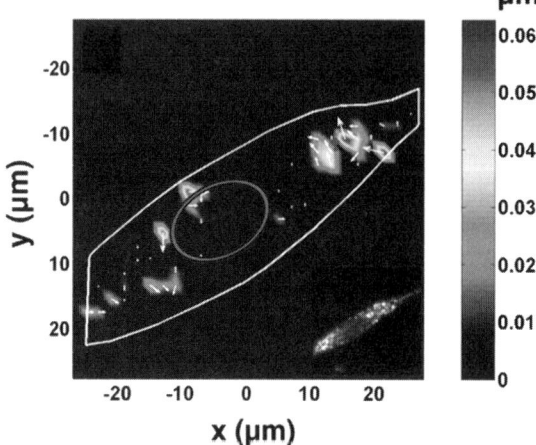

Fig. 6-5. Evidence of the action-at-a-distance effect. Displacement map in living human airway smooth muscle cells obtained using the intracellular tomography technique (Hu et al., 2003). Load is applied to the cell by twisting a ferromagnetic bead bound to integrin receptors on the cell apical surface. The bead position is shown on the phase-contrast image of the cell (inset), the black dot on the image is the bead. The white arrows indicate the direction of the displacement field and the gray-scale map represents its magnitude. Displacements do not decay quickly away from the bead center. Appreciable "lumps" of displacement concentration could be seen at distances more than 20 μm from the bead, consistent with the action-at-a-distance effect. The inner elliptical contour indicates the position of the nucleus. Adapted with permission from Hu et al., 2003.

such disturbances produced only local deformations. However, the authors did not confirm formation of focal adhesions at points of application of external loading, which is essential for load transfer between cell surface and the interior cytoskeleton (Ingber et al., 2000). Thus, their results remain controversial.

Do microtubules carry compression?

Microscopic visualization of green fluorescent protein-labeled microtubules of living cells (see Fig. 6-2) shows that microtubules buckle as they oppose contraction of the actin network (Waterman-Storer and Salmon, 1997; Wang et al., 2001). It was not known, however, whether the compression that causes this buckling could balance a substantial fraction of the contractile prestress. To investigate this possibility, an energetic analysis of buckling of microtubules was carried out (Stamenović et al., 2002a). The assumption was that energy stored in microtubules during compression was transferred to a flexible substrate upon disruption of microtubules. Thus, measurement of an increase in elastic energy of the substrate following disruption of microtubules should indicate compression energy stored in microtubules prior to their disruption. Elastic energy stored in the substrate was obtained from traction microscopy measurements as a work done by traction forces during cell contraction. It was found in highly stimulated and spread human airway smooth muscle cells that following disruption of microtubules by colchicine, the work of traction increases on average by ~30 percent relative to the state before disruption, and equals 0.13 pJ (Stamenović et al., 2002a). This result was then utilized in the energetic analysis. Based on the model of Brodland and Gordon (1990), the microtubules were assumed as slender elastic rods laterally supported by intermediate filaments. Using the post-buckling equilibrium theory of Euler struts (Timoshenko and Gere, 1988), the energy stored during buckling of microtubules was estimated as ~0.18 pJ, which is close to the measured value of ~0.13 pJ (Stamenović et al., 2002a). This is further evidence in support of the idea that microtubules are intracellular compression-bearing elements. Potential concerns are that disruption of microtubules may activate myosin light-chain phosphorylation (Kolodney and Elson, 1995) or could cause a release of intracellular calcium (Paul et al., 2000). Thus, the observed increase in traction and work of traction following disruption of microtubules could be due entirely to chemical mechanisms rather than through mechanical load transfer. These concerns are alleviated by observations indicating that microtubule disruption results in an increase of traction even when the level of myosin light-chain phosphorylation and the level of calcium do not change (Wang et al., 2001; Stamenović et al., 2002a).

From the same experimental data used in the energetic analysis, the contribution of microtubules to balancing the prestress was obtained as follows (Wang et al., 2001; Stamenović et al., 2002a). An increase in traction following microtubule disruption indicates the part of the prestress balanced by microtubules that is transferred to the substrate (see Fig. 6-3b). It was found that this increase ranges from ~5–30 percent, depending on the cell, and is on average ~14 percent, suggesting that microtubules balance only a small fraction of the cytoskeletal prestress and that the substrate balances the bulk of it (Stamenović et al., 2002a). An increase in traction in the response to disruption of microtubules had been observed previously, in different cell types, by other investigators, but has not been quantified (Kolodney and Wysolmersky, 1992; Kolodney and Elson, 1995). More recently, Hu et al. (2004) showed that the contribution of microtubules to balancing the prestress and to the energy budget of the cell depends on the extent of cell spreading. Using the traction cytometry technique, these investigators found that in airway smooth muscle cells, changes in traction and the substrate energy following disruption of microtubules decrease with increasing cell spreading. For example, as the cell projected area increases from 500 to 1800 μm^2, the percent increase in traction following disruption of microtubules decreases from 80 percent to a very small percent. Because in their natural habitat cells seldom exhibit highly spread forms, the above results suggest that the contribution of microtubules in balancing the prestress cannot be overlooked.

The role of intermediate filaments

Cytoskeletal-based intermediate filaments also carry prestress and link the nucleus to the cell surface and the cytoskeleton (Ingber, 1993; 2003a). In support of this view, vimentin-deficient fibroblasts were found to exhibit reduced contractility and reduced traction on the substrate in comparison to the wild-type cells (Eckes et al., 1998). Also, it was observed that the intermediate filament network alone is sufficient to transfer mechanical load from cell surface to the nucleus in cells in which the actin and microtubule networks are chemically disrupted (Maniotis et al., 1997). Taken together, these observations suggest that intermediate filaments play a role in transferring the contractile prestress to the substrate and in long-distance load transfer within the cytoskeleton. Both are key features of the cellular tensegrity model. In addition, inhibition of intermediate filaments causes a decrease in cell stiffness (Wang et al., 1993; Eckes et al., 1998; Wang and Stamenović, 2000), as well as cytoplasmic tearing in response to high applied strains (Maniotis et al., 1997; Eckes et al., 1998). In fact, it appears that the intermediate filaments' contribution to a cell's resistance to shape distortion is substantial only at relatively large strains (Wang and Stamenović, 2000). Another role of intermediate filaments is suggested by Brodland and Gordon (1990). According to these authors, intermediate filaments provide a lateral stabilizing support to microtubules as they buckle while opposing contractile forces transmitted by the cytoskeletal actin lattice. This description is consistent with experimental data (Stamenović et al., 2002a).

Summary

Results from experimental measurements on living adherent cells indicate that their behavior is consistent with the cellular tensegrity model. It was found that cell stiffness increases directly proportionally with increasing contractile stress. It was also found that microtubules carry compression that, in turn, balances a substantial portion of the cytoskeletal prestress. This contribution of microtubules is much smaller in highly spread cells, roughly a few percent, whereas in poorly spread cells it can be as high as ~50 percent. The majority of data from measurements of the action-at-a-distance phenomenon indicate that cells exhibit this type of behavior when the force is applied through integrin receptors at the cell surface and focal adhesions were formed at the site of force application. Intermediate filaments appear to be important contributors to cell contractility and thus to supporting the prestress. They serve as molecular "guy wires" that facilitate transfer of mechanical loads between the cell surface and the nucleus. Finally, intermediate filaments appear to stabilize microtubules as the latter balance the cytoskeletal prestress. Taken together, these observations provide strong evidence in support of the cellular tensegrity model. Although they can have alternative interpretations, there is no single model other than tensegrity that can explain all these data together.

Examples of mathematical models of the cytoskeleton based on tensegrity

Despite its geometric complexity, its dynamic nature, and its inelastic properties, the cytoskeleton is often modeled as a static, elastic, isotropic, and homogeneous network of idealized geometry. The idea is that if the mechanisms by which such an idealized model develops mechanical stress are indeed embodied within the cytoskeleton, then, despite all simplifications, the model should be able to capture key features that characterize mechanical behavior of cells under the steady-state. With the tensegrity model, however, each element is individually taken into account for a discrete formulation of the model. This section describes three types of prestressed structures that have been commonly used as models of cellular mechanics: the cortical membrane

Models of cytoskeletal mechanics based on tensegrity

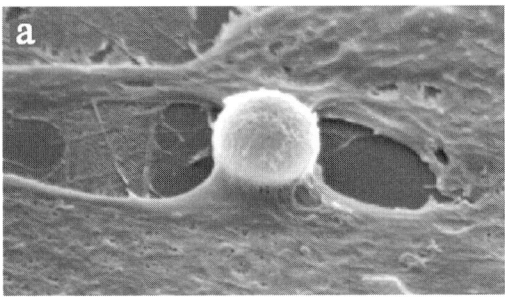 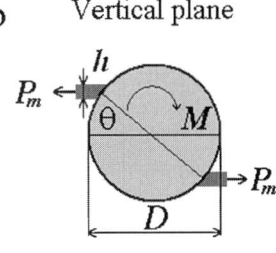

Fig. 6-6. (a) Ferromagnetic beads bound to the apical surface of cultured human airway smooth muscle cells (unpublished data kindly supplied by Dr. B. Fabry). (b) A free-body diagram of a magnetic bead of diameter D half embedded into an elastic membrane of thickness h. The bead is rotated in a vertical plane by specific torque M through angle θ. The rotation is resisted by the membrane tension (prestress) (P_m).

model; the tensed cable net model; and the cable-and-strut model. All three models are stabilized by the prestress. They differ from each other in their topological and structural organization, and in the manner by which they balance the prestress. Results obtained from the models are compared with data from living cells.

The cortical membrane model

This model assumes that the main force-bearing elements of the cytoskeleton are confined either within a thin (~100 nm) cortical layer (Zhelev et al., 1994) or several distinct layers (Heidemann et al., 1999). The cortical layer is under sustained tension (that is, prestress) that is either entirely balanced by the pressurized cytoplasm in suspended cells, or balanced partly by the cytoplasmic pressure and partly by traction at the extracellular adhesions in adherent cells. This model has been successful in describing mechanical features of various suspended cells (Evans and Yeung, 1989; Zhelev et al., 1994; Discher et al., 1998). However, in the case of adherent cells, this model has enjoyed limited success (Fung and Liu, 1993; Schmid-Schönbein et al., 1995; Coughlin and Stamenović, 2003). To illustrate the usefulness of this model, a simulation of a magnetic twisting cytometry measurement is described below (Stamenović and Ingber, 2002).

In the magnetic twisting cytometry technique, small ferromagnetic beads (4.5-μm diameter) bound to integrin receptors on the apical surface of an adherent cell are twisted by a magnetic field, as shown in Fig. 6-6a. Because integrins are physically linked to the cytoskeleton, twisting of the bead is resisted by restoring forces of the cytoskeleton. Using the cortical membrane model, magnetic twisting measurements are simulated as follows.

A rigid spherical bead of diameter D is half-embedded in an initially tensed (prestressed) membrane of thickness h. A twisting torque (M) is applied to the bead in the vertical plane (Fig. 6-6b). Rotation of the bead is impeded by the prestress (P_m) in the membrane. By considering mechanical balance between M and P_m it was found (Stamenović and Ingber, 2002) that

$$M = D^2 P_m h \sin\theta, \qquad (6.1)$$

where θ is the angle of bead rotation. In magnetic twisting measurements, a scale for the applied shear stress (T) is defined as the ratio of M and 6 times bead volume, where δ is the shape factor, and shear stiffness (G) as the ratio of T and θ (Wang et al., 1993; Wang and Ingber, 1994). Thus it follows from Eq. 6.1 that

$$G = \frac{1}{\pi} P_m \frac{h}{D} \frac{\sin\theta}{\theta}. \qquad (6.2)$$

In the limit of $\theta \to 0$, $G \to (1/\pi)P_m(h/D)$ and represents the shear modulus of Hookean elasticity. It follows from Eq. 6.2 that G increases in direct proportion with P_m, a feature consistent with the behavior observed in living cells during magnetic twisting measurements (see Fig. 6-4). Taking into account experimentally based values for $h = 0.1$ μm, $D = 4.5$ μm, and $P_m = O(10^4-10^5)$ Pa it follows from Eq. 6.2 that $G = O(10^2-10^3)$ Pa, which is consistent with experimentally obtained values for G (see Fig. 6-4). [P_m was estimated as follows. It scales with the cytoskeletal prestress P as the ratio of cell radius R to membrane thickness h. Experimental data show that $P = O(10^2-10^3)$ Pa (Fig. 6-4), $R = O(10^1)$ μm and $h = O(10^{-1})$ μm, thus $P_m = O(10^4-10^5)$ Pa.]

Despite this agreement, several aspects of this model are not consistent with experimental results. First, Eq. 6.2 predicts that G decreases with increasing angular strain θ, in other words, softening behavior, whereas magnetic twisting measurements show stress hardening (Wang et al., 1993) or constant stiffness (Fabry et al., 2001). Second, Eq. 6.3 predicts that G decreases with increasing bead diameter D, whereas experiments on cultured endothelial cells show the opposite trend (Wang and Ingber, 1994). One reason for these discrepancies could be the assumption that the cortical layer is a membrane that carries only tensile force. In reality, the cortical layer can support bending, for example in red-blood cells (Evans, 1983; Fung, 1993), and hence a more appropriate model may be a shell-like rather than a membrane-like structure. Regardless, the assumption that the cytoskeleton is confined within a thin cortical layer that surrounds liquid cytoplasm contradicts observations in adherent cells that mechanical perturbations applied to the cell surfaces are transmitted deep into the cytoplasmic domain (Maniotis et al., 1997; Wang et al., 2001; Hu et al., 2003). These observations suggest that mechanical force transmission through the cell is facilitated through the molecular connectivity of the intracellular solid-state cytoskeletal lattice. Taken together, these inconsistencies lower our enthusiasm for the cortical-membrane model as an adequate depiction of the mechanics of adherent cells. However, it remains a good mechanical model for suspended cells where the cytoskeleton appears to be organized within a thin cortical membrane (Bray et al., 1986).

Tensed cable nets

These are reticulated networks comprised entirely of tensile cable elements (Volokh and Vilnay, 1997). Because cables do not support compression, they need to carry initial tension to prevent their buckling and subsequent collapse in the presence of externally applied load. This initial tension defines prestress that is balanced externally (for example, by attachment to the extracellular matrix), and/or internally (such as, by cytoplasmic swelling). A simple illustration of key features of tensed cable nets can be obtained by using the affine network model. A key premise of such a model is that local strains follow the macroscopic (continuum) strain field. (This assumption is known as the affine approximation.) Using this approach and assuming that initially all cable orientations in the network are equally probable, one can obtain that the shear modulus (G) (Stamenović, 2005) is

$$G = (0.8 + 0.2B) P \qquad (6.3)$$

where P is the prestress, $B \equiv (dF/dL)/(F/L)$ is nondimensional cable stiffness, and the F vs. L dependence represents the cable tension-length characteristsic (Budiansky and Kimmel, 1987). In general, B may depend on the level of cable tension (that is, on P). In that case, according to Eq. 6.3, the G vs. P relationship is nonlinear. If, however, B is constant, then G is directly proportional to P. The first term on the right-hand side of Eq. 6.3 represents the sum of the contributions of changes in spacing and orientation of the cables ($0.5P + 0.3P$) to G, whereas the second term ($0.2BP$) is the contribution of the lengthening of the cables to G.

To test whether the prediction of Eq. 6.3 is quantitatively consistent with experimental data from living cells (see Fig. 6-4), we estimate B from measurements of force-extension properties of isolated acto-myosin interactions (Ishijima et al., 1996). Based on these measurements, $(dF/dL)/F = 0.024 \text{ nm}^{-1}$ for a wide range of F. Thus, for a 100-nm long actin filament $B = 2.4$. The choice of filament length of $L = 100$ nm is based on the observation of the average pore-size of the actin cytoskeletal network of endothelial cells (Satcher et al., 1997). By substituting this value into Eq. 6.3, it follows that $G = 1.28P$. This is a modest overestimate of the experimentally obtained result $G = 1.04P$ (Fig. 6-4).

The most favorable aspect of this model is that it provides a mathematically transparent insight into mechanisms that may determine cytoskeletal deformability; G is primarily determined by P through change in spacing and orientation of the cable elements, and to a lesser extent by their stiffness. The model can also provide a reasonably good quantitative correspondence to experimental G vs. P data. The latter is obtained under the crude assumptions of the affine strain approximation and of equally probable distribution of cable orientations. These assumptions are known to lead toward an overestimate of G (Stamenović, 1990). The model also assumes a homogeneous distribution of the prestress throughout the cytoskeleton, although measurements show that the prestress is greatest near the cell edges and decreases toward the nuclear region (Tolić-Nørrelykke et al., 2002). However, in the experimental data for the G vs. P relationship (Fig. 6-4), P represents the mean value of the prestress distribution throughout the cell, and thus the model assumption of uniform prestress is reasonable. The model focuses only on the contribution of the actin network and ignores potential contributions of other components of the cytoskeleton. These contributions will be considered shortly. Nevertheless, the model provides a reasonably good prediction of the G vs. P relationship suggesting that the tensed actin network plays a major role in determining cell mechanical properties. The model also describes the cytoskeleton as a static, elastic network, whereas the cytoskeleton is a dynamic and inelastic structure. This issue is discussed in the section on tensegrity and cellular dynamics.

It is noteworthy that two-dimensional cable nets also have been used to model the cortical membrane. In those models the cortical membrane has been depicted as a two-dimensional network of triangles (Boey et al., 1998) and hexagons (Coughlin and Stamenović, 2003). In the case of suspended cells, this model provides very good correspondence to experimental data. For example, the model of the spectrin lattice successfully describes the behavior of red blood cells during micropipette aspiration measurements (Discher et al., 1998). However, in the case of adherent cells, the model of the actin cortical lattice has enjoyed only moderate success. While it provides a reasonably good correspondence to data from cell poking measurements, it exhibits only some qualitative features of the cell response to twisting and pulling of magnetic beads bound to integrin receptors (Coughlin and Stamenović, 2003). Taken together, the above results show that the two-dimensional cable net model is incomplete to describe mechanical behavior of adherent cells; however the results also show that prestress is a key determinant of the model response.

Cable-and-strut model

This is a cable net model in which the prestress in the cables is balanced by internal compression-supporting struts rather than by inflating pressure. At each free node, one strut meets several cables (see Fig. 6-1). Cables carry initial tension that is balanced by compression of the strut. Together, cables and struts form a self-equilibrated and stable form in the space. This structure may also be attached to the substrate (Fig. 6-1). In this case, the anchoring forces of the substrate also contribute to the balance of tension in the cables. The main difference between these structures and the cable nets is that in the former, the struts directly contribute to the structure's resistance to shape distortion, whereas in the latter this contribution does not exist.

The shear modulus (G) of the cable and strut model can be also obtained using the affine network approach, as in the case of the tensed cable net model. It was found (Stamenović, 2005) that

$$G = 0.8(P - P_Q) + 0.2(BP + B_Q P_Q) \tag{6.4}$$

where P is the prestress carried by the cables and P_Q is the portion of P balanced by the struts, $B \equiv (dF/dL)/(F/L)$ is the nondimensional cable stiffness and $B_Q \equiv (dQ/dl)/(Q/l)$ is the nondimensional strut stiffness. The difference $P - P_Q$ represents the portion of P transmitted to and balanced by the substrate and is denoted by P_S (see Fig. 6-2b). It is this P_S that can be directly measured using the traction microscopy technique (Wang et al., 2002).

It was shown in the section on tensed cable nets that $B = 2.4$. The quantity B_Q is determined based on the buckling behavior of microtubules (Stamenović et al., 2002a). It is found that $B_Q \approx 0.54$. By substituting this value and $B = 2.4$ into Eq. 6.4 and taking into account that in well-spread smooth muscle cells, microtubules balance on average ~14 percent of P_S, that is, $P_Q = 0.14 P_S$ (Stamenović et al., 2002a), it is obtained that $G = 1.19 P$, which is close to the experimental data of $G = 1.04 P$ (Fig. 6-4).

It is noteworthy that if $P_Q = 0$, for example, in a case where microtubules are disrupted, Eq. 6.4 reduces to Eq. 6.3. If disruption of microtubules would not affect P, then according to Eqs. 6.3 and 6.4, for a given P, the shear modulus G would be ~8 percent lower in the case of intact microtubules than in the case of disrupted microtubules. In reality, such conditions in cells are hard to achieve. An experimental

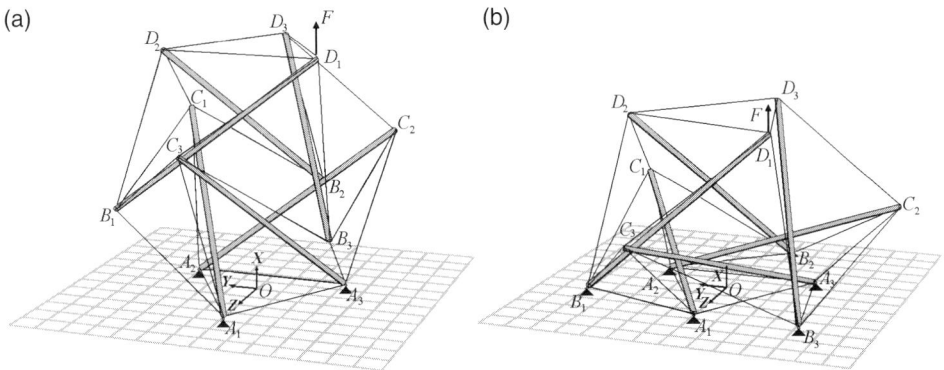

Fig. 6-7. Six-strut tensegrity model in the round (a) and spread (b) configurations anchored to the substrate. Anchoring nodes A_1, A_2 and A_3 (round) and A_1, A_2, A_3, B_1, B_2, and B_3 (spread) are indicated by solid triangles. Pulling force F (thick arrow) is applied at node D_1. Reprinted with permission from Coughlin and Stamenović, 1998.

condition that comes close to this occurs in airway smooth muscle cells stimulated by a saturated dose of histamine (10 µM). In those cells the level of prestress was maintained constant prior to and after disruption of microtubules by colchicine (Wang et al., 2001; 2002). It was found that disruption of microtubules causes a small (∼10 percent), but not significant, increase in cell stiffness (Stamenović et al., 2002b), which is close to the predicted value of ∼8 percent. On the other hand, in nonstimulated endothelial cells, disruption of microtubules causes a significant (∼20 percent) decrease in cell stiffness (Wang et al., 1993; Wang, 1998), which is opposite from the model prediction. A possible reason for this decrease in stiffness in endothelial cells is that in the absence of compression-supporting microtubules, cytoskeletal prestress in those cells decreased, and consequently the cytoskeletal lattice became more compliant.

Most of the criticism for the cable net model also applies to the cable-and-strut model. However, the ability of the model to predict the G vs. P relationship as well as the mechanical role of cytoskeleton-based microtubules such that they are consistent with corresponding experimental data, suggests that the model has captured the basic mechanisms by which the cytoskeleton resists shape distortion.

Consider next an application of a so-called six-strut tensegrity model to study the effect of cell spreading on cell deformability (Coughlin and Stamenović, 1998). This particular model has been frequently used in studies of cytoskeletal mechanics (Ingber, 1993; Stamenović et al., 1996; Coughlin and Stamenović, 1998; Volokh et al., 2000; Wang and Stamenović, 2000; Wendling et al., 1999). It is comprised of six struts interconnected with twenty-four cables (see Fig. 6-7). Although this model represents a gross oversimplification of cytoskeletal architecture, surprisingly it has provided good predictions and simulations of various mechanical behaviors observed in living cells, suggesting that it embodies key mechanisms that determine cytoskeleton mechanics.

In the six-strut tensegrity model, the struts are viewed as slender bars that support no lateral load. Initially, the cables are under tension balanced entirely by compression

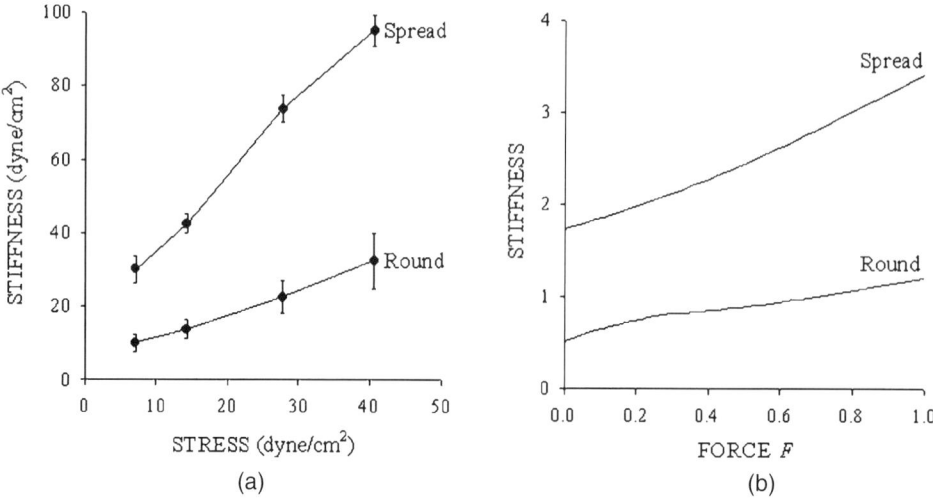

Fig. 6-8. (a) Data for stiffness vs. applied stress in round and spread cultured endothelial cells measured by magnetic twisting cytometry; points means ± SE ($n = 3$ wells, 20,000 cells/well). Both configurations exhibit stress-hardening behavior with greater hardening in the spread than in the round configuration. (Adapted with permission from Wang and Ingber (1994).) (b) Simulations of stiffness vs. applied force (F) in spread and round configurations of the six-strut model (Fig. 6-7) are qualitatively consistent with the data in panel (a). The force is given in the unit of force and the stiffness in the unit of force/length. Adapted with permission from Coughlin and Stamenović, 1998.

of the struts. The structure is then attached to a rigid substrate at three nodes through frictionless ball-joint connections (Fig. 6-7a). The initial force distribution within the structure is not affected by this attachment. This is referred to as a 'round configuration.' To mimic cell spreading, three additional nodes are also anchored to the substrate (Fig. 6-7b). This is referred to as a 'spread configuration.' As a consequence of spreading, force distribution is altered from the one in the round configuration. Tension in the cables is now partly balanced by the struts and partly by reaction forces at the anchoring nodes. In both spread and round configurations, a vertical pulling force (F) is applied at a node distal from the substrate (Fig. 6-7). The corresponding vertical displacement (Δx) is calculated and the structural stiffness as $G = F/\Delta x$. Two cases were considered, one where struts are rigid and cables linearly elastic, and the other where both struts and cables are elastic and struts buckle under compression.

Here we present results from the case with rigid struts; corresponding results obtained with buckling struts are qualitatively similar (Coughlin and Stamenović, 1998). The model predicts that stiffness increases with spreading (Fig. 6-8b). The reason is that tension (prestress) in the cables increases with spreading. The model also predicts approximately linear stress-hardening behavior and predicts that this dependence is greater in the spread than in the round configuration (Fig. 6-8b). All these predictions are consistent (Fig. 6-8a) with the corresponding behavior in round and in spread endothelial cells (Wang and Ingber, 1994). Further attachments of the nodes to the substrate, that is, further spreading, would gradually eliminate the struts from the force balance scheme and their role will be taken over by the substrate.

Taken together, the above results indicate that the cable-and-strut model provides a good and plausible description of cytoskeletal mechanics. It reiterates the central role of cytoskeletal prestress in cell deformability. The cable-and-strut model also reveals the potential contribution of microtubules; they balance a fraction of the prestress, and their deformability (buckling) contributes to the overall deformability of the cytoskeleton. This contribution decreases as cell spreading increases.

In all of the above considerations, intermediate filaments are viewed only as a stabilizing support during buckling of microtubules. To investigate their contribution to cytoskeletal mechanics as stress-bearing members, elastic cables that connect the nodes of the six-strut tensegrity model with its geometric center are added to the model (Wang and Stamenović, 2000). This was based on the observed role of intermediate filaments as "guy wires" between the cell surface and the nucleus (Maniotis et al., 1997). It was shown that by including those cable members in the six-strut model, the model can account for the observed difference in the stress-strain behavior measured by magnetic twisting cytometry between normal cells and cells in which intermediate filaments were inhibited (Wang and Stamenović, 2000).

Summary

Mathematical descriptions of standard tensegrity models of cellular mechanics provide insight into how the cytoskeletal prestress determines cell deformability. Three key mechanisms through which the prestress secures shape stability of the cytoskeleton are changes in spacing, orientation, and length of structural members of the cytoskeleton. Importantly, these mechanisms are not tied to the manner by which the cytoskeletal prestress is balanced. This, in turn, implies that the close association between cell stiffness and the cytoskeletal prestress is a common characteristic of all prestressed structures. Quantitatively, however, this relationship does depend on the architectural organization of the cytoskeletal lattice, including the manner in which the prestress is balanced. The cable-and-strut model shows that in highly spread cells, where virtually the entire prestress is balanced by the substrate, the contribution of microtubules to deformability of the cytoskeleton is negligible. In less-spread cells, however, where the contribution of internal compression members to balancing the prestress increases at the expense of the substrate, deformability of microtubules importantly contributes to the overall lattice deformability. Thus, which of the three models would be appropriate to describe mechanical behavior of a cell would depend upon the cell type and the extent of cell spreading.

Tensegrity and cellular dynamics

In previous sections it was shown how tensegrity-based models could account for static elastic behavior of cells. However, cells are known to exhibit time- and rate-of-deformation-dependent viscoelastic behavior (Petersen et al., 1982; Evans and Yeung, 1989; Sato et al., 1990; Wang and Ingber, 1994; Bausch et al., 1998; Fabry et al., 2001). Because in their natural habitat cells are often exposed to dynamic loads (for example, pulsatile blood flow in vascular endothelial cells, periodic stretching of the extracellular matrix in various pulmonary adherent cells), their viscoelastic properties

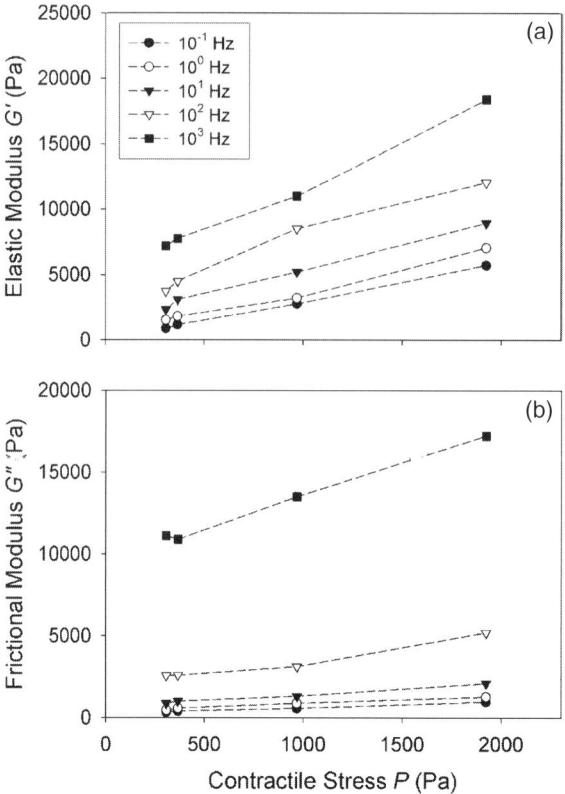

Fig. 6-9. (a) For a given frequency of loading (ω), the storage (elastic) modulus (G') increases with increasing cytoskeletal contractile prestress (P) at all frequencies. (b) The loss (viscous) modulus (G'') also increased with P at all frequencies. Cell contractility was modulated by histamine and isoproterenol. P was measured by traction cytometry and G' and G'' by magnetic oscillatory cytometry. Data are means ± SE. Adapted with permission from Stamenović et al., 2004.

are important determinants of their mechanical behavior. As the tensegrity-based models have provided a reasonably good description of elastic behavior of adherent cells, it is of considerable interest to investigate whether these models can be extended to describe viscoelastic cell behavior.

Recent oscillatory measurements on cultured airway smooth muscle cells indicate that the cytoskeletal prestress may play an important role in determining cell dynamics. It was found (see Fig. 6-9) that the cell dynamic modulus (G^*) is systematically altered in response to modulations of cell contractility; at a given frequency, the real and imaginary components of G^* – the storage (elastic) modulus (G') and loss (viscous) modulus (G''), respectively – increase with increasing contractile prestress P (Stamenović et al., 2002b; 2004). These prestress-dependences of G' and G'' suggest the possibility that cells may utilize similar mechanisms to resist dynamical loads as they do in the case of static loads. Whereas it is clear how geometrical rearrangements of cytoskeletal filaments may come into play in determining the dependence of G' on P, it is not that obvious how they could explain the dependence of G'' on P. A possible explanation for the latter is as follows. In a purely elastic prestressed structure that is subjected to a harmonic strain excitation, all three mechanisms are in phase with the applied strain as long as the structural response is approximately linear and inertial effects are negligible. Consequently, $G'' \equiv 0$. However, in a structure affected by linear damping, the three mechanisms may not all be in phase with the applied strain. As these mechanisms depend on P, phase lags associated with each of them will also depend on P. Consequently, $G'' \neq 0$ and depends on P. The mathematical description of this argument is as follows.

There have been several attempts to model cell viscoelastic behavior using the cable-and-strut model. Cañadas et al. (2002) and Sultan et al. (2004) used the six-strut tensegrity model (Fig. 6-7a) with viscoelastic Voigt elements instead of elastic cables and with rigid struts to study the creep and the oscillatory responses of the cell, respectively. Their models predicted prestress-dependent viscoelastic properties that are qualitatively consistent with experiments. Sultan et al. (2004) also attempted to quantitatively match model predictions with experimental data. They showed that with a suitable choice of model parameters one can provide a very good quantitative correspondence to the observed dependences of G' and G'' on P (Fig. 6-9). However, this could be accomplished only with a very high degree of inhomogeneity in model parameters (variation of several orders of magnitudes), which is not physically realistic.

The specific issue of time- and rate-of-deformation-dependence in explaining the viscoelastic behavior of cells is covered in Chapters 3, 4 and 5 of this book. However, it will be addressed briefly here in the context of the tensegrity idea. A growing body of evidence indicates that the oscillatory response of various cell types follows a weak power-law dependence on frequency, ω^k where $0 \leq k \leq 1$, over several orders of magnitude of ω (Goldmann and Ezzel, 1996; Fabry et al., 2001; Alcaraz et al., 2003). In the limits when $k = 0$, rheological behavior is Hookean elastic solid-like, and when $k = 1$ it is Newtonian viscous fluid-like. A power-law behavior implies the absence of an internal time scale in the structure. Thus, it rules out the Voigt model, the Maxwell model, the standard linear solid model, and other models with a discrete number of time constants (see for example, Sato et al., 1990; Baush et al., 1998). The power-law behavior observed in cells persists even after cell contractility is altered. The only parameter that changes is the power-law exponent k; in contracted airway smooth muscle cells k decreases, whereas in relaxed cells it increases relative to the baseline (Fabry et al., 2001; Stamenović et al., 2004). Based on these observations, an empirical relationship between k and P has been established (Stamenović et al., 2004). It was found that k decreases approximately logarithmically with increasing P. This result suggests that the cytoskeletal contractile stress regulates the transition between solid-like and fluid-like cell behavior.

The observed relationship between k and P appears not to be an a priori prediction of the tensegrity-based models. Rather, it depends on rheological properties of individual structural members and is rooted in the dynamics (thermal fluctuations) of molecules of the cytoskeleton. These dynamics can lead to a power-law behavior of the entire network (Suki et al., 1994). It is feasible, however, that tensile force carried by prestressed cytoskeletal filaments may influence their molecular dynamics, which in turn may explain why P affects the exponent k in the power-law behavior of cells (Stamenović et al., 2004). This has yet to be shown. Another possibility is that molecules of the cytoskeleton exhibit highly nonhomogeneous properties that would lead to a wide distribution of time constants, and thereby to a power-law behavior, as shown by Sultan et al. (2004).

In summary, the basic mechanisms of the tensegrity model can explain the dependence of cell viscoelastic properties on the cytoskeletal prestress. These mechanisms cannot completely explain the frequency response of cells, however, which conforms to a power-law. This power-law behavior seems to be primarily determined by rheology of individual cytoskeletal filaments and their own dynamics (thermal fluctuations, and so forth), rather than by structural dynamics of the cytoskeleton.

Conclusion

This chapter has shown that the tensegrity model is a useful approach for studying mechanics of living cells starting from first principles. This approach elucidates how simple structural models naturally come to express many seemingly complex behaviors observed in cells. This does not preclude the numerous chemically and genetically mediated mechanisms (such as, cytoskeletal remodeling, acto-myosin motor kinetics, cross-linking) that are known to regulate cytoskeletal filament assembly and force generation. Rather, it elucidates a higher level of organization in which these events function and may be regulated.

Taken together, results presented in this chapter can be summarized as follows. First, the cytoskeletal prestress is a key determinant of cell deformability. This feature is consistent with all forms of cellular tensegrity models: the cortical membrane model; the cable net model; and the cable-and-strut model. As a consequence, cell stiffness increases with increasing prestress in nearly direct proportion. Second, depending on the cell type and the extent of cell spreading, one may invoke accordingly different types of tensegrity models in order to describe the effect of the prestress on cellular mechanics. Clearly, various types of ad hoc models unrelated to tensegrity may also provide very useful descriptions of cell mechanical behavior under certain experimental conditions (compare Theret et al., 1988; Sato et al., 1990; Forgacs, 1995; Satcher and Dewey, 1996; Bausch et al., 1998; Fabry et al., 2001). However, the studies described here show that the current formulation of the cellular tensegrity model, although highly simplified, embodies many of the key behaviors of cells. Third, the tensegrity model can explain some aspects of cell viscoelastic behavior, but not all. The behavior appears to be primarily related to rheology and molecular dynamics of individual cytoskeletal filaments. Nevertheless, the observed relationship between viscoelastic properties of the cell and cytoskeletal prestress suggests that rheology of individual filaments may be modulated by the prestress through the mechanisms of tensegrity. This is a subject of future studies that will show whether the tensegrity model is useful only in describing and understanding static elastic behavior of cells, or whether it is also useful for describing and understanding cell dynamic viscoelastic behavior.

A long-term goal is to use the tensegrity idea as a mathematical framework to help understand and predict how mechanical and chemical signals interplay to regulate cell function as well as gene expression. In addition, this model may reveal how cytoskeletal structure, prestress, and the extracellular matrix come into play in the control of cellular information processes (Ingber, 2003b). The biological ground for these applications has already been laid (Ingber 2003a; 2003b). It is the task of bioengineers to carry on this work further.

Acknowledgement

I thank Drs. D. E. Ingber, N. Wang, M. F. Coughlin, and J. J. Fredberg for their collaboration and support in the course of my research of cellular mechanics. Special thanks go to Drs. Ingber and Wang for critically reviewing this chapter.

This work was supported by National Heart, Lung, and Blood Institute Grant HL-33009.

References

Albert-Buehler G (1987) Role of cortical tension in fibroblast shape and movement. *Cell Motil. Cytoskel.*, 7: 54–67.
Alcaraz J, Buscemi L, Grabulosa M, Trepat X, Fabry B, Farre R, Navajas D. (2003) Microrheology of human lung epithelial cells measured by atomic force microscopy. *Biophys. J.*, 84: 2071–2079.
An SS, Laudadio RE, Lai J, Rogers RA, Fredberg JJ (2002) Stiffness changes in cultured airway smooth muscle cells. *Am. J. Physiol. Cell Physiol.*, 283: C792–C801.
Bausch A, Ziemann F, Boulbitch AA, Jacobson K, Sackmann E (1998) Local measurements of viscoelastic parameters of adherent cell surfaces by magnetic bead microrheometry. *Biophys. J.*, 75: 2038–2049.
Boey SK, Boal DH, Discher DE (1998) Simulations of the erythrocyte cytoskeleton at large deformation. I. Microscopic models. *Biophys. J.*, 75: 1573–1583.
Bray D, Heath J, Moss D (1986) The membrane-associated "cortex" of animal cells: Its structure and mechanical properties. *J. Cell Sci. Suppl.*, 4: 71–88.
Brodland GW, Gordon R (1990) Intermediate filaments may prevent buckling of compressively loaded microtubules. *ASME J. Biomech. Eng.*, 112: 319–321.
Budiansky B, Kimmel E (1987) Elastic moduli of lungs. *ASME J. Appl. Mech.*, 54: 351–358.
Burkhardt R (2004) A technology for designing tensegrity domes and spheres. http:www.intergate.com/~bobwb/ts/prospect/prospect/htm.
Butler JP, Tolić-Nørrelykke IM, Fredberg JJ (2002) Estimating traction fields, moments, and strain energy that cells exert on their surroundings. *Am. J. Physiol. Cell Physiol.*, 282: C595–C605.
Cai S, Pestic-Dragovich L, O'Donnell ME, Wang N, Ingber DE, Elson E, Lanorelle P (1998) Regulation of cytoskeletal mechanics and cell growth by myosin light chain phosphorylation. *Am. J. Physiol., Cell Physiol.*, 275: C1349–C1356.
Cañadas P, Laurent VM, Oddou C, Isabey D, Wendling S (2002) A cellular tensegrity model to analyze the structural viscoelasticity of the cytoskeleton. *J. Theor. Biol.*, 218: 155–173.
Coughlin MF, Stamenović D (1998) A tensegrity model of the cytoskeleton in spread and round cells. *ASME J. Biomech. Eng.*, 120: 770–777.
Coughlin MF, Stamenović D (2003) A prestressed cable network model of adherent cell cytoskeleton. *Biophys. J.*, 84: 1328–1336.
Discher DE, Boal DH, Boey SK (1998) Stimulations of the erythrocyte cytoskeleton at large deformation. II. Micropipette aspiration. *Biophys. J.*, 75: 1584–1597.
Eckes B, Dogic D, Colucci-Guyon E, Wang N, Maniotis A, Ingber D, Merckling A, Langa F, Aumailley M, Delouvée A, Koteliansky V Babinet C, Krieg T (1998) Impaired mechanical stability, migration and contractile capacity in vimentin-deficient fibroblasts. *J. Cell Sci.*, 111: 1897–1907.
Evans E (1983) Bending elastic modulus of red blood cell membrane derived from buckling instability in micropipet aspiration tests. *Biophys. J.*, 43: 27–30.

Evans E, Yeung A (1989) Apparent viscosity and cortical tension of blood granulocytes determined by micropipet aspiration. *Biophys. J.*, 56: 151–160.

Evans E, Leung A, Zhelev D (1993) Synchrony of cell spreading and contraction force as phagocytes engulf large pathogens. *J. Cell Biol.*, 122: 1295–1300.

Fabry B, Maksym GN, Hubmayr RD, Butler JP, Fredberg JJ (1999) Implications of heterogeneous bead behavior on cell mechanical properties measured with magnetic twisting cytometry. *J. Magnetism Magnetic Materials* 194: 120–125.

Fabry B, Maksym GN, Butler JP, Glogauer M, Navajas D, Fredberg JJ (2001) Scaling the microrheology of living cells. *Phys. Rev. Lett.*, 87: 148102(1–4).

Forgacs G (1995) On the possible role of cytoskeletal filamentous networks in intracellular signaling: An approach based on percolation. *J. Cell Sci.*, 108: 2131–2143.

Fredberg JJ, Jones KA, Nathan M, Raboudi S, Prakash YS, Shore SA, Butler JP, Sieck GC (1996) Friction in airway smooth muscle: Mechanism, latch, and implications in asthma. *J. Appl. Physiol.*, 81: 2703–2712.

Fuller B (1961) Tensegrity. *Portfolio Artnews Annual* 4: 112–127.

Fung YC (1993) *Biomechanics – Mechanical Properties of Living Tissues*, 2nd edition. New York, Springer.

Fung YC, Liu SQ (1993) Elementary mechanics of the endothelium of blood vessels. *ASME J. Biomech. Eng.*, 115: 1–12.

Gittes F, Mickey B, Nettleton J, Howard J (1993) Flexural rigidity of microtubules and actin filaments measured from thermal fluctuations in shape. *J. Cell Biol.*, 120: 923–934.

Goldmann, WH, Ezzel, RM (1996) Viscoelasticity of wild-type and vinculin deficient (5.51) mouse F9 embryonic carcinoma cells examined by atomic force microscopy and rheology. *Exp. Cell Res.*, 226: 234–237.

Harris AK, Wild P, Stopak D (1980) Silicon rubber substrata: A new wrinkle in the study of cell locomotion. *Science*, 208: 177–179.

Heidemann SR, Buxbaum RE (1990) Tension as a regulator and integrator of axonal growth. *Cell Motil. Cytoskel.*, 17: 6–10.

Heidemann SR, Kaech S, Buxbaum RE, Matus A (1999) Direct observations of the mechanical behavior of the cytoskeleton in living fibroblasts. *J. Cell Biol.*, 145: 109–122.

Helmke BP, Rosen AB, Davies PF (2003) Mapping mechanical strain of an endogenous cytoskeletal network in living endothelial cells. *Biophys. J.*, 84: 2691–2699.

Hu S, Chen J, Fabry B, Namaguchi Y, Gouldstone A, Ingber DE, Fredberg JJ, Butler JP, Wang N (2003) Intracellular stress tomography reveals stress and structural anisotropy in the cytoskeleton of living cells. *Am. J. Physiol. Cell Physiol.*, 285: C1082–C1090.

Hu S, Chen, J, Wang N (2004) Cell spreading controls balance of prestress by microtubules and extracellular matrix. *Frontiers in Bioscience*, 9: 2177–2182.

Hubmayr RD, Shore SA, Fredberg JJ, Planus E, Panettieri Jr, RA, Moller W, Heyder J, Wang N (1996) Pharmacological activation changes stiffness of cultured human airway smooth muscle cells. *Am. J. Physiol. Cell Physiol.*, 271: C1660–C1668.

Ingber DE (1993) Cellular tensegrity: Defining new rules of biological design that govern the cytoskeleton. *J. Cell Sci.*, 104: 613–627.

Ingber DE (2003a) Cellular tensegrity revisited I. Cell structure and hierarchical systems biology. *J. Cell Sci.*, 116: 1157–1173.

Ingber DE (2003b) Tensegrity II. How structural networks influence cellular information-processing networks. *J. Cell Sci.*, 116: 1397–1408.

Ingber DE, Jameison JD (1985) Cells as tensegrity structures: Architectural regulation of histodifferentiation by physical forces transduced over basement membrane. In: *Gene Expression during Normal and Malignant Differentiation*. (eds. Anderson LC, Gahmberg GC, Ekblom P), Orlando, FL: Academic Press, pp. 13–32.

Ingber DE, Madri, JA, Jameison JD (1981) Role of basal lamina in the neoplastic disorganization of tissue architecture. *Proc. Nat. Acad. Sci. USA*, 78: 3901–3905.

Ingber DE, Heidemann SR, Lamoroux P, Buxbaum RE (2000) Opposing views on tensegrity as a structural framework for understanding cell mechanics. *J. Appl. Physiol.*, 89: 1663–1670.

Ishijima A, Kojima H, Higuchi H, Harada Y, Funatsu T, Yanagida T (1996) Multiple- and single-molecule analysis of the actomyosin motor by nanometer-piconewton manipulation with a microneedle: Unitary steps and forces. *Biophys. J.*, 70: 383–400.

Kaech S, Ludin B., Matus A (1996) Cytoskeletal plasticity in cells expressing neuronal microtubule-expressing proteins. *Neuron*, 17: 1189–1199.

Kolodney MS, Wysolmerski RB (1992) Isometric contraction by fibroblasts and endothelial cells in tissue culture: A quantitative study. *J. Cell Biol.*, 117: 73–82.

Kolodney MS, Elson EL (1995) Contraction due to microtubule disruption is associated with increased phosphorylation of myosin regulatory light chain. *Proc. Natl. Acad. Sci. USA*, 92: 10252–10256.

Kurachi M, Masuki H, Tashiro H (1995) Buckling of single microtubule by optical trapping forces: Direct measurement of microtubule rigidity. *Cell Motil. Cytoskel.*, 30: 221–228.

MacKintosh FC, Käs J, Janmey PA (1995) Elasticity of semiflexible biopolymer networks. *Phys. Rev. Lett.*, 75: 4425–4428.

Maniotis AJ, Chen CS, Ingber DE (1997) Demonstration of mechanical connectivity between integrins, cytoskeletal filaments, and nucleoplasm that stabilize nuclear structure. *Proc. Natl. Acad. Sci. USA*, 94: 849–854.

Mehta D, Gunst SJ (1999) Actin polymerization stimulated by contractile activation regulates force development in canine tracheal smooth muscle. *J. Physiol. (Lund.)*, 519: 829–840.

Mijailovich SM, Butler JP, Fredberg JJ (2000) Perturbed equilibrium of myosin binding in airway smooth muscle: Bond-length distributions, mechanics, and ATP metabolism. *Biophys. J.*, 79: 2667–2681.

Mijailovich SM, Kojic M, Zivkovic M, Fabry B, Fredberg JJ (2002) A finite element model of cell deformation during magnetic bead twisting. *J. Appl. Physiol.*, 93: 1429–1436.

Paul RJ, Bowman P, Kolodney MS (2000) Effects of microtubule disruption on force, velocity, stiffness and $[Ca^{2+}]_i$ in porcine coronary arteries. *Am. J. Physiol. Heart Circ. Physiol.*, 279: H2493–H2501.

Pelham RJ, Wang YL (1997) Cell locomotion and focal adhesions are regulated by substrate flexibility. *Proc. Natl. Acad. Sci. USA*, 94: 13661–13665.

Petersen NO, McConnaughey WB, Elson EL (1982) Dependence of locally measured cellular deformability on position on the cell, temperature, and cytochalasin B. *Proc. Natl. Acad. Sci. USA*, 79: 5327–5331.

Porter KR (1984) The cytomatrix: A short history of its study. *J. Cell Biol.*, 99: 3s–12s.

Pourati J, Maniotis A, Spiegel D, Schaffer JL, Butler JP, Fredberg JJ, Ingber DE, Stamenović D, Wang N (1998) Is cytoskeletal tension a major determinant of cell deformability in adherent endothelial cells? *Am J. Physiol. Cell Physiol.*, 274: C1283–C1289.

Satcher RL Jr, Dewey CF Jr (1996) Theoretical estimates of mechanical properties of the endothelial cell cytoskeleton. *Biophys. J.*, 71: 109–118.

Satcher R, Dewey CF Jr, Hartwig JH (1997) Mechanical remodeling of the endothelial surface and actin cytoskeleton induced by fluid flow. *Microcirculation*, 4: 439–453.

Sato M, Theret DP, Wheeler LT, Ohshima N, Nerem RM (1990) Application of the micropipette technique to the measurements of cultured porcine aortic endothelial cell viscoelastic properties. *ASME J. Biomech. Eng.*, 112: 263–268.

Schmid-Schönbein GW, Kosawada T, Skalak R, Chien S (1995) Membrane model of endothelial cell and leukocytes. A proposal for the origin of cortical stress. *ASME J. Biomech. Eng.*, 117: 171–178.

Sims JR, Karp S, Ingber DE (1992) Altering the cellular mechanical force balance results in integrated changes in cell, cytoskeletal and nuclear shape. *J. Cell Sci.*, 103: 1215–1222.

Stamenović D (1990) Micromechanical foundations of pulmonary elasticity. *Physiol. Rev.*, 70: 1117–1134.

Stamenović D (2005) Microtubules may harden or soften cells, depending on the extent of cell distension. *J. Biomech.*, 38:1728–1732.

Stamenović D, Ingber DE (2002) Models of cytoskeletal mechanics of adherent cells. *Biomech. Model Mechanobiol.*, 1: 95–108.

Stamenović D, Fredberg JJ, Wang N, Butler JP, Ingber DE (1996) A microstructural approach to cytoskeletal mechanics based on tensegrity. *J. Theor. Biol.*, 181: 125–136.

Stamenović D, Mijailovich SM, Tolić-Nørrelykke IM, Chen J, Wang N (2002a) Cell prestress. II. Contribution of microtubules. *Am. J. Physiol. Cell Physiol.*, 282: C617–C624.

Stamenović D, Liang Z, Chen J, Wang N (2002b) The effect of cytoskeletal prestress on the mechanical impedance of cultured airway smooth muscle cells. *J. Appl. Physiol.*, 92: 1443–1450.

Stamenović D, Suki B, Fabry B, Wang N, Fredberg JJ (2004) Rheology of airway smooth muscle cells is associated with cytoskeletal contractile stress. *J. Appl. Physiol.*, 96: 1600–1605.

Suki B, Barabási A-L, Lutchen KR (1994) Lung tissue viscoelasticity: A mathematical framework and its molecular basis. *J. Appl. Physiol.*, 76: 2749–2759.

Sultan C, Stamenović D, Ingber DE (2004) A computational tensegrity model predicts dynamic rheological behaviors in living cells. *Ann. Biomed. Eng.*, 32: 520–530.

Tang D, Mehta D, Gunst SJ (1999) Mechanosensitive tyrosine phosphorylation of paxillin and focal adhesion kinase in tracheal smooth muscle. *Am. J. Physiol. Cell Physiol.*, 276: C250–C258.

Theret DP, Levesque MJ, Sato M, Nerem RM, Wheeler LT (1988) The application of a homogeneous half-space model in the analysis of endothelial cell micropipette measurements. *ASME J. Biomech. Eng.*, 110: 190–199

Timoshenko SP, Gere JM (1988) *Theory of Elastic Stability*. New York: McGraw-Hill.

Tolić-Nørrelykke IM, Butler JP, Chen J, and Wang N (2002) Spatial and temporal traction response in human airway smooth muscle cells. *Am J Physiol. Cell Physiol.*, 283, C1254–C1266.

Volokh, KY, Vilnay O (1997) New cases of reticulated underconstrained structures. *Int J Solids Structures* 34: 1093–1104.

Volokh KY, Vilnay O, Belsky M (2000), Tensegrity architecture explains linear stiffening and predicts softening of living cells. *J. Biomech.*, 33: 1543–1549.

Wang N (1998) Mechanical interactions among cytoskeletal filaments. *Hypertension*, 32: 162–165.

Wang N, Ingber DE (1994) Control of the cytoskeletal mechanics by extracellular matrix, cell shape, and mechanical tension. *Biophys. J.*, 66: 2181–2189.

Wang N, Stamenović D (2000) Contribution of intermediate filaments to cell stiffness, stiffening and growth. *Am. J. Physiol. Cell Physiol.*, 279: C188–C194.

Wang N, Butler JP, Ingber DE (1993) Mechanotransduction across cell surface and through the cytoskeleton. *Science*, 26: 1124–1127.

Wang N, Naruse K, Stamenović D, Fredberg JJ, Mijailovich SM, Tolić-Nørrelykke IM, Polte T, Mannix R, Ingber DE (2001) Mechanical behavior in living cells consistent with the tensegrity model. *Proc. Natl. Acad. Sci. USA*, 98: 7765–7770.

Wang N, Tolić-Nørrelykke IM, Chen J, Mijailovich SM, Butler JP, Fredberg JJ, Stamenović D (2002) Cell prestress. I. Stiffness and prestress are closely associated in adherent contractile cells. *Am. J. Physiol. Cell Physiol.*, 282: C606–C616.

Waterman-Storer CM, Salmon ED (1997) Actomyosin-based retrograde flow of microtubules in lamella of migrating epithelial cells influences microtubule dynamic instability and turnover and is associated with microtubule breakage and treadmilling. *J. Cell Biol.*, 139: 417–434.

Wendling S, Oddou C, Isabey D (1999) Stiffening response of a cellular tensegrity model. *J. Theor. Biol.*, 196: 309–325.

Yanagida T, Nakase M, Nishiyama K, Oosawa F (1984) Direct observations of motion of single F-actin filaments in the presence of myosin. *Nature*, 307: 58–60.

Zhelev DV, Needham D, Hochmuth RM (1994) Role of the membrane cortex in neutrophil deformation in small pipettes. *Biophys. J.*, 67: 696–705.

7 Cells, gels, and mechanics

Gerald H. Pollack

ABSTRACT: The cell is known to be a gel. If so, then a logical approach to the understanding of cell function may be through an understanding of gel function. Great strides have been made recently in understanding the principles of gel dynamics. It has become clear that a central mechanism in biology is the polymer-gel phase-transition – a major structural change prompted by a subtle change of environment. Phase-transitions are capable of doing work, and such mechanisms could be responsible for much of the work of the cell. Here, we consider this approach. We set up a polymer-gel-based foundation for cell function, and explore the extent to which this foundation explains how the cell generates various types of mechanical motion.

Introduction

The cell is a network of biopolymers, including proteins, nucleic acids, and sugars, whose interaction with solvent (water) confers a gel-like consistency. This revelation is anything but new. Even before the classic book by Frey-Wyssling a half-century ago (Frey-Wyssling, 1953), the cytoplasm's gel-like consistency had been perfectly evident to any who ventured to crack open a raw egg. The "gel-sol" transition as a central biological mechanism is increasingly debated (Jones, 1999; Berry, et al., 2000), as are other consequences of the cytoplasm's gel-like consistency (Janmey, et al., 2001; Hochachka, 1999). Such phenomena are well studied by engineers, surface scientists, and polymer scientists, but the fruits of their understanding have made little headway into the biological arena.

Perhaps it is for this reason that virtually all cell biological mechanisms build on the notion of an aqueous solution – or, more specifically, on free diffusion of solutes in aqueous solution. One merely needs to peruse representative textbooks to note the many diffusional steps required in proceeding from stimulus to action. These steps invariably include: ions diffusing into and out of membrane channels; ions diffusing into and out of membrane pumps; ions diffusing through the cytoplasm; proteins diffusing toward other proteins; and substrates diffusing toward enzymes, among others. A cascade of diffusional steps underlies virtually every intracellular process, notwithstanding the cytoplasm's character as a gel, where diffusion can be slow enough to be biologically irrelevant. This odd dichotomy between theory and evidence has grown unchecked, in large part because modern cell biology has been pioneered by those with limited familiarity with gel function. The gel-like consistency of the cytoplasm has been largely ignored. What havoc has such misunderstanding wrought?

Problems with the aqueous-solution-based paradigm

Consider the consequences of assuming that the cytoplasm is an aqueous solution. To keep this "solution" and its solutes constrained, this liquid-like milieu is surrounded by a membrane, which is presumed impervious to most solutes. But solutes need to pass into and out of the cell – to nourish the cell, to effect communication between cells, to exude waste products, etc. In order for solutes to pass into and out of the cell, the membrane requires openable pores. Well over 100 solute-specific channels have been identified, with new ones emerging regularly.

The same goes for membrane pumps: Because ion concentrations inside and outside the cell are rarely in electrochemical equilibrium, the observed concentration gradients are thought to be maintained by active pumping mediated by specific entities lodged within the membrane. The text by Alberts et al. (1994) provides a detailed review of this foundational paradigm, along with the manner in which this paradigm accounts for many basics of cell function. In essence, solute partitioning between the inside and the outside of the cell is assumed to be a product of an impermeable membrane, membrane pumps, and membrane channels.

How can this foundational paradigm be evaluated?

If partitioning requires a continuous, impermeant barrier, then violating the barrier should collapse the gradients. Metabolic processes should grind to a halt, enzymes and fuel should dissipate as they diffuse out of the cytoplasm, and the cell should be quickly brought to the edge of death.

Does this really happen?

To disrupt the membrane experimentally, scientists have concocted an array of implements not unlike lances, swords, and guns:

- *Microelectrodes*. These are plunged into cells in order to measure electrical potentials between inside and outside or to pass substances into the cell cytoplasm. The microelectrode tip may seem diminutive by conventional standards, but to the 10-µm cell, invasion by a 1-µm probe is roughly akin to the reader being invaded by a fence post.
- *Electroporation* is a widely used method of effecting material transfer into a cell. By shotgunning the cell with a barrage of high-voltage pulses, the membrane becomes riddled with orifices large enough to pass genes, proteins, and other macromolecules – and certainly large enough to pass ions and small molecules easily.
- The *patch-clamp* method involves the plucking of a 1-µm patch of membrane from the cell for electrophysiological investigation; the cell membrane is grossly violated.

Although such insults may sometimes inflict fatal injuries they are not, in fact, necessarily consequential. Consider the microelectrode plunge and subsequent withdrawal. The anticipated surge of ions, proteins, and metabolites might be thwarted if the hole could be kept plugged by the microelectrode shank – but this is not always the case. Micropipettes used to microinject calcium-sensitive dyes at multiple sites along muscle cells require repeated withdrawals and penetrations, each withdrawal leaving multiple micron-sized injuries. Yet, normal function is observed for up to several days (Taylor et al., 1975). The results of patch removal are similar. Here again, the hole in question is more than a million times the cross-section of the potassium ion. Yet, following removal of the 1-µm patch, the 10 µm isolated heart cell is commonly found to live on and continue beating (L. Tung and G. Vassort, personal communication).

Similarly innocuous is the insult of electroporation. Entry of large molecules into the cell is demonstrable even when molecules are introduced into the bath up to many hours after the end of the electrical barrage that creates the holes (Xie et al., 1990; Klenchin et al., 1991; Prausnitz et al., 1994; Schwister and Deuticke, 1985; Serpersu et al., 1985). Hence, the pores must remain open for such long periods without resealing. Nor do structural studies in muscle and nerve cells reveal any evidence of resealing following membrane disruption (Cameron, 1988; Krause et al., 1984). Thus, notwithstanding long-term membrane orifices of macromolecular size – with attendant leakage of critical-for-life molecules anticipated – the cell does not perish.

Now consider the common alga *Caulerpa*, a single cell whose length can grow to several meters. This giant cell contains stem, roots, and leaves in one cellular unit undivided by any internal walls or membranes (Jacobs, 1994). Although battered by pounding waves and gnawed on by hungry fish, such breaches of integrity do not impair survival. In fact, deliberately cut sections of stem or leaf will grow back into entire cells. Severing of the membrane is devoid of serious consequence.

Yet another example of major insult lies within the domain of experimental genetics, where cells are routinely sectioned in order to monitor the fates of the respective fragments. When cultured epithelial cells are sectioned by a sharp micropipette, the nonnucleated fraction survives for one to two days, while the nucleated, centrosome-containing fraction survives indefinitely and can go on to produce progeny (Maniotis and Schliwa, 1991). Sectioned muscle and nerve cells similarly survive (Yawo and Kuno, 1985; Casademont et al., 1988; Krause et al., 1984), despite the absence of membrane resealing (Cameron, 1988; Krause et al., 1984).

Finally, and perhaps not surprisingly in light of all that has been said, ordinary cells in the body are continually in a "membrane-wounded" state. Cells that suffer mechanical abrasion in particular – such as skin cells, gut endothelial cells, and muscle cells – are especially prone to membrane wounds, as confirmed by passive entry into the cell of large tracers that ordinarily fail to enter. Yet such cells appear structurally and functionally normal (McNeil and Ito, 1990; McNeil and Steinhardt, 1997). The fraction of wounded cells in different tissues is variable. In cardiac muscle cells it is ~20 percent, but the fraction rises to 60 percent in the presence of certain kinds of performance-enhancing drugs (Clarke et al., 1995). Thus, tears in the cell membrane occur commonly and frequently even in normal, functioning tissue, possibly due to surface abrasion.

Evidently, punching holes in the membrane does not wreak havoc with the cell, even though the holes may be monumental in size relative to an ion. It appears we are stuck on the horns of a dilemma. If a continuous barrier envelops the cell and is consequential for function, one needs to explain why breaching the barrier is not more consequential than the evidence seems to indicate. On the other hand, if we entertain the possibility that the barrier may be noncontinuous, so that creating yet another opening makes little difference, we then challenge the dogma on which all mechanisms of cell biological function rest, for the continuous barrier concept has become axiomatic.

Is there an escape?

If the cytoplasm is not an aqueous solution after all, then the need for a continuous barrier (with pumps and channels) becomes less acute. If the cytoplasm is a gel, for example, the membrane could be far less consequential. This argument does not imply that the membrane is *absent* – only that its continuity may not be essential for function. Such an approach could go a long way toward explaining the membrane-breach anomalies described above, for gels can be sliced with relative impunity. Major insults might or might not be tolerable by the gel-like cell, depending on the nature of the insult and the degree of cytoplasmic damage inflicted. Death is not obligatory. A continuous barrier is not required for gel integrity, just as we have seen that a continuous barrier is not required for cell integrity. A critical feature of the cytoplasm, then, may be its gel-like consistency.

Cells as gels

Gels are built around a scaffold of long-chain polymers, often cross-linked to one another and invested with solvent. The cytoplasm is much the same. Cellular polymers such as proteins, polysaccharides, and nucleic acids are long-chained molecules, frequently cross-linked to one another to form a matrix. The matrix holds the solvent (water) – which is retained even when the cell is demembranated. "Skinned" muscle cells, for example, retain water in the same way as gels. Very much, then, the cytoplasm resembles an ordinary gel – as textbooks assert.

How the gel/cell matrix holds water is a matter of some importance (Rand et al., 2000), and there are at least two mechanistic possibilities. The first is osmotic: charged surfaces attract counter-ions, which draw in water. In the second mechanism, water-molecule dipoles adsorb onto charged surfaces, and subsequently onto one another to form multilayers. The first mechanism is unlikely to be the prevailing one because: (1) gels placed in a water bath of sufficient size should eventually be depleted of the counter-ions on which water retention depends, yet, the hydrated gel state is retained; and (2) cytoplasm placed under high-speed centrifugation loses ions well before it loses water (Ling and Walton, 1976).

The second hypothesis, that charged surfaces attract water dipoles in multilayers, is an old one (Ling, 1965). The thesis is that water can build layer upon layer (Fig. 7-1). This view had been controversial at one time, but it has been given support by several groundbreaking observations. The first is the now-classical observation by Pashley and Kitchener that polished quartz surfaces placed in a humid atmosphere will adsorb films of water up to 600 molecular layers thick (Pashley and Kitchener, 1979); this implies adsorption of a substantial number of layers, one upon another. The second set of observations are those of Israelachvili and colleagues, who measured the force required to displace solvents sandwiched between closely spaced parallel mica surfaces

Cells, gels, and mechanics

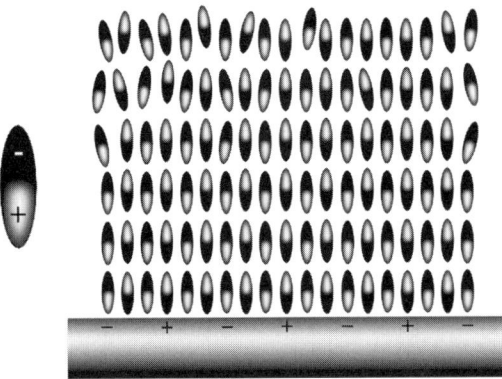

Fig. 7-1. Organization of water molecules adjacent to charged surface.

(Horn and Israelachvili, 1981; Israelachvili and McGuiggan, 1988; Israelachvili and Wennerström, 1996). The overall behavior was largely classical, following DLVO theory. However, superimposed on the anticipated monotonic response was a series of regularly spaced peaks and valleys (Fig. 7-2). The spacing between peaks was always equal to the molecular diameter of the sandwiched fluid. Thus, the force oscillations appeared to arise from a layering of molecules between the surfaces.

Although the Israelachvili experiments confirm molecular layering near charged surfaces, they do not prove that the molecules are linked to one another in the manner implied in Fig. 7-1. However, more recent experiments using carbon-nanotube-tipped AFM probes approaching flexible monolayer surfaces in water show similar layering (Jarvis et al., 2000), implying that the ordering does not arise merely from packing constraints; and, the Pashley/Kitchener experiment implies that many layers are

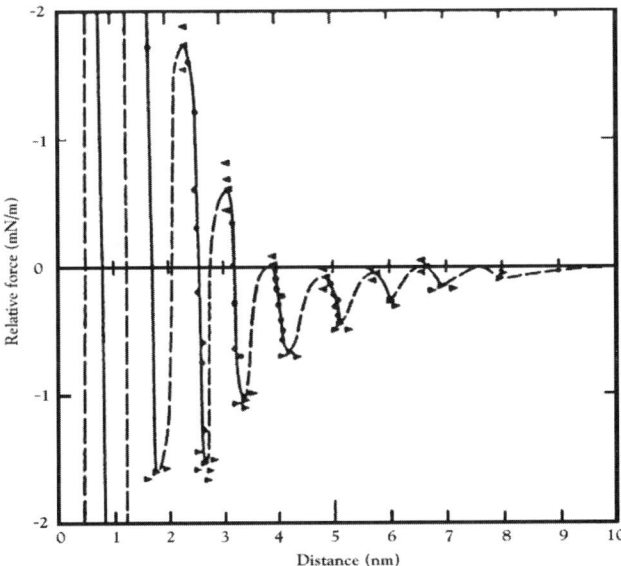

Fig. 7-2. Effect of separation on force between closely spaced mica plates. Only the oscillatory part of the response is shown. After Horn and Israelachvili, 1981.

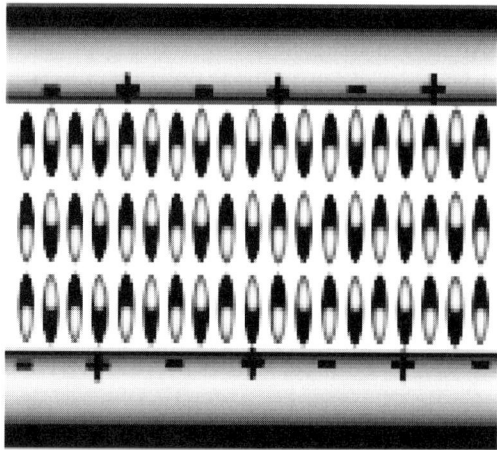

Fig. 7-3. Structured water dipoles effectively "glue" charged surfaces to one another.

possible. Hence, the kind of layering diagrammed in Fig. 7-1 is collectively implied by these experiments.

When two charged polymeric surfaces lie in proximity of one another, the interfacial water layers can bond the surfaces much like glue (Fig. 7-3). This is revealed in common experience. Separating two glass slides stacked face-to-face is no problem; when the slides are wet, however, separation is formidable: sandwiched water molecules cling tenaciously to the glass surfaces and to one another, preventing separation. A similar principle holds in sand. A foot will ordinarily sink deeply into dry sand at the beach, leaving a large imprint, but in wet sand, the imprint is shallow. Water clings to the sand particles, bonding them together with enough strength to support one's full weight.

The picture that emerges, then, is that of a cytoplasmic matrix very much resembling a gel matrix. Water molecules are retained in both cases because of their affinity for the charged (hydrophilic) surfaces and their affinity for one another. The polymer matrix and adsorbed water largely make up the gel. This explains why demembranated cells retain their integrity.

Embodied in this gel-like construct are many features that have relevance for cell function. One important one is ion partitioning. The prevailing explanation for the ion gradients found between extracellular and intracellular compartments lies in a balance between passive flow through channels and active transport by pumps. Thus, the low sodium concentration inside the cell relative to outside is presumed to arise from the activity of sodium pumps, which transport sodium ions against their concentration gradient from the cytoplasm across the cell membrane. The gel construct invites an alternative explanation. It looks toward differences of solubility between extracellular bulk water and intracellular layered, or "structured" water, as well as differences of affinity of various ions for the cell's charged polymeric surfaces (Ling, 1992). Na^+ has a larger hydrated diameter than K^+, and is therefore more profoundly excluded from the cytoplasm than K^+; and, because the hydration layers require more energy to remove from Na^+ than from K^+, the latter has higher affinity for the cell's negatively charged polymeric surfaces. Hence, the cytoplasm has considerably more potassium

Cells, gels, and mechanics

than sodium. A fuller treatment of this fundamental biological feature is given in the recent book by the author (Pollack, 2001).

Similarly, the gel construct provides an explanation for the cell potential. The cell is filled with negatively charged polymers. These polymers attract cations. The number of cations that can enter the cell (gel) is restricted by the cations' low solubility in structured water. Those cations that do enter compete with water dipoles for the cell's fixed anionic charges. Hence, the negative charge in the cell is not fully balanced by cations. The residual charge amounts to approximately 0.3 mol/kg (Wiggins, 1990). With net negative charge, the cytoplasm will have a net negative potential. Indeed, depending on conditions, membrane-free cells can show potentials as large as 50 mV (Collins and Edwards, 1971). And gels made of negatively charged polymers show comparable or larger negative potentials, while gels built of positively charged polymers show equivalent positive potentials (Fig. 7-4). Membranes, pumps, and channels evidently play no role.

Hence, the gel paradigm can go quite far in explaining the cell's most fundamental attributes – distribution of ions and the presence of a cell potential. These are equilibrium processes; they require no energy for maintenance.

Cell dynamics

The cell is evidently not a static structure, but a machine designed to carry out a multitude of tasks. Such tasks are currently described by a broad variety of mechanisms, apparently lacking any single identifiable underlying theme – at least to this author. For virtually every process there appears to be another mechanism.

Whether a common underlying theme might govern the cell's many operational tasks is a question worth asking. After all, the cell began as a simple gel and evolved from there. As it specialized, gel structure and processes gained in intricacy. Given such lineage, the potential for a simple, common, underlying, gel-based theme should not necessarily be remote. Finding a common underlying theme has been a long-term quest in other fields. In physics, for example, protégés of Einstein continue the search for a unifying force. That nature works in a parsimonious manner, employing variations of a few simple principles to carry out multitudinous actions, is an attractive notion, one I do not believe has yet been seriously pursued in the realm of cell function, although simplicity is a guiding principle in engineering.

If the cell is a gel, then a logical approach to the question of a common underlying principle of cell function is to ask whether a common underlying principle governs gel function. Gels do "function." They undergo transition from one state to another. The process is known as a polymer-gel phase-transition – much like the transition from ice to water – a small change of environment causing a huge change in structure.

Such change can generate work. Just as ice formation has sufficient power to fracture hardened concrete, gel expansion or contraction is capable of many types of work, ranging from solute/solvent separation to force generation (Fig. 7-5). Common examples of useful phase-transitions are the time-release capsule (in which a gel-sol transition releases bioactive drugs), the disposable diaper (where a condensed gel undergoes enormous hydration and expansion to capture the "load"), and various

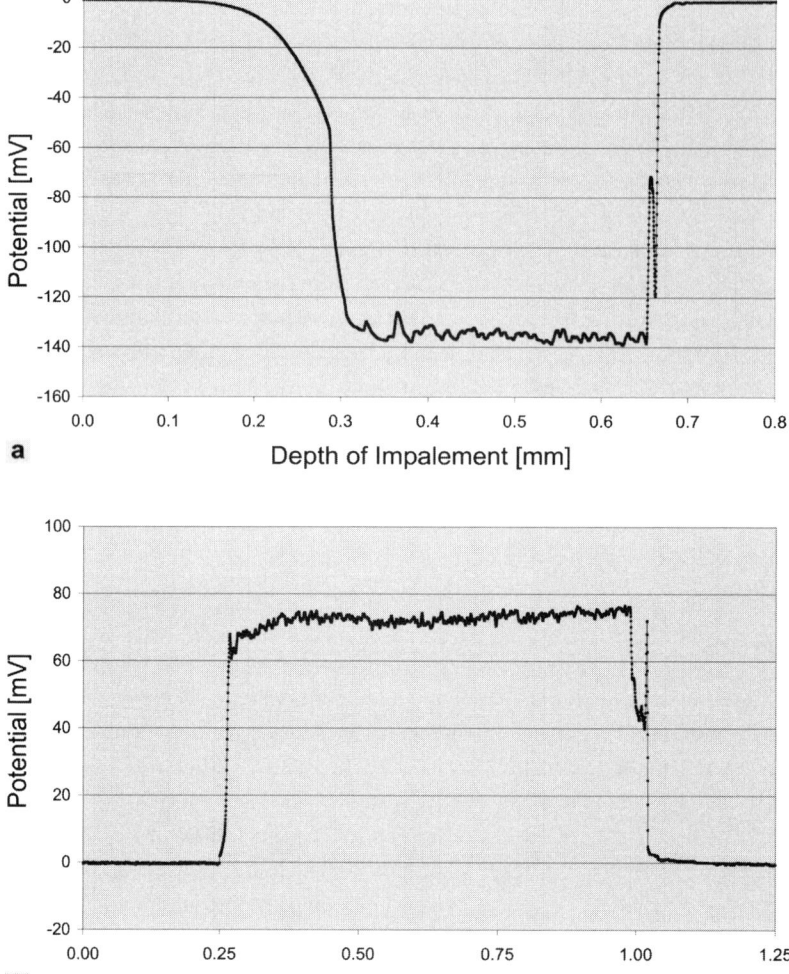

Fig. 7-4. KCl-filled microelectrodes stuck into gel strips at slow constant velocity, and then withdrawn. (a) Typical anionic gel, polyacrylamide/polypotassiumacrylate, shows negative potential. (b) Typical cationic gel, polyacrylamide/polydialyldimethylammonium chloride, shows positive potential. Courtesy of R. Gülch.

artificial muscles. Such behaviors are attractive in that a large change of structure can be induced by a subtle change of environment (Fig. 7-6).

Like synthetic gels, the natural gel of the cell may have the capacity to undergo similarly useful transitions. The question is whether they do. This question is perhaps more aptly stated a bit differently, for the cell is not a homogeneous gel but a collection of gel-like organelles, each of which is assigned a specific task. The more relevant question, then, is whether any or all such organelles carry out their functions by undergoing phase-transition.

The short answer is yes: it appears that this is the case. Pursuing so extensive a theme in a meaningful way in the short space of a review article is challenging; for

Cells, gels, and mechanics

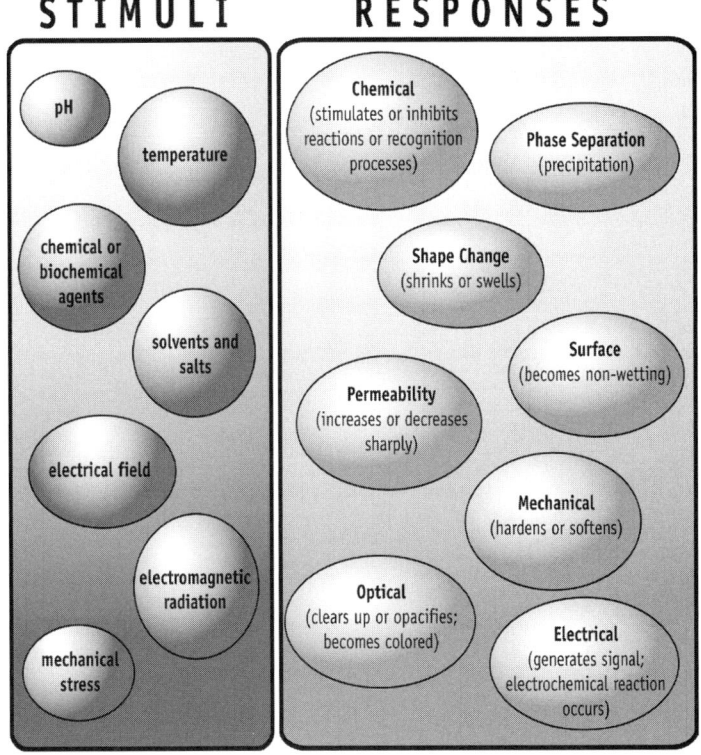

Fig. 7-5. Typical stimuli and responses of artificial polymer hydrogels. After Hoffman, 1991.

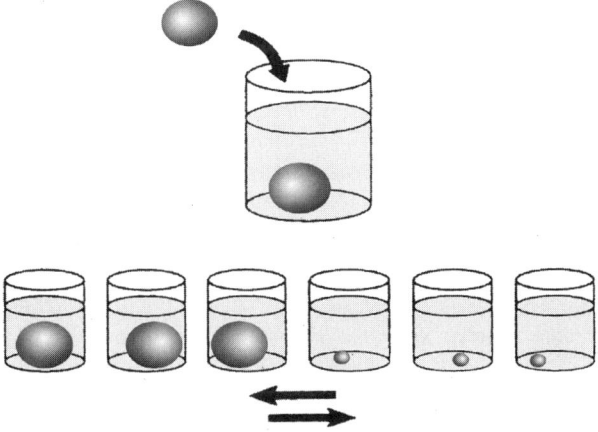

Temperature, solvent composition, pH, ions electric field, UV, light, specific molecules, or chemicals

Fig. 7-6. Phase-transitions are triggered by subtle shifts of environment. After Tanaka et al., 1992.

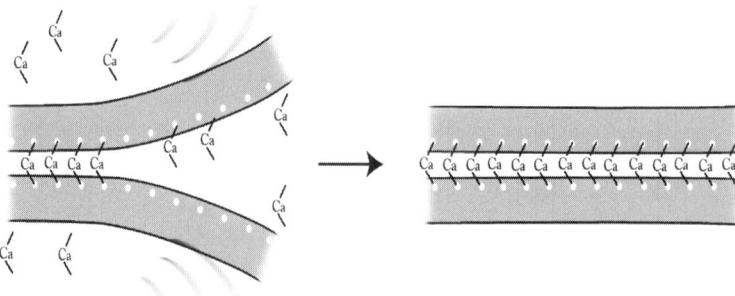

Fig. 7-7. Calcium and other divalent cations can bridge the gap between negatively charged sites, resulting in zipper-like condensation.

a fuller development I refer the reader to Pollack (2001). In this venue I focus on a single aspect: the relevance of phase-transitions in the production of motion.

Gels and motion

The classes of motion produced by phase-transitions fall largely into two categories, isotropic and linear. In isotropic gels, polymers are randomly arranged and sometimes cross-linked. Water is held largely by its affinity to polymers (or proteins, in the case of the cell). The gel is thus well hydrated – and may in the extreme contain as much as 99.97 percent water (Osada and Gong, 1993). In the transitioned state, the dominant polymer-water affinity gives way to a higher polymer-polymer affinity, condensing the gel into a compact mass and expelling solvent. Thus, water moves, and polymer moves.

Linear polymers also undergo transition – from extended to shortened states. The extended state is stable because it maximizes the number of polymer-water contacts and therefore minimizes the system's energy. Water builds layer upon layer. In the shortened state the affinity of polymer for itself exceeds the affinity of polymer for water, and the polymer folds. It may fold entirely, or it may fold regionally, along a fraction of its length. As it folds, polymer and water both move. And, if a load is placed at the end of the shortening filament, the load can move as well.

Phase-transitions are inevitably cooperative: once triggered, they go to completion. The reason lies in the transition's razor-edge behavior. Once the polymer-polymer affinity (or the polymer-water affinity) begins to prevail, its prevalence increases; hence the transition goes to completion. An example is illustrated in Fig. 7-7. In this example, the divalent ion, calcium, cross-links the polymer strands. Its presence thereby shifts the predominant affinity from polymer-water to polymer-polymer. Once a portion of the strand is bridged, flanking segments of the polymer are brought closer together, increasing the proclivity for additional calcium bridging. Thus, local action enhances the proclivity for action in a neighboring segment, ensuring that the reaction proceeds to completion. In this way, transitions propagate toward completion.

Evidently, the polymer-gel phase-transition can produce different classes of motion. If the cell were to exploit this principle, it could have a simple way of producing a broad array of motions, depending on the nature and arrangement of constituent polymers.

Cells, gels, and mechanics

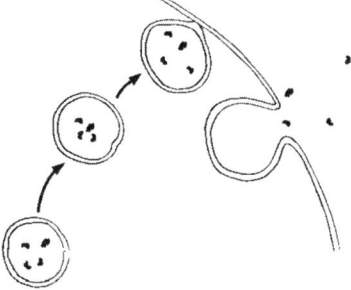

Fig. 7-8. Textbook view of secretion. Chemicals are packed in vesicles, which work their way to the cell surface, poised for discharge.

In all cases, a small shift of some environmental variable such as pH or chemical content, for example, could give rise to a cooperative, all-or-none response, which could produce massive mechanical action.

As representative examples of such action, two fundamental cellular processes are considered – secretion and contraction. The first involves the propulsive motion of small molecules, the second the motion of protein filaments. (Additional details on these and other mechanisms can be found in Pollack (2001)).

Secretion

Secretion is the mechanism by which the cell exports molecules. The molecules are packed into small spherical vesicles, which lie just within the cell boundary, awaiting export (Fig. 7-8). According to prevailing views, the vesicle is a kind of "soup" surrounded by a membrane – a miniature of the prevailing view of the cell itself. For discharge, the vesicle docks with the cell membrane; cell and vesicle membranes fuse, opening the interior of the vesicle to the extracellular space and allowing the vesicle's contents to escape by diffusion. Although attractive in its apparent simplicity, this mechanism does not easily reconcile with several lines of evidence.

The first is that the vesicle is by no means a clear broth. It is a dense matrix of tangled polymers, invested with the molecules to be secreted. Getting those molecules to diffuse through this entwining thicket and leave the cell is as implausible as envisioning a school of fish escaping from an impossibly tangled net.

A second concern is the response to solvents. Demembranated vesicle matrices can be expanded and recondensed again and again by exposure to various solutions, but these solutions are not the ones expected from classical theory. When condensed matrices from mast cells or goblet cells (whose matrices hydrate to produce mucus) are exposed to low osmolarity solutions – even distilled water – they remain condensed even though the osmotic draw for water ought to be enormous (Fernandez et al., 1991; Verdugo et al., 1992). Something keeps the network condensed, and it appears to be multivalent cations, in some cases calcium and in other cases the molecule to be secreted, which is commonly a multivalent cation. These multivalent cations cross-link the negatively charged matrix and keep it condensed, even in the face of solutions of extremely low osmolarity (Fig. 7-9).

A third issue is that discharge does not appear to be a passive event. It is often accompanied by dramatic vesicle expansion. Isolated mucin-producing secretory

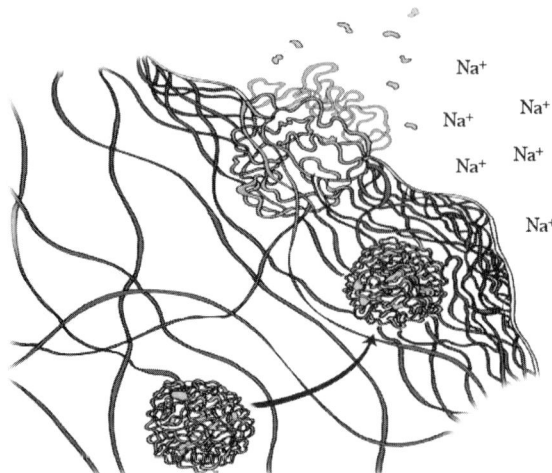

Fig. 7-9. Phase-transition model of secretion. The phase-transition is triggered as extracellular Na^+ replaces the multivalent cation holding the anionic network condensed.

vesicles, for example, undergo a 600-fold volume expansion within 40 ms (Verdugo et al., 1992). Vesicles of nematocysts (aquatic stinging cells) are capable of linear expansion rates of 2,000 μm/ms (Holstein and Tardent, 1984). Such phenomenal expansion rates imply something beyond mere passive diffusion of solutes and water.

Given these features, it is no surprise that investigators have begun looking for mechanistic clues within the realm of the phase-transition, where expansion can be large and rapid. A feature of secretory discharge consistent with this mechanism is that discharge happens or doesn't happen depending on a critical shift of environment – the very hallmark of the polymer-gel phase-transition. Goblet-cell and mast-cell matrices condense or expand abruptly as the solvent ratio (either glycerol/water or acetone/water) is edged just past a threshold or the temperature edges past a threshold, the transition thresholds in both cases lying within a window as narrow as 1 percent of the critical value (Verdugo et al., 1992). Hence, the phase-transition's signature criterion is satisfied. The abrupt expansion and hydration would allow the relevant molecules to escape into the extracellular fluid.

Such a system might work as follows. When the condensed matrix is exposed to the extracellular space, sodium displaces the divalent cross-linker. No longer cross-linked, the polymer can satisfy its intense thirst for hydration, imbibing water and expanding explosively, in a manner described as a jack-in-the-box (Verdugo et al., 1992). Meanwhile, the messenger molecules are discharged. Diffusion may play some role in release, but the principal role is played by convective forces, for multivalent ions are relatively insoluble in the layered water surrounding the charged polymers (Vogler, 1998), and will therefore be forcefully ejected. Hence, discharge into the extracellular space occurs by explosive convection.

Muscle contraction

As is now well known, muscle sarcomeres, or contractile units, contain three filament types (Fig. 7-10): thick, thin, and connecting – the latter interconnecting the ends of

Cells, gels, and mechanics

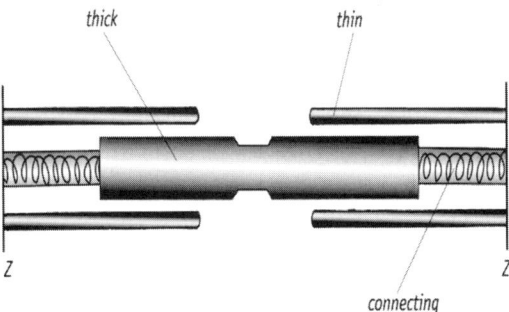

Fig. 7-10. Muscle sarcomere contains three filament types, bounded by Z-lines.

the thick filament with respective Z-lines and behaving as a molecular spring. All three filaments are polymers: thin filaments consist largely of repeats of monomeric actin; thick filaments are built around multiple repeats of myosin; and connecting filaments are built of titin (also known as connectin), a huge protein containing repeating immunoglobulin-like (Ig) and other domains. Together with water, which is held with extreme tenacity by these proteins (Ling and Walton, 1976), this array of polymers forms a gel-like lattice.

Until the mid-1950s, muscle contraction was held to occur by a mechanism not much different from the phase-transition mechanism to be considered. All major research groups subscribed to this view. With the discovery of interdigitating filaments in the mid-1950s, it was tempting to dump this notion and suppose instead that contraction arose out of pure filament sliding. This supposition led Sir Andrew Huxley and Hugh Huxley to examine independently whether filaments remained at constant length during contraction. Back-to-back papers in *Nature*, using the optical microscope (Huxley and Niedergerke, 1954; Huxley and Hanson, 1954) appeared to confirm this supposition. The constant-filament-length paradigm took hold, and has held remarkably firm ever since – notwithstanding more than thirty subsequent reports of thick filament or A-band shortening (Pollack, 1983; 1990) – a remarkable disparity of theory and evidence. The motivated reader is invited to check the cited papers and make an independent judgment.

With the emerging notion of sliding filaments, the central issue became the nature of the driving force; the model that came to the forefront was the so-called swinging cross-bridge mechanism (Huxley, 1957). In this model, filament translation is driven by oar-like elements protruding like bristles of a brush from the thick filaments, attaching transiently to thin filaments, swinging, and propelling the thin filaments to slide along the thick. This mechanism explains many known features of contraction, and has therefore become broadly accepted (Spudich, 1994; Huxley, 1996; Block, 1996; Howard, 1997; Cooke, 1997).

On the other hand, contradictory evidence abounds. In addition to the conflicting evidence on the constancy of filament length, which contradicts the pure sliding model, a serious problem is the absence of compelling evidence for cross-bridge swinging (Thomas, 1987). Electron-spin resonance, X-ray diffraction, and fluorescence-polarization methods have produced largely negative results, as has high-resolution electron microscopy (Katayama, 1998). The most positive of these results has been an angle change of $3°$ measured on a myosin light chain (Irving et al., 1995) – far short of the anticipated 45°. Other concerns run the gamut from instability (Irving et al., 1995), to mechanics (Fernandez et al., 1991), structure (Schutt and Lindberg, 1993; 1998), and chemistry (Oplatka, 1996; 1997). A glance at these reviews conveys a picture different from the one in textbooks.

An alternative approach considers the possibility that the driving mechanism does not lie in cross-bridge rotation, but in a paradigm in which all three filaments shorten. If contiguous filaments shorten synchronously, the event is global, and may qualify as a phase-transition. We consider the three filaments one at a time.

First consider the connecting filament. Shortening of the connecting filament returns the extended, unactivated sarcomere to its unstrained length. Conversely, applied stress lengthens the connecting filament. Shortening may involve a sequential folding of domains along the molecule, whereas stretch includes domain unfoldings – the measured length change being stepwise (Rief et al., 1997; Tskhovrebova et al., 1997). Similarly in the intact sarcomere, passive length changes also occur in steps (Blyakhman et al., 1999), implying that each discrete event is synchronized in parallel over many filaments.

Next, consider the thick filament. Thick filament shortening could transmit force to the ends of the sarcomere through the thin filaments, thereby contributing to active sarcomere shortening. Evidence for thick filament shortening was mentioned above. Although rarely discussed in contemporary muscle literature, the observations of thick filament length changes are extensive: they have been carried out in more than fifteen laboratories worldwide and have employed electron and light microscopic techniques on specimens ranging from crustaceans and insects to mammalian heart and skeletal muscle – even human muscle. Evidence to the contrary is relatively rare (Sosa et al., 1994). These extensive observations cannot be summarily dismissed merely because they are not often discussed.

Thick filament shortening cannot be the sole mechanism underlying contraction. If it were, the *in vitro* motility assay in which thin filaments translate over individual myosin molecules planted on a substrate could not work, for it contains no filaments that could shorten. On the other hand, filament shortening cannot be dismissed as irrelevant. Thick filament shortening could contribute directly to sarcomere shortening. It could be mediated by an alpha-helix to random-coil transition along the myosin rod, which lies within the thick filament backbone (Pollack, 1990). The helix-coil transition is a classical phase-transition well known to biochemists – and also to those who have put a wool sweater into a hot clothes dryer and watched it shrink.

The thin filament may also shorten. There is extensive evidence that some structural change takes place along the thin filament (Dos Remedios and Moens, 1995; Pollack, 1996; Käs et al., 1994). Crystallographic evidence shows that monomers of actin can pack interchangeably in either of two configurations along the filament – a "long" configuration and a shorter one (Schutt and Lindberg, 1993). The difference leads to a filament length change of 10–15 percent. The change in actin is worth dwelling on, for although it may be more subtle than the ones cited above, it may be more universal, as actin filaments are contained in all eukaryotic cells.

Isolated actin filaments show prominent undulations. Known as "reptation" because of its snake-like character, the constituent undulations are broadly observed: in filaments suspended in solution (Yanagida et al., 1984); embedded in a gel (Käs

Cells, gels, and mechanics

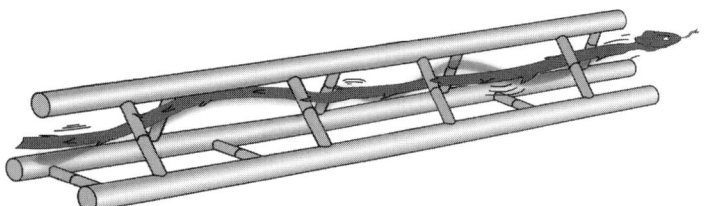

Fig. 7-11. Reptation model. An actin filament snakes its way toward the center of the sarcomere, past myosin cross-bridges, which may well interconnect adjacent thick filaments. From Vogler, 1998.

et al., 1994); and gliding on a myosin-coated surface (Kellermayer and Pollack, 1996). Such undulations had been presumed to be of thermal origin, but that notion is challenged by the observation that they can be substantially intensified by exposure to myosin (Yanagida et al., 1984) or ATP (Hatori et al., 1996). These effects imply a specific structural change rather than a thermally induced change.

In fact, structural change in actin is implied by a long history of evidence. Molecular transitions had first been noted in the 1960s and 1970s (Asakura et al., 1963; Hatano et al., 1967; Oosawa et al., 1972). On exposure to myosin, actin monomers underwent a 10° rotation (Yanagida and Oosawa, 1978). Conformational changes have since been confirmed not only in probe studies, but also in X-ray diffraction studies, phosphorescence-anisotropy studies, and fluorescence-energy transfer studies, the latter showing a myosin-triggered actin-subdomain-spacing change of 17 percent (Miki and Koyama, 1994).

That such structural change propagates along the filament is shown in several experimental studies. Gelsolin is a protein that binds to one end (the so-called "barbed" end) of the actin filament, yet the impact of binding is felt along the entire filament: molecular orientations shift by 10°, and there is a three-fold decrease of the filament's overall torsional rigidity (Prochniewicz et al., 1996). Thus, structural change induced by point binding propagates over the entire filament. Such propagated action may account for the propagated waves seen traveling along single actin filaments – observable either by cross-correlation of point displacements (deBeer et al., 1998) or by tracking fluorescence markers distributed along the filament (Hatori et al., 1996; 1998). In the latter, waves of shortening can be seen propagating along the filament, much like a caterpillar.

Could such a propagating structural transition drive the thin filament to slide along the thick? A possible vehicle for such action is the inchworm mechanism (Fig. 7-11). By propagating along the thin filament, a shortening transition could propel the thin filament to reptate past the thick filament, each propagation cycle advancing the filament incrementally toward the center of the sarcomere.

Perhaps the most critical prediction of such a mechanism is the anticipated quantal advance of the thin filament. With each propagation cycle, the filament advances by a step (Fig. 7-12). The advance begins as an actin monomer unbinds from a myosin cross-bridge; it ends as the myosin bridge rebinds an actin momomer farther along the thin filament. Hence, the filament-translation step size must be an integer multiple of the actin-repeat spacing (see Fig. 7-12). The translation step could be one, two, ... or n times the actin-repeat spacing along the thin filament.

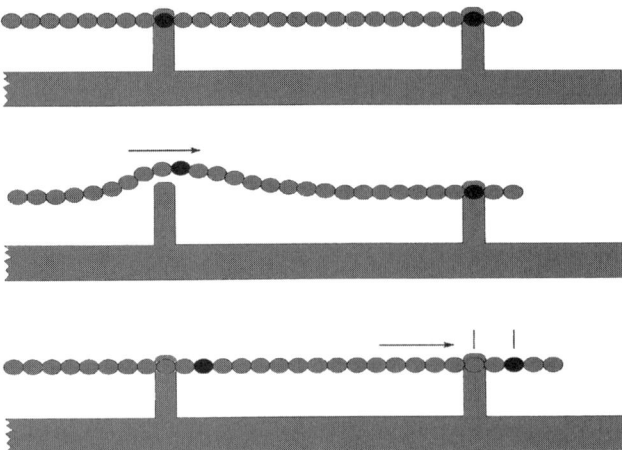

Fig. 7-12. Reptation model predicts that each advance of the thin filament will be an integer multiple of the actin-monomer repeat along the filament.

This signature-like prediction is confirmed in several types of experiment. The thin filament advances in steps; and, step size is an integer multiple of the actin-monomer spacing (Figs. 7-13–7-16). This is true in the isolated molecular system, where the single myosin molecule translates along actin (Kitamura et al., 1999); in the intact sarcomere, where thin filaments translate past thick filaments (Blyakhman et al., 1999; Yakovenko et al., 2002); and in isolated actin and myosin filaments sliding past one another (Liu and Pollack, 2004).

In the sarcomere experiments, the striated image of a single myofibril is projected onto a photodiode array. The array is scanned repeatedly, producing successive traces

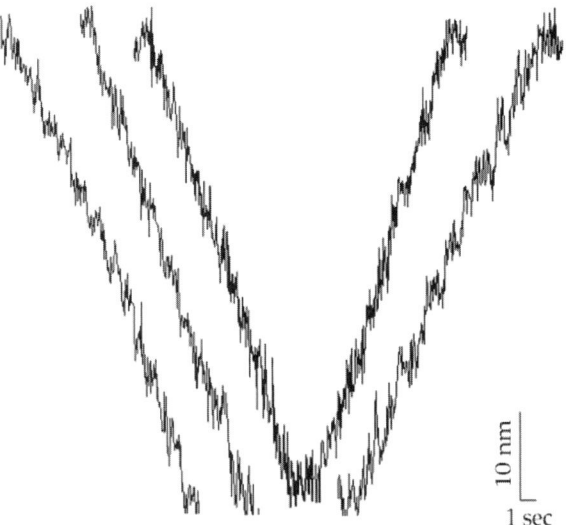

Fig. 7-13. Time course of single sarcomere shortening in single activated myofibrils. From Yakovenko et al. (2002).

Cells, gels, and mechanics

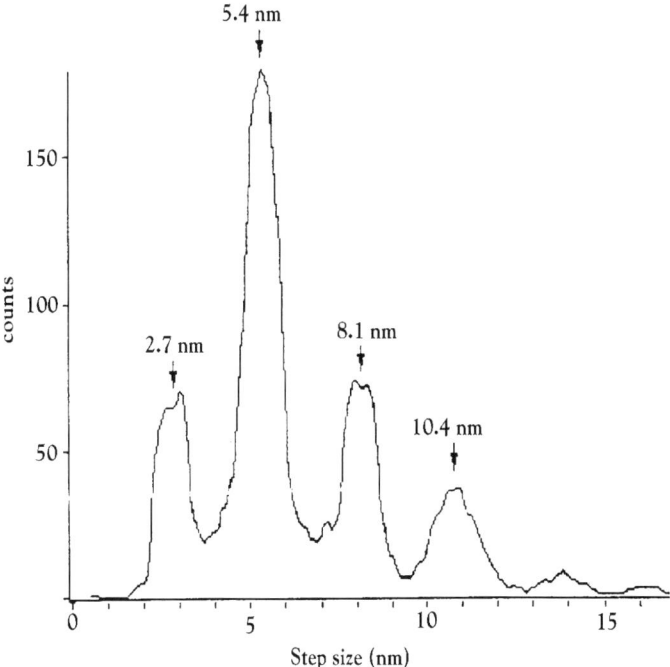

Fig. 7-14. Analysis of steps in records such as those of Fig. 7-13. Steps are integer multiples of 2.7 nm.

of intensity along the myofibril axis. Hence, single sarcomeres can be tracked. The sarcomere-length change is consistently stepwise (Fig. 7-13). Analysis of many steps showed that their size is an integer multiple of 2.7 nm, the actin-monomer spacing projected on the filament axis (Fig. 7-14).

Similar results are obtained when a single actin filament slides past a single thick filament (Liu and Pollack, 2004). Here the trailing end of the actin filament is attached to the tip of a deflectable nanolever (Fauver et al., 1998), normal to its long axis. On addition of ATP, the actin filament slides, bending the nanolever. Displacement of the actin filament is monitored by tracking the position of the nanolever tip, which is projected onto a photodiode array. Representative traces are shown in Fig. 7-15. They reveal the stepwise nature of filament translation. Step size was measured by applying an algorithm that determined the least-squares linear fit to each pause. The vertical displacement between pauses gave the step. Fig. 7-16 shows a histogram obtained from a large number of steps. The histogram is similar to that of Fig. 7-14. It shows, once again, steps that are integer multiples of 2.7 nm, both when the actin filament slides forward during contraction and when the actin filament is forcibly pulled in the sarcomere-lengthening direction. Hence, the results obtained with single isolated filaments and single sarcomeres are virtually indistinguishable.

Agreement between these results and the model's prediction lends support to the proposed thesis. Conventional mechanisms might generate a step advance during each cross-bridge stroke; and, with a fortuitously sized cross-bridge swinging arc, the step could have the appropriate size. But it is not at all clear how integer multiples of the

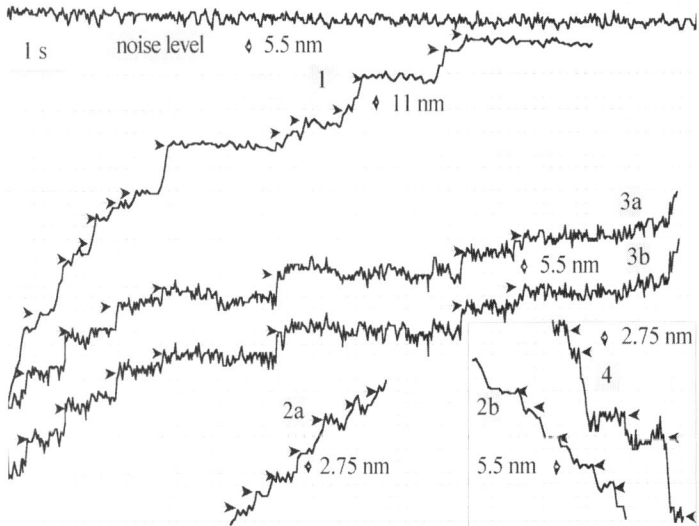

Fig. 7-15. Step-wise interaction between single actin and single thick filament. Horizontal bar represents 1 s except in traces 1 and 2, where it equals 0.5 s. The two traces in the box show backward steps. Arrows indicate positions of some pauses. Top trace shows noise level, and has same scale as trace 2.

fundamental size might be generated in a simple way, although they are observed frequently (Figs. 7-14, 7-16). By contrast, the detailed quantitative observations described above are direct predictions of the reptation mechanism.

In sum, contraction of the sarcomere could well arise out of contraction of each of the three filaments – connecting, thick, and thin. Connecting and thick filaments appear to shorten by local phase-transitions, each condensation shortening the

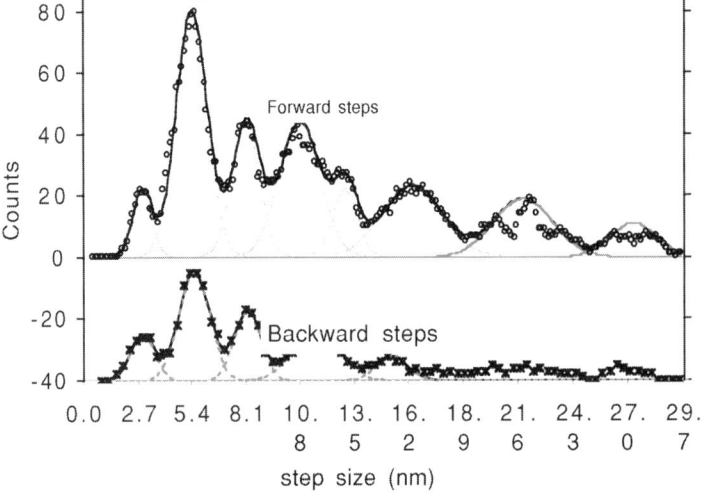

Fig. 7-16. Continuous histogram of step-size distribution from data similar to those shown on Fig. 7-15, with bin width of 1.0 nm and increments of 0.1 nm.

Cells, gels, and mechanics

respective filament by an incremental step. Because these two filaments lie in series, filament shortening leads directly to sarcomere shortening. The thin filament appears to undergo a local, propagating transition, each snake-like cycle advancing the thin filament past the thick by an increment. Repeated cycles produce large-scale translation. (A similar process may occur in the *in vitro* motility assay, where the myosins are firmly planted on a substrate rather than in the lattice of filaments; the filament may "snake" its way along.)

The incremental steps anticipated from these transitions are observable at various levels of organization. These levels range from the single filament pair (Liu and Pollack, 2004) and single myofibrillar sarcomere (Blyakhman et al., 1999; Yakovenko et al., 2002), to bundles of myofibrils (Jacobson et al., 1983), to segments of whole muscle fibers (Granzier et al., 1987). Hence, the transitions are global, as the phase-transition anticipates. It is perhaps not surprising that phase-transitions arise in all three filamentary elements. This endows the system with a versatile array of features that makes muscle the effective machine that it is. Indeed, muscle is frequently referred to as the jewel in mother nature's crown of achievements.

Conclusion

Two examples of biological motion have been presented, each plausibly driven by phase-transitions and each producing a different type of motion. Isotropic structures such as secretory vesicles undergo condensations and expansions, whereas filamentary bundles such as actin and myosin produce linear contraction. Linear contraction can also occur in microtubules, another of nature's linear polymers: when cross-linked into a bundle, microtubules along one edge of the bundle are often observed to shorten (McIntosh, 1973); this may mediate bending, as occurs for example in a bimetal strip. Hence, diverse motions are possible.

Given such mechanistic versatility, it would not be surprising if the phase-transition were a generic mechanism for motion production, extending well beyond the examples considered here. Phase-transitions are simple and powerful. They can bring about large-scale motions induced by subtle changes of environment. This results in a kind of switch-like action with huge amplification. Such features seem attractive enough to imply that if nature has chosen the phase-transition as a common denominator of cell motion (and perhaps sundry other processes) it has made a wise choice.

Acknowledgement

The consent of *Ebner and Sons* to reprint figures from Pollack, *Cells, Gels and the Engines of Life: A New, Unifying Approach to Cell Function* (2001), is gratefully acknowledged.

References

Alberts, B., Bray, D., Lewis, J., Raff, M., Roberts, K., and Watson, J.D. (1994). Molecular Biology of the Cell. Third Edition. New York: Garland.

Asakura, A., Taniguchi, M., and Oosawa. F. (1963). Mechano-chemical behavior of F-actin. *J. Mol. Biol.*, 7, 55–63.

Berry, H., Pelta, J., Lairez, D., and Larreta-Garde, V. (2000). Gel-sol transition can describe the proteolysis of extracellular matrix gels. *Biochim. Biophys. Acta.*, 1524(2–3), 110–117.

Block, S.M. (1996). Fifty ways to love your lever: Myosin motors. *Cell*, 87, 151–157.

Blyakhman, F., Shklyar, T., and Pollack, G.H. (1999). Quantal length changes in single contracting sarcomeres. *J. Mus. Res. Cell Motil.*, 20, 529–538.

Cameron, I. (1988). Ultrastructural observations on the transectioned end of frog skeletal muscles. *Physiol. Chem. Phys. Med.* NMR, 20, 221–225.

Casademont, J., Carpenter, S., and Karpati, G. (1988). Vacuolation of muscle fibers near sarcolemmal breaks represents T tubule dilation secondary to enhanced sodium pump activity. *J. Neuropath. Exp. Neurol.*, 47, 618–628.

Clarke, M.S.F., Caldwell, R.W., Miyake, K., and McNeil, P.L. (1995). Contraction-induced cell wounding and release of fibroblast growth factor in heart. *Circ. Res.*, 76, 927–934.

Collins, E.W., Jr., and Edwards, C. (1971). Role of Donnan equilibrium in the resting potentials in glycerol-extracted muscle. *Am. J. Physiol.*, 22(4), 1130–1133.

Cooke, R. (1997). Actomyosin interaction in striated muscle. *Physiol. Rev.*, 77(3), 671–679.

deBeer, E.L., Sontrop, A., Kellermayer, M.S.Z., and Pollack, G.H. (1998). Actin-filament motion in the in vitro motility assay is periodic. *Cell Motil. and Cytoskel.*, 38, 341–350.

Dos Remedios, C.G., and Moens, P.D. (1995). Actin and the actomyosin interface: a review. *Biochim. Biophys. Acta.*, 1228(2–3), 99–124.

Fauver, M., Dunaway, D., Lilienfield, D., Craighead, H., and Pollack, G.H. (1998). Microfabricated cantilevers for measurement of subcellular and molecular forces. IEEE *Trans. Biomed. Engr.*, 45, 891–898.

Fernandez, J.M., Villalon, M., and Vedugo, P. (1991). Reversible condensation of mast cell secretory products in vitro. *Biophys. J.*, 59, 1022–1027.

Frey-Wyssling, A. (1953). Submicroscopic Morphology of Protoplasm. Amsterdam: Elsevier.

Granzier, H.L.M., Myers, J.A., and Pollack, G.H. (1987). Stepwise shortening of muscle fiber segments. *J. Mus. Res. & Cell Motility*, 8, 242–251.

Hatano, S., Totsuka, T., and Oosawa, F. (1967). Polymerization of plasmodium actin. *Biochem. Biophys. Acta.*, 140, 109–122.

Hatori, K., Honda, H., and Matsuno, K. (1996). Communicative interaction of myosins along an actin filament in the presence of ATP. *Biophys. Chem.*, 60, 149–152.

Hatori, K., Honda, H., Shimada, K., and Matsuno, K. (1998). Propagation of a signal coordinating force generation along an actin filament in actomyosin complexes. *Biophys. Chem.*, 75, 81–85.

Hochachka, P.W. (1999). The metabolic implications of intracellular circulation. *Proc. Natl. Acad. Sci. USA*, 96(22), 12233–12239.

Hoffman, A. S. (1991). Conventionally and environmentally sensitive hydrogels for medical and industrial use: a review paper. *Polymer Gels*, 268(5), 82–87.

Holstein, T., and Tardent, P. (1984). An ultra high-speed analysis of exocytosis: nematocyst discharge. *Science*, 223, 830–833.

Horn, R.G., and Israelachvili, J.N. (1981). Direct measurement of astructural forces between two surfaces in a nonpolar liquid. *J. Chem. Phys.*, 75(3), 1400–1411.

Howard, J. (1997). Molecular motors: structural adaptations to cellular functions. *Nature*, 389, 561–567.

Huxley, A.F. (1957). Muscle structure and theories of contraction. *Prog. Biophys Biophys. Chem.*, 7, 255–318.

Huxley, A.F., and Niedergerke, R. (1954). Structural changes in muscle during contraction: Interference microscopy of living muscle fibres. *Nature*, 173, 971–973.

Huxley, H.E. (1996). A personal view of muscle and motility mechanisms. *Ann. Rev. Physiol.*, 58, 1–19.

Huxley, H.E., and Hanson, J. (1954). Changes in the cross striations of muscle during contraction and stretch and their structural interpretation. *Nature*, 173, 973–976.

Irving, M., Allen, T. St.-C., Sabido-David, C., Craik, J.S., Brandmeler, B., Kendrick-Jones, J., Corrie, J.E.T., Trentham, D.R., and Goldman, Y.E. (1995). Tilting the light-chain region of myosin during step length changes and active force generation in skeletal muscle. *Nature*, 375, 688–691.

Israelachvili, J.N., and McGuiggan, P.M. (1988). Forces between surfaces in liquids. *Science*, 241, 795–800.

Israelachvili, J.N., and Wennerström, H. (1996). Role of hydration and water structure in biological and colloidal interactions. *Nature*, 379, 219–225.

Iwazumi, (1970). A New Field Theory of Muscle Contraction. Ph.D Thesis, University of Pennsylvania.

Jacobs, W.P. (1994). Caulerpa. *Sci. Amer.*, 100–105.

Jacobson, R.C., Tirosh, R., Delay, M.J., and Pollack, G.H. (1983). Quantized nature of sarcomere shortening steps. *J. Mus. Res. Cell Motility*, 4, 529–542.

Janmey, P.A., Shah, J.V., Tang, J.X., and Stossel, T.P. (2001). Actin filament networks. *Results Probl. Cell Differ.*, 32, 181–99.

Jarvis, S.P., et al., (2000). Local solvation shell measurement in water using a carbon nanotube probe. *J. Phys. Chem. B*, 104, 6091–6097.

Jones, D. S. (1999). Dynamic mechanical analysis of polymeric systems of pharmaceutical and biomedical significance. *Int. J. Pharm.*, 179(2), 167–178.

Käs, J., Strey, H., and Sackmann, E. (1994). Direct imaging of reptation for semi-flexible actin filaments. *Nature*, 368, 226–229.

Katayama, E. (1998). Quick-freeze deep-etch electron microscopy of the actin-heavy meromyosin complex during the in vitro motility assay. *J. Mol. Biol.*, 278, 349–367.

Kellermayer, M.S.Z., and Pollack, G.H. (1996). Rescue of in vitro actin motility halted at high ionic strength by reduction of ATP to submicromolar levels. *Biochim. Biophys. Acta.*, 1277, 107–114.

Kitamura, K., Tokunaga, M., Iwane, A., and Yanagida. T. (1999). A single myosin head moves along an actin filament with regular steps of 5.3 nanometers. *Nature*, 397(6715), 129–134.

Klenchin, V.A., Sukharev, S.I., Serov, S.M., Chernomordik, L.V., and Chizmadzhev, Y.A. (1991). Electrically induced DNA uptake by cells is a fast process involving DNA electrophoresis. *Biophys. J.*, 60(4), 804–811.

Krause, T.L., Fishman, H.M., Ballinger, M.L., Ballinger, G.D., and Bittner, G.D. (1984). Extent mechanism of sealing in transected giant axons of squid and earthworms. *J. Neurosci.*, 14, 6638–6651.

Ling, G.N. (1965). The physical state of water in living cell and model systems. *Ann. N. Y. Acad. Sci.*, 125, 401.

Ling, G.N. (1992). A Revolution in the Physiology of the Living Cell. Malabar, FL: Krieger.

Ling, G.N., and Walton, C.L. (1976). What retains water in living cells? *Science*, 191, 293–295.

Liu, X., and Pollack, G.H. (2004). Stepwise sliding of single actin and myosin filaments. *Biophys. J.*, 86, 353–358.

Maniotis, A., and Schliwa, M. (1991). Microsurgical removal of centrosomes blocks cell reproduction and centriole generation in BSC-1 cells. *Cell*, 67, 495–504.

McIntosh, J.R. (1973). The axostyle of Saccinobaculus. II. Motion of the microtubule bundle and a structural comparison of straight and bent axostyles. *J. Cell Bio.*, 56, 324–339.

McNeil, P.L., and Ito, S. (1990). Molecular traffic through plasma membrane disruptions of cells in vivo. *J. Cell Sci.*, 67, 495–504.

McNeil, P.L., and Steinhardt, R.A. (1997). Loss, restoration, and maintenance of plasma membrane integrity. *J. Cell Bio.*, 137(1), 1–4.

Miki, M., and Koyama, T. (1994). Domain motion in actin observed by fluorescence resonance energy transfer. *Biochem.*, 33, 10171–10177.

Oosawa, F., Fujime, S., Ishiwata S., and Mihashi, K. (1972). Dynamic property of F-actin and thin filament. *CSH Symposia on Quant. Biol.*, XXXVII, 277–285.

Oplatka, A. (1996). The rise, decline, and fall of the swinging crossbridge dogma. *Chemtracts Bioch. Mol. Biol.*, 6, 18–60.

Oplatka, A. (1997). Critical review of the swinging crossbridge theory and of the cardinal active role of water in muscle contraction. *Crit. Rev. Biochem. Mol. Biol.*, 32(4), 307–360.

Osada, Y., and Gong, J. (1993). Stimuli-responsive polymer gels and their application to chemomechanical systems. *Prog. Polym. Sci.*, 18, 187–226.

Pashley, R.M., and Kitchener, J.A. (1979). Surface forces in adsorbed multilayers of water on quartz. *J. Colloid Interface Sci.*, 71, 491–500.

Pollack, G.H. (1983). The sliding filament/cross-bridge theory. *Physiol. Rev.*, 63, 1049–1113.

Pollack, G.H. (1990). Muscle & Molecules: Uncovering the Principles of Biological Motion. Seattle: Ebner and Sons.

Pollack, G.H. (1996). Phase-transitions and the molecular mechanism of contraction. *Biophys. Chem.*, 59, 315–328.

Pollack, G.H. (2001). Cells, Gels and the Engines of Life: A New, Unifying Approach to Cell Function. Seattle: Ebner and Sons.

Prausnitz, M.R., Milano, C.D., Gimm, J.A., Langer, R., and Weaver, J.C. (1994). Quantitative study of molecular transport due to electroporation: uptake of bovine serum albumin by erythrocyte ghosts. *Biophys. J.*, 66(5), 1522–1530.

Prochniewicz, E., Zhang, Q., Janmey, P.A., and Thomas, D.D. (1996). Cooperativity in F-actin: binding of gelsolin at the barbed end affects structure and dynamics of the whole filament. *J. Mol. Biol.*, 260(5), 756–766.

Rand, R.P., Parsegian V.A., Rau, D.C. (2000). Intracellular osmotic action. *Cell Mol. Life Sci.*, 57(7), 1018–1032.

Rief, M., Gautel, M., Oesterhelt, F., Fernandez, J.M., and Gaub, H.E. (1997). Reversible unfolding of individual titin immunoglobin domains by AFM. *Science*, 276, 1109–1112.

Schutt, C.E., and Lindberg, U. (1993). A new perspective on muscle contraction. *FEBS*, 325, 59–62.

Schutt, C.E., and Lindberg, U. (1998). Muscle contraction as a Markhov process I: energetics of the process. *Acta. Physiol. Scan.*, 163, 307–324.

Schwister, K., and Deuticke, B. (1985). Formation and properties of aqueous leaks induced in human erythrocytes by electrical breakdown. *Biophys. Acta.*, 816(2), 332–348.

Serpersu, E.H., Kinosita, K., Jr., and Tsong, T.Y. (1985). Reversible and irreversible modification of erythrocyte membrane permeability by electric field. *Biochim. Biophys. Acta.*, 812(3), 779–85.

Sosa, H., Popp, D., Ouyang, G., and Huxley, H.E. (1994). Ultrastructure of skeletal muscle fibers studied by a plunge quick freezing method: myofilament lengths. *Biophys. J.*, 67, 283–292.

Spudich, J.A. (1994). How molecular motors work. *Nature*, 372, 515–518.

Tanaka, T., Anaka, M., et al. (1992). Phase transitions in gels: Mechanics of swelling. NATO ASI Series Vol. H64, Berlin: Springer Verlag.

Taylor, S.R., Shlevin, H.H., and Lopez, J.R. (1975). Calcium in excitation-contraction coupling of skeletal muscle. *Biochem. Soc. Transact.*, 7, 759–764.

Thomas, D.D. (1987). Spectroscopic probes of muscle cross-bridge rotaion. *Ann. Rev. Physiol.*, 49, 641–709.

Tskhovrebova, L., Trinick, J., Sleep, J.A., and Simmons, R.M. (1997). Elasticity and unfolding of single molecules of the giant muscle protein titin. *Nature*, 387, 308–312.

Tung, L., and Vassort, G. Personal communication.

Verdugo, P., Deyrup-Olsen, I., Martin, A.W., and Luchtel, D.L. (1992). Polymer gel-phase transition: the molecular mechanism of product release in mucin secretion? NATO ASI Series Vol. H64 Mechanics of Swelling T.K. Karalis, Springer, Berlin:

Vogler, E. (1998). Structure and reactivity of water at biomaterial surfaces. *Adv. Colloid and Interface Sci.*, 74, 69–117.

Wiggins, P.M. (1990). Role of water in some biological processes. *Microbiol. Rev.*, 54(4), 432–449.

Xie, T.D., Sun, L., and Tsong, T.Y. (1990). Studies of mechanisms of electric field-induced DNA transfection. *Biophys. J.*, 58, 13–19.

Yakovenko, O., Blyakhman, F., and Pollack, G.H. (2002). Fundamental step size in single cardiac and skeletal sarcomeres. *Am J. Physiol (Cell)*, 283(9), C735–C743.

Yanagida, T., and Oosawa, F. (1978). Polarized fluorescence from epsilon-ADP incorporated into F-actin in a myosin-free single fibre. *J. Mol. Biol.*, 126, 507–524.

Yanagida, T., Nakase, M., Nishiyama K., and Oosawa, F. (1984). Direct observation of motion of single F-actin filaments in the presence of myosin. *Nature*, 307, 58–60.

Yawo, H., and Kuno, M. (1985). Calcium dependence of membrane sealing at the cut end of the cockroach giant axon. *J. Neurosci.*, 5, 1626–1632.

8 Polymer-based models of cytoskeletal networks

F.C. MacKintosh

ABSTRACT: Most plant and animal cells possess a complex structure of filamentous proteins and associated proteins and enzymes for bundling, cross-linking, and active force generation. This *cytoskeleton* is largely responsible for cell elasticity and mechanical stability. It can also play a key role in cell locomotion. Over the last few years, the single-molecule micromechanics of many of the important constituents of the cytoskeleton have been studied in great detail by biophysical techniques such as high-resolution microscopy, scanning force microscopy, and optical tweezers. At the same time, numerous *in vitro* experiments aimed at understanding some of the unique mechanical and dynamic properties of solutions and networks of cytoskeletal filaments have been performed. In parallel with these experiments, theoretical models have emerged that have both served to explain many of the essential material properties of these networks, as well as to motivate quantitative experiments to determine, for example concentration dependence of shear moduli and the effects of cross-links. This chapter is devoted to theoretical models of the cytoskeleton based on polymer physics at both the level of single protein filaments and the level of solutions and networks of cross-linked or entangled filaments. We begin with a derivation of the static and dynamic properties of single cytoskeletal filaments. We then proceed to build up models of solutions and cross-linked gels of cytoskeletal filaments and we discuss the comparison of these models with a variety of experiments on *in vitro* model systems.

Introduction

Understanding the mechanical properties of cells and even whole tissues continues to pose significant challenges. Cells experience a variety of external stresses and forces, and they exert forces on their surroundings – for instance, in cell locomotion. The mechanical interaction of cells with their surroundings depends on structures such as cell membranes and complex networks of filamentous proteins. Although these cellular components have been known for many years, important outstanding problems remain concerning the origins and regulation of cell mechanical properties (Pollard and Cooper, 1986; Alberts et al., 1994; Boal, 2002). These mechanical factors determine how a cell maintains and modifies its shape, how it moves, and even how cells adhere to one another. Mechanical stimulus of cells can also result in changes in gene expression.

Cells exhibit rich composite structures ranging from the nanometer to the micrometer scale. These structures combine soft membranes and rather rigid filamentous

Polymer-based models of cytoskeletal networks

proteins or biopolymers, among other components. Most plant and animal cells, in fact, possess a complex network structure of biopolymers and associated proteins and enzymes for bundling, cross-linking, and active force generation. This *cytoskeleton* is often the principal determinant of cell elasticity and mechanical stability.

Over the last few years, the single-molecule properties of many of the important building blocks of the cytoskeleton have been studied in great detail by biophysical techniques such as high-resolution microscopy, scanning force microscopy, and optical tweezers. At the same time, numerous *in vitro* experiments have aimed to understand some of the unique mechanical and dynamic properties of solutions and networks of cytoskeletal filaments. In parallel with these experiments, theoretical models have emerged that have served both to explain many of the essential material properties of these networks, as well as to motivate quantitative experiments to determine the way material properties are regulated by, for example, cross-linking and bundling proteins. Here, we focus on recent theoretical modeling of cytoskeletal solutions and networks.

One of the principal components of the cytoskeleton, and even one of the most prevalent proteins in the cell, is actin. This exists in both monomeric or globular (G-actin) and polymeric or filamentary (F-actin) forms. Actin filaments can form a network of entangled, branched, and/or cross-linked filaments known as the actin cortex, which is frequently found near the periphery of cells. *In vivo*, this network is far from passive, with both active motion and (contractile) force generation during cell locomotion, and with a strong coupling to membrane proteins that appears to play a key role in the ability of cells to sense and respond to external stresses.

In order to understand these complex structures, quantitative models are needed for the structure, interactions, and mechanical response of networks such as the actin cortex. Unlike networks and gels of most synthetic polymers, however, these networks have been clearly shown to possess properties that cannot be modeled by existing polymer theories. These properties include rather large shear moduli (compared with synthetic polymers under similar conditions), strong signatures of nonlinear response (in which, for example, the shear modulus can increase by a full factor of ten or more under modest strains of only 10 percent or so) (Janmey et al., 1994), and unique dynamics. In a very close and active collaboration between theory and experiment over the past few years, a standard model of sorts for the material properties of *semiflexible* polymer networks has emerged, which can explain many of the observed properties of F-actin networks, at least *in vitro*. Central to these models has been the semiflexible nature of the constituent filaments, which is both a fundamental property of almost any filamentous protein, as well as a clear departure from conventional polymer physics, which has focused on flexible or rod-like limits. In contrast, biopolymers such as F-actin are nearly rigid on the scale of a micrometer, while at the same time showing significant thermal fluctuations on the cellular scale of a few microns.

This chapter begins with an introduction to models of single-filament response and dynamics, and in fact, spends most of its time on a detailed understanding of these single-filament properties. Because cytoskeletal filaments are the most important structural components in cells, a quantitative understanding of their mechanical response to bending, stretching, and compression is crucial for any model of the mechanics of networks of these filaments. We shall see how these fundamental

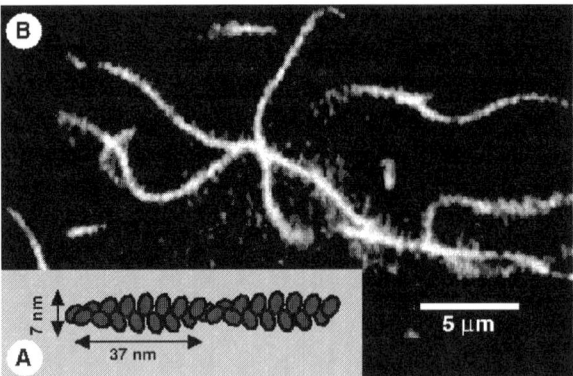

Fig. 8-1. Entangled solution of semiflexible actin filaments. (A) In physiological conditions, individual monomeric actin proteins (G-actin) polymerize to form double-stranded helical filaments known as F-actin. These filaments exhibit a polydisperse length distribution of up to 70 μm in length. (B) A solution of 1.0 mg/ml actin filaments, approximately 0.03% of which have been labeled with rhodamine-phalloidin in order to visualize them by florescence microscopy. The average distance ξ between chains in this figure is approximately 0.3 μm. (Reprinted with permission from MacKintosh F C, Käs J, and Janmey P A, *Physical Review Letters*, 75 4425 (1995). Copyright 1995 by the American Physical Society.

properties of the individual filaments can explain many of the properties of solutions and networks.

Single-filament properties

The biopolymers that make up the cytoskeleton consist of aggregates of large globular proteins that are bound together rather weakly, as compared with most synthetic, covalently bonded polymers. Nevertheless, they can be surprisingly strong. The most rigid of these are *microtubules*, which are hollow tube-like filaments that have a diameter of approximately 20 nm. The most basic aspect determining the mechanical behavior of cytoskeletal polymers on the cellular scale is their bending rigidity.

Even with this mechanical resistance to bending, however, cytoskeletal filaments can still exhibit significant thermally induced bending fluctuations because of Brownian motion in a liquid. Thus such filaments are said to be *semiflexible* or *wormlike*. This is illustrated in Fig. 8-1, showing fluorescently labeled F-actin filaments on the micrometer scale. The effect of the Brownian forces on the filament leads to increasingly contorted shapes over larger-length segments. The length at which significant bending fluctuations occur actually provides a simple yet quantitative characterization of the mechanical stiffness of such polymers. This thermal bending length, or *persistence length* ℓ_p, is defined in terms of the the angular correlations (for example, of the local orientation along the polymer backbone), which decay exponentially with a characteristic length ℓ_p. In simple terms, however, this just says that a typical filament in thermal equilibrium in a liquid will appear rather straight over lengths that are short compared with this persistence length, while it will begin to exhibit a random, contorted shape only on longer-length scales. The persistence lengths of a few important biopolymers are given in Table 8-1, along with their approximate diameter and length.

Table 8-1. *Persistence lengths and other parameters of various biopolymers (Howard, 2001; Gittes et al., 1993)*

Type	Approximate diameter	Persistence length	Contour length
DNA	2 nm	50 nm	≲1 m
F-actin	7 nm	17 μm	≲50 μm
Microtubule	25 nm	~1–5 mm	10s of μm

The worm-like chain model

Rigid polymers can be thought of as elastic rods, except on a small scale. The mechanical description of these is essentially the same as for a macroscopic rod with quantitative differences in parameters. The important role of thermal fluctuations, however, introduces a qualitative difference from the macroscopic case. Because the diameter of a filamentous protein is so much smaller than other length scales of interest – and especially the cellular scale – it is often sufficient to think of a filament as an idealized curve that resists bending. This is the essence of the so-called worm-like chain model. This can be described by an energy of the form,

$$H_{\text{bend}} = \frac{\kappa}{2} \int ds \left| \frac{\partial \vec{t}}{\partial s} \right|^2, \qquad (8.1)$$

where κ is the *bending modulus* and $\vec{t}$ is a (unit) tangent vector along the chain. The variation (derivative) of the tangent is a measure of curvature, which appears here quadratically because it is assumed that there is no preferred direction of curvature. Here, the chain position $\vec{r}(s)$ is described in terms of a coordinate s corresponding to the length along the chain backbone. Hence, the tangent vector

$$\vec{t} = \frac{\partial \vec{r}}{\partial s}.$$

These quantities are illustrated in Fig. 8-2.

The bending modulus κ has units of energy times length. A natural energy scale for a rod subject to Brownian fluctuations is kT, where T is the temperature and k is Boltzmann's constant. This is the typical kinetic energy of a molecule or particle. The persistence length described above is simply given by $\ell_p = \kappa/(kT)$, because the fluctuations tend to decrease with stiffness κ and increase with temperature. As noted, this is the typical length scale over which the polymer forgets its orientation, due to the constant Brownian forces it experiences in a medium at finite temperature.

More precisely, for a homogeneous rod of diameter $2a$ consisting of a homogeneous elastic material, the bending modulus should be proportional to the Young's modulus E. The Young's modulus, or the stiffness of the material, has units of energy per volume. Thus, on dimensional grounds, we expect that $\kappa \sim Ea^4$. In fact (Landau and Lifshitz 1986),

$$\kappa = \frac{\pi}{4} E a^4.$$

The prefactor in front of Ea^4 depends on the geometry of the rod (in other words, its cross-section). The factor $\pi a^4/4$ is for a solid rod of radius a. For a hollow tube, such

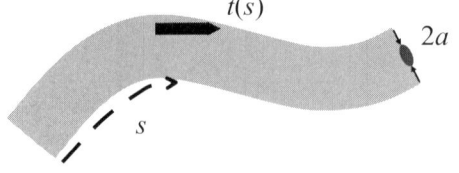

Fig. 8-2. A filamentous protein can be regarded as an elastic rod of radius a. Provided the length of the rod is very long compared with the monomeric dimension a, and that the rigidity is high (specifically, the persistence length $\ell_p \gg a$), this can be treated as an abstract line or curve, characterized by the length s along its backbone. A unit vector $\vec{t}$ tangent to the filament defines the local orientation of the filament. Curvature is present when this orientation varies with s. For bending in a plane, it is sufficient to consider the angle $\theta(s)$ that the filament makes with respect to some fixed axis. The curvature is then $\partial \theta / \partial s$.

as one might use to model a microtubule, the prefactor would be different, but still of order a^4, where a is the (outer) radius. This is often expressed as $\kappa = EI$, where I is the moment of inertia of the cross-section (Howard, 2001).

In general, for bending in 3D, there are two independent directions for deflections of the rod or polymer transverse to its local axis. It is often instructive, however, to consider a simpler case of a single transverse degree of freedom, in other words, motion confined to a plane, as illustrated by Fig. 8-2. Here, the integrand in Eq. 8.1 becomes $(\partial \theta / \partial s)^2$, where $\theta(s)$ is simply the local angle that the chain axis makes at point s, relative to any fixed axis. Using basic principles of statistical mechanics (Grosberg and Khokhlov, 1994), one can calculate the thermal average angular correlation between distant points along the chain, for which

$$\langle \cos[\theta(s) - \theta(s')] \rangle \simeq \langle \cos(\Delta\theta) \rangle^{|s-s'|/\Delta s} \simeq e^{-|s-s'|/2\ell_p}. \tag{8.2}$$

As noted at the outset, so far this is all for motion confined to a plane. In three dimensions, there is another direction perpendicular to the plane that the filament can move in. This increases the rate of decay of the angular correlations by a factor of two relative to the result above:

$$\langle \vec{t}(s) \cdot \vec{t}(s') \rangle = e^{-|s-s'|/\ell_p}, \tag{8.3}$$

where ℓ_p is the same persistence length defined above. This is a general definition of the persistence length, which also provides a purely *geometric* measure of the *mechanical* stiffness of the rod, provided that it is in equilibrium at temperature T. In principle, this means that one can measure the stiffness of a biopolymer by simply examining its bending fluctuations in a microscope. In practice, however, it is usually better to measure the amplitudes of a number of different bending modes (that is, different wavelengths) in order to ensure that thermal equilibrium is established (Gittes et al., 1993).

Force-extension of single chains

In order to understand how a network of filaments responds to mechanical loading, we need to understand at least two things: the way a single filament responds to stress; and the way in which the individual filaments are connected or otherwise interact with each other. We address the single-filament properties here, and reserve the characterization of the way filaments interact for later.

A single filament can respond to forces in at least two ways. It can respond to both transverse and longitudinal forces by either bending or stretching/compressing. On length scales shorter than the persistence length, bending can be described in mechanical terms, as for elastic rods. By contrast, stretching and compression can involve both a purely elastic or mechanical response (again, as in the stretching, compression, or even buckling of macroscopic elastic rods), as well as an entropic response. The latter comes from the thermal fluctuations of the filament. Perhaps surprisingly, as will be shown, the longitudinal response can be dominated by entropy even on length scales small compared with the persistence length. Thus, it is incorrect to think of a filament as truly rod-like, even on length scales short compared with ℓ_p.

The longitudinal single-filament response is often described in terms of a so-called force-extension relationship. Here, the force required to extend the filament is measured or calculated in terms of the degree of extension along a line. At any finite temperature, there is a resistance to such extension due to the presence of thermal fluctuations that make the polymer deviate from a straight conformation. This has been the basis of mechanical studies, for example, of long DNA (Bustamante et al., 1994). In the limit of large persistence length, this can be calculated as follows (MacKintosh et al., 1995).

We consider a filament segment of length ℓ that is short compared with the persistence length ℓ_p. It is then nearly straight, with small transverse fluctuations. We let the x-axis define the average orientation of the chain segment, and let u and v represent the two independent transverse degrees of freedom. These can then be thought of as functions of x and time t in general. For simplicity, we shall mostly consider just one of these coordinates, $u(x, t)$. The bending energy is then

$$H_{\text{bend}} = \frac{\kappa}{2} \int dx \left(\frac{\partial^2 u}{\partial x^2} \right)^2 = \frac{\ell}{4} \sum_q \kappa q^4 u_q^2, \qquad (8.4)$$

where we have represented $u(x)$ by a Fourier series

$$u(x, t) = \sum_q u_q \sin(qx). \qquad (8.5)$$

As illustrated in Fig. 8-3, the local orientation of the filament is given by the slope $\partial u/\partial x$, while the local curvature is given by the second derivative $\partial^2 u/\partial x^2$. Such a description is appropriate for the case of a nearly straight filament with fixed boundary conditions $u = 0$ at the ends, $x = 0, \ell$. For this case, the wave vectors $q = n\pi/\ell$, where $n = 1, 2, 3, \ldots$.

We assume that the chain has no compliance in its contour length, in other words, that the total arc length $\int ds$ is unchanged by the fluctuations. As illustrated in Fig. 8-3, for a nearly straight filament, the arc length ds of a short segment is approximately given by $\sqrt{(dx)^2 + (du)^2} = dx\sqrt{1 + |\partial u/\partial x|^2}$. The contraction of the chain relative to its full contour length in the presence of thermal fluctuations in u is then

$$\Delta \ell = \int dx \left(\sqrt{1 + |\partial u/\partial x|^2} - 1 \right) \simeq \frac{1}{2} \int dx \, |\partial u/\partial x|^2. \qquad (8.6)$$

Fig. 8-3. From one fixed end, a filament tends to wander in a way that can be characterized by $u(x)$, the transverse displacement from an initial straight line (dashed). If the arc length of the filament is unchanged, then the transverse thermal fluctuations result in a contraction of the end-to-end distance, which is denoted by $\Delta \ell$. In fact, this contraction is actually distributed about a thermal average value $\langle \Delta \ell \rangle$. The mean-square (longitudinal) fluctuations about this average are denoted by $\langle \delta \ell^2 \rangle$, while the mean-square lateral fluctuations (that is, with respect to the dashed line) are denoted by $\langle u^2 \rangle$.

The integration here is actually over the projected length of the chain. But, to leading (quadratic) order in the transverse displacements, we make no distinction between projected and contour lengths here, and above in H_{bend}.

Thus, the contraction

$$\Delta \ell = \frac{\ell}{4} \sum_q q^2 u_q^2. \tag{8.7}$$

Conjugate to this variable is the tension τ in the chain. Thus, we consider the effective energy

$$H = \frac{1}{2} \int dx \left[\kappa \left(\frac{\partial^2 u}{\partial x^2} \right)^2 + \tau \left(\frac{\partial u}{\partial x} \right)^2 \right] = \frac{\ell}{4} \sum_q (\kappa q^4 + \tau q^2) u_q^2. \tag{8.8}$$

Under a constant tension τ, therefore, the equilibrium amplitudes u_q must satisfy

$$\langle |u_q|^2 \rangle = \frac{2kT}{\ell (\kappa q^4 + \tau q^2)}, \tag{8.9}$$

and the contraction

$$\langle \Delta \ell \rangle = kT \sum_q \frac{1}{(\kappa q^2 + \tau)}. \tag{8.10}$$

There are, of course, two transverse degrees of freedom, and so this last answer incorporates a factor of two appropriate for a chain fluctuating in 3D.

Semiflexible filaments exhibit a strong suppression of bending fluctuations for modes of wavelength less than the persistence length ℓ_p. More precisely, as we see from Eq. 8.9 the mean-square amplitude of shorter wavelength modes are increasingly suppressed as the fourth power of the wavelength. This has important consequences for many of the thermal properties of such filaments. In particular, it means that the longest unconstrained wavelengths tend to be dominant in most cases. This allows us, for instance, to anticipate the scaling form of the end-to-end contraction $\Delta \ell$ between points separated by arc length ℓ in the absence of an applied tension. We note that it is a length and it must vary inversely with stiffness κ and must increase with temperature. Thus, as the dominant mode of fluctuations is that of the maximum wavelength, ℓ, we expect the contraction to be of the form $\langle \Delta \ell \rangle_0 \sim \ell^2/\ell_p$. More precisely, for $\tau = 0$,

$$\langle \Delta \ell \rangle_0 = \frac{kT \ell^2}{\kappa \pi^2} \sum_{n=1}^{\infty} \frac{1}{n^2} = \frac{\ell^2}{6 \ell_p}. \tag{8.11}$$

Polymer-based models of cytoskeletal networks

Similar scaling arguments to those above lead us to expect that the typical transverse amplitude of a segment of length ℓ is approximately given by

$$\langle u^2 \rangle \sim \frac{\ell^3}{\ell_p} \tag{8.12}$$

in the absence of applied tension. The precise coefficient for the mean-square amplitude of the midpoint between ends separated by ℓ (with vanishing deflection at the ends) is $1/24$.

For a finite tension τ, however, there is an extension of the chain (toward full extension) by an amount

$$\delta\ell = \langle \Delta\ell \rangle_0 - \langle \Delta\ell \rangle_\tau = \frac{kT\ell^2}{\kappa\pi^2} \sum_n \frac{\phi}{n^2(n^2 + \phi)}, \tag{8.13}$$

where $\phi = \tau\ell^2/(\kappa\pi^2)$ is a dimensionless force. The characteristic force $\kappa\pi^2/\ell^2$ that enters here is the critical force in the classical Euler buckling problem (Landau and Lifshitz, 1986). Thus, the force-extension curve can be found by inverting this relationship. In the linear regime, this becomes

$$\delta\ell = \frac{\ell^2}{\ell_p \pi^2} \phi \sum_n \frac{1}{n^4} = \frac{\ell^4}{90\ell_p \kappa} \tau, \tag{8.14}$$

that is, the effective spring constant for longitudinal extension of the chain segment is $90\kappa\ell_p/\ell^4$. The scaling form of this could also have been anticipated, based on very simple physical arguments similar to those above. In particular, given the expected dominance of the longest wavelength mode (ℓ), we expect that the end-to-end contraction scales as $\delta\ell \sim \int (\partial u/\partial x)^2 \sim u^2/\ell$. Thus, $\langle \delta\ell^2 \rangle \sim \ell^{-2}\langle u^4 \rangle \sim \ell^{-2}\langle u^2 \rangle^2 \sim \ell^4/\ell_p^2$, which is consistent with the effective (linear) spring constant derived above. The full nonlinear force-extension curve can be calculated numerically by inversion of the expression above. This is shown in Fig. 8-4. Here, one can see both the linear regime for small forces, with the effective spring constant given above, as well as a divergent force near full extension. In fact, the force diverges in a characteristic way, as the inverse square of the distance from full extension: $\tau \sim |\delta\ell - \Delta\ell|^{-2}$ (Fixman and Kovac, 1973).

We have calculated only the longitudinal response of semiflexible polymers that arises from their thermal fluctuations. It is also possible that such filaments will actually lengthen (in arc length) when pulled on. This we can think of as a *zero-temperature* or *purely mechanical* response. After all, we are treating semiflexible polymers as little bendable rods. To the extent that they behave as rigid rods, we might expect them to respond to longitudinal stresses by stretching as a rod. Based on the arguments above, it seems that the persistence length ℓ_p determines the length below which filaments behave like rods, and above which they behave like flexible polymers with significant thermal fluctuations. Perhaps surprisingly, however, even for semiflexible filament segments as short as $\ell \sim \sqrt[3]{a^2\ell_p}$, which is much shorter than the persistence length, their longitudinal response can be dominated by the entropic force-extension described above, that is, in which the response is due to transverse thermal fluctuations (Head et al., 2003b).

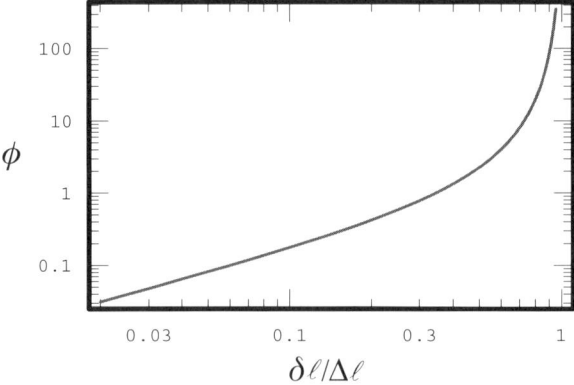

Fig. 8-4. The dimensionless force ϕ as a function of extension $\delta\ell$, relative to maximum extension $\Delta\ell$. For small extension, the response is linear.

Dynamics of single chains

The same Brownian forces that give rise to the bent shapes of filaments such as in Fig. 8.1 also govern the dynamics of these fluctuating filaments. Both the relaxation dynamics of bent filaments, as well as the dynamic fluctuations of individual chains exhibit rich behavior that can have important consequences even at the level of bulk solutions and networks. The principal dynamic modes come from the transverse motion, that is, the degrees of freedom u and v above. Thus, we must consider time dependence of these quantities. The transverse equation of motion of the chain can be found from H_{bend} above, together with the hydrodynamic drag of the filaments through the solvent. This is done via a Langevin equation describing the net force per unit length on the chain at position x,

$$0 = -\zeta \frac{\partial}{\partial t} u(x,t) - \kappa \frac{\partial^4}{\partial x^4} u(x,t) + \xi_\perp(x,t), \qquad (8.15)$$

which is, of course, zero within linearized, inertia-free (low Reynolds number) hydrodynamics that we assume here.

Here, the first term represents the hydrodynamic drag per unit length of the filament. We have assumed a constant transverse drag coefficient that is independent of wavelength. In fact, given that the actual drag per unit length on a rod of length L is $\zeta = 4\pi\eta/\ln(AL/a)$, where L/a is the aspect ratio of the rod, and A is a constant of order unity that depends on the precise geometry of the rod. For a filament fluctuating freely in solution, a weak logarithmic dependence on wavelength is thus expected. In practice, the presence of other chains in solution gives rise to an effective screening of the long-range hydrodynamics beyond a length of order the separation between chains, which can then be taken in place of L above. The second term in the Langevin equation above is the restoring force per unit length due to bending. It has been calculated from $-\delta H_{\text{bend}}/\delta u(x,t)$ with the help of integration by parts. Finally, we include a random force $\xi_\perp$ that accounts for the motion of the surrounding fluid particles.

A simple force balance in the Langevin equation above leads us to conclude that the characteristic relaxation rate of a mode of wavevector q is (Farge and Maggs, 1993)

$$\omega(q) = \kappa q^4/\zeta. \tag{8.16}$$

The fourth-order dependence of this rate on q is to be expected from the appearance of a single time derivative along with four spatial derivatives in Eq. 8.15. This relaxation rate determines, among other things, the correlation time for the fluctuating bending modes. Specifically, in the absence of an applied tension,

$$\langle u_q(t) u_q(0) \rangle = \frac{2kT}{\ell \kappa q^4} e^{-\omega(q)t}. \tag{8.17}$$

That the relaxation rate varies as the fourth power of the wavevector q has important consequences. For example, while the time it takes for an actin filament bending mode of wavelength 1 μm to relax is of order 10 ms, it takes about 100 s for a mode of wavelength 10 μm. This has implications, for instance, for imaging of the thermal fluctuations of filaments, as is done in order to measure ℓ_p and the filament stiffness (Gittes et al., 1993). This is the basis, in fact, of most measurements to date of the stiffness of DNA, F-actin, and other biopolymers. Using Eq. 8.17, for instance, one can both confirm thermal equilibrium and determine ℓ_p by measuring the mean-square amplitude of the thermal modes of various wavelengths. However, in order both to resolve the various modes as well as to establish that they behave according to the thermal distribution, one must sample over times long compared with $1/\omega(q)$ for the longest wavelengths $\lambda \sim 1/q$. At the same time, one must be able to resolve fast motion on times of order $1/\omega(q)$ for the shortest wavelengths. Given the strong dependence of these relaxation times on the corresponding wavelengths, for instance, a range of order a factor of 10 in the wavelengths of the modes corresponds to a range of 10^4 in observation times.

Another way to look at the result of Eq. 8.16 is that a bending mode of wavelength λ relaxes (that is, fully explores its equilibrium conformations) in a time of order $\zeta \lambda^4/\kappa$. Because it is also true that the longest (unconstrained) wavelength bending mode has by far the largest amplitude, and thus dominates the typical conformations of any filament (see Eqs. 8.10 and 8.17), we can see that in a time t, the *typical* or dominant mode that relaxes is one of wavelength $\ell_\perp(t) \sim (\kappa t/\zeta)^{1/4}$. As we have seen above in Eq. 8.12, the mean-square amplitude of transverse fluctuations increases with filament length ℓ as $\langle u^2 \rangle \sim \ell^3/\ell_p$. Thus, in a time t, the expected mean-square transverse motion is given by (Farge and Maggs, 1993; Amblard et al., 1996)

$$\langle u^2(t) \rangle \sim (\ell_\perp(t))^3/\ell_p \sim t^{3/4}, \tag{8.18}$$

because the typical and dominant mode contributing to the motion at time t is of wavelength $\ell_\perp(t)$. Equation 8.18 represents what can be called *subdiffusive* motion because the mean-square displacement grows less strongly with time than for diffusion or Brownian motion. Motion consistent with Eq. 8.18 has been observed in living cells, by tracking small particles attached to microtubules (Caspi et al., 2000). Thus, in some cases, the dynamics of cytoskeletal filaments in living cells appear to follow the expected motion for transverse equilibrium thermal fluctuations in viscous fluids.

The dynamics of longitudinal motion can be calculated similarly. It is found that the means-square amplitude of longitudinal fluctuations of filament of length ℓ are also governed by (Granek, 1997; Gittes and MacKintosh, 1998)

$$\langle \delta\ell(t)^2 \rangle \sim t^{3/4}, \tag{8.19}$$

where this mean-square amplitude is smaller than for the transverse motion by a factor of order ℓ/ℓ_p. Thus, both for the short-time fluctuations as well as for the static fluctuations of a filament segment of length ℓ, a filament end explores a disk-like region with longitudinal motion smaller than perpendicular motion by this factor. Although the amplitude of longitudinal motion is smaller than for transverse, the longitudinal motion of Eq. 8.19 can explain the observed high-frequency viscoelastic response of solutions and networks of biopolymers, as discussed below.

Solutions of semiflexible polymer

Because of their inherent rigidity, semiflexible polymers interact with each other in very different ways than flexible polymers would, for example, in solutions of the same concentration. In addition to the important characteristic lengths of the molecular dimension (say, the filament diameter $2a$), the material parameter ℓ_p, and the contour length of the chains, there is another important new length scale in a solution – the *mesh size*, or typical spacing between polymers in solution, ξ. This can be estimated as follows in terms of the molecular size a and the polymer volume fraction ϕ (Schmidt et al., 1989). In the limit that the persistence length ℓ_p is large compared with ξ, we can approximate the solution on the scale of the mesh as one of rigid rods. Hence, within a cubical volume of size ξ, there is of order one polymer segment of length ξ and cross-section a^2, which corresponds to a volume fraction ϕ of order $(a^2\xi)/\xi^3$. Thus,

$$\xi \sim a/\sqrt{\phi}. \tag{8.20}$$

This mesh size, or spacing between filaments, does not completely characterize the way in which filaments interact, even sterically with each other. For a dilute solution of rigid rods, it is not hard to imagine that one can embed a long rigid rod rather far into such a solution before touching another filament. A true estimate of the distance between typical interactions (points of contact) of semiflexible polymers must account for their thermal fluctuations (Odijk, 1983). As we have seen, the transverse range of fluctuations δu a distance ℓ away from a fixed point grows according to $\delta u^2 \sim \ell^3/\ell_p$. Along this length, such a fluctuating filament explores a narrow cone-like volume of order $\ell \delta u^2$. An entanglement that leads to a constraint of the fluctuations of such a filament occurs when another filament crosses through this volume, in which case it will occupy a volume of order $a^2 \delta u$, as $\delta u \ll \ell$. Thus, the volume fraction and the contour length ℓ between constraints are related by $\phi \sim a^2/(\ell \delta u)$. Taking the corresponding length as an entanglement length, and using the result above for $\delta u = \sqrt{\delta u^2}$, we find that

$$\ell_e \sim \left(a^4 \ell_p\right)^{1/5} \phi^{-2/5}, \tag{8.21}$$

which is larger than the mesh size ξ in the semiflexible limit $\ell_p \gg \xi$.

These transverse entanglements, separated by a typical length ℓ_e, govern the elastic response of solutions, in a way first outlined in Isambert and Maggs (1996). A more complete discussion of the rheology of such solutions can be found in Morse (1998b) and Hinner et al. (1998). The basic result for the rubber-like plateau shear modulus for such solutions can be obtained by noting that the number density of entropic constraints (entanglements) is thus $n\ell/\ell_c \sim 1/(\xi^2 \ell_e)$, where $n = \phi/(a^2 \ell)$ is the number density of chains of contour length ℓ. In the absence of other energetic contributions to the modulus, the entropy associated with these constraints results in a shear modulus of order $G \sim kT/(\xi^2 \ell_e) \sim \phi^{7/5}$. This has been well established in experiments such as those of Hinner et al. (1998).

With increasing frequency, or for short times, the macroscopic shear response of solutions is expected to show the underlying dynamics of individual filaments. One of the main signatures of the frequency response of polymer solutions in general is an increase in the shear modulus with increasing frequency. This is simply because the individual filaments are not able to fully relax or explore their conformations on short times. In practice, for high molecular weight F-actin solutions of approximately 1 mg/ml, this frequency dependence is seen for frequencies above a few Hertz. Initial experiments measuring this response by imaging the dynamics of small probe particles have shown that the shear modulus increases as $G(\omega) \sim \omega^{3/4}$ (Gittes et al., 1997; Schnurr et al., 1997), which has since been confirmed in other experiments and by other techniques (for example, Gisler and Weitz, 1999).

If, as noted above, this increase in stiffness with frequency is due to the fact that filaments are not able to fully fluctuate on the correspondingly shorter times, then we should be able to understand this more quantitatively in terms of the dynamics described in the previous section. In particular, this behavior can be understood in terms of the longitudinal dynamics of single filaments (Morse, 1998a; Gittes and MacKintosh, 1998). Much as the static longitudinal fluctuations $\langle \delta \ell^2 \rangle \sim \ell^4/\ell_p^2$ correspond to an effective longitudinal spring constant $\sim kT\ell_p^2/\ell^4$, the time-dependent longitudinal fluctuations shown above in Eq. 8.19 correspond to a time- or frequency-dependent compliance or stiffness, in which the effective spring constant increases with increasing frequency. This is because, on shorter time scales, fewer bending modes can relax, which makes the filament less compliant. Accounting for the random orientations of filaments in solution results in a frequency-dependent shear modulus

$$G(\omega) = \frac{1}{15}\rho\kappa\ell_p\left(-2i\zeta/\kappa\right)^{3/4}\omega^{3/4} - i\omega\eta, \tag{8.22}$$

where ρ is the polymer concentration measured in length per unit volume.

Network elasticity

In a living cell, there are many different specialized proteins for binding, bundling, and otherwise modifying the network of filamentous proteins. Many tens of actin-associated proteins alone have been identified and studied. Not only is it important to understand the mechanical roles of, for example, cross-linking proteins, but as we shall see, these can have a much more dramatic effect on the network properties than is the case for flexible polymer solutions and networks.

The introduction of cross-linking agents into a solution of semiflexible filaments introduces yet another important and distinct length scale, which we shall call the cross-link distance ℓ_c. As we have just seen, in the limit that $\ell_p \gg \xi$, individual filaments may interact with each other only infrequently. That is to say, in contrast with flexible polymers, the distance between interactions of one polymer with its neighbors (ℓ_e in the case of solutions) may be much larger than the typical spacing between polymers. Thus, if there are biochemical cross-links between filaments, these may result in significant variation of network properties even when ℓ_c is larger than ξ.

Given a network of filaments connected to each other by cross-links spaced an average distance ℓ_c apart along each filament, the response of the network to macroscopic strains and stresses may involve two distinct single-filament responses: (1) bending of filaments; and (2) stretching/compression of filaments. Models based on both of these effects have been proposed and analyzed. Bending-dominated behavior has been suggested both for ordered (Satcher and Dewey, 1996) and disordered (Kroy and Frey, 1996) networks. That individual filaments bend under network strain is perhaps not surprising, unless one thinks of the case of uniform shear. In this case, only rotation and stretching or compression of individual rod-like filaments are possible. This is the basis of so-called *affine* network models (MacKintosh et al., 1995), in which the macroscopic strain falls uniformly across the sample. In contrast, bending of constituents involves (*non-affine* deformations, in which the state of strain varies from one region to another within the sample.

We shall focus mostly on random networks, such as those studied *in vitro*. It has recently been shown (Head et al., 2003a; Wilhelm and Frey, 2003; Head et al., 2003b) that which of the *affine* or *non-affine* behaviors is expected depends, for instance, on filament length and cross-link concentration. Non-affine behavior is expected either at low concentrations or for short filaments, while the deformation is increasingly affine at high concentration or for long filaments. For the first of these responses, the network shear modulus (*Non-Affine*) is expected to be of the form

$$G_{\mathrm{NA}} \sim \kappa/\xi^4 \sim \phi^2 \quad (8.23)$$

when the density of cross-links is high (Kroy and Frey 1996). This quadratic dependence on filament concentration c is also predicted for more ordered networks (Satcher and Dewey 1996).

For affine deformations, the modulus can be estimated using the effective single-filament longitudinal spring constant for a filament segment of length ℓ_c between cross-links, $\sim \kappa \ell_p/\ell_c^4$, as derived above. Given an area density of $1/\xi^2$ such chains passing through any shear plane (see Fig. 8-5), together with the effective tension of order $(\kappa \ell_p/\ell_c^3)\epsilon$, where ϵ is the strain, the shear modulus is expected to be

$$G_{\mathrm{AT}} \sim \frac{\kappa \ell_p}{\xi^2 \ell_c^3}. \quad (8.24)$$

This shows that the shear modulus is expected to be strongly dependent on the density of cross-links. Recent experiments on *in vitro* model gels consisting of F-actin with permanent cross-links, for instance, have shown that the shear modulus can vary from less than 1 Pa to over 100 Pa at the same concentration of F-actin, by varying the cross-link concentration (Gardel et al., 2004).

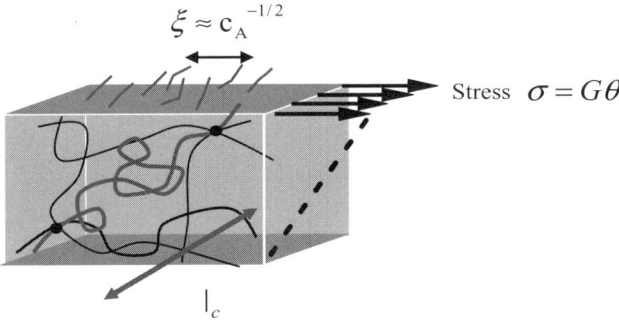

Fig. 8-5. The macroscopic shear stress σ depends on the mean tension in each filament, and on the area density of such filaments passing any plane. There are on average $1/\xi^2$ such filaments per unit area. This gives rise to the factor ξ^{-2} in both Eqs. 8.24 and 8.25. The macroscopic response can also depend strongly on the typical distance ℓ_c between cross-links, as discussed below.

In the preceding derivation we have assumed a thermal/entropic (*A*ffine and *T*hermal) response of filaments to longitudinal forces. As we have seen, however, for shorter filament segments (that is, for small enough ℓ_c), one may expect a mechanical response characteristic of rigid rods that can stretch and compress (with a modulus μ). This would lead to a different expression (Affine, Mechanical) for the shear modulus

$$G_{AM} \sim \frac{\mu}{\xi^2} \sim \phi, \qquad (8.25)$$

which is proportional to concentration. The expectations for the various mechanical regimes is shown in Fig. 8.6 (Head et al., 2003b).

Nonlinear response

In contrast with most polymeric materials (such as gels and rubber), most biological materials, from the cells to whole tissues, stiffen as they are strained even by a few percent. This nonlinear behavior is also quite well established by *in vitro* studies of a wide range of biopolymers, including networks composed of F-actin, collagen, fibrin, and a variety of intermediate filaments (Janmey et al., 1994; Storm et al., 2005). In particular, these networks have been shown to exhibit approximately ten-fold

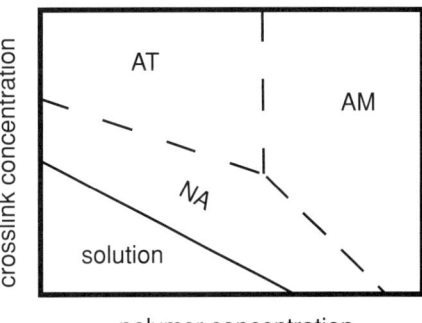

Fig. 8-6. A sketch of the expected diagram showing the various elastic regimes in terms of cross-link density and polymer concentration. The solid line represents the rigidity percolation transition where rigidity first develops from a solution at a macroscopic level. The other, dashed lines indicate crossovers (not thermodynamic transitions). NA indicates the non-affine regime, while AT and AM refer to affine thermal (or entropic) and mechanical, respectively.

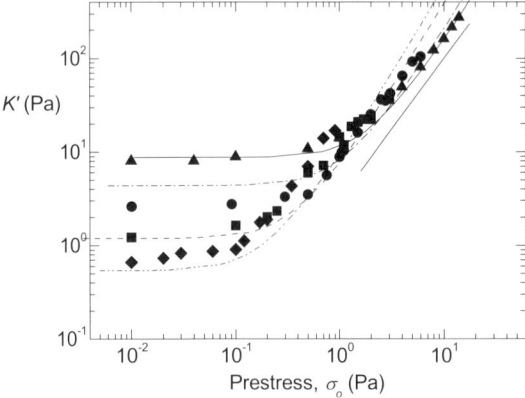

Fig. 8-7. The differential modulus $K' = d\sigma/d\gamma$ describes the increase in the stress σ with strain γ in the nonlinear regime. This was measured for cross-linked actin networks by small-amplitude oscillations at low frequency, corresponding to a nearly purely elastic response, after applying a constant prestress σ_0. This was measured for four different concentrations represented by the various symbols. For small prestress σ_0, the differential modulus K' is nearly constant, corresponding to a linear response for the network. With increasing σ_0, the network stiffens, in a way consistent with theoretical predictions (MacKintosh et al., 1995; Gardel et al., 2004), as illustrated by the various theoretical curves. Specifically, it is expected that in the strongly nonlinear regime, the stiffening increases according to the straight line, corresponding to $d\sigma/d\gamma \sim \sigma^{3/2}$. Data taken from Gardel et al., 2004.

stiffening under strain. Thus these materials are compliant, while being able to withstand a wide range of shear stresses.

This strain-stiffening behavior can be understood in simple terms by looking at the characteristic force-extension behavior of individual semiflexible filaments, as described above. As can be seen in Fig. 8-4, for small extensions or strains, there is a linear increase in the force. As the strain increases, however, the force is seen to grow more rapidly. In fact, in the absence of any compliance in the arc length of the filament, the force strictly diverges at a finite extension. This suggests that for a network, the macroscopic stress should diverge, while in the presence of high stress, the macroscopic shear strain is bounded and ceases to increase. In other words, after being compliant at low stress, such a material will be seen to stop responding, even under high applied stress.

This can be made more quantitative by calculating the macroscopic shear stress of a strained network, including random orientations of the constituent filaments (MacKintosh et al., 1995; Kroy and Frey, 1996; Gardel et al., 2004; Storm et al., 2005). Specifically, for a given shear strain γ, the tension in a filament segment of length ℓ_c is calculated, based on the force-extension relation above. This is done within the (affine) approximation of uniform strain, in which the microscopic strain on any such filament segment is determined precisely by the macroscopic strain and the filament's orientation with respect to the shear. The contribution of such a filament's tension to the macroscopic stress, in other words, in a horizontal plane in Fig. 8.5, also depends on its orientation in space. Finally, the concentration or number density of such filaments crossing this horizontal plane is a function of the overall polymer concentration, and the filament orientation.

The full nonlinear shear stress is calculated as a function of γ, the polymer concentration, and ℓ_c, by adding all such contributions from all (assumed random) orientations of filaments. This can then be compared with macroscopic rheological studies of cross-linked networks, such as done recently by Gardel et al. (2004). These experiments measured the differential modulus, $d\sigma/d\gamma$ versus applied stress σ, and found good agreement with the predicted increase in this modulus with increasing stress (Fig. 8-7). In particular, given the quadratic divergence of the single-filament tension shown above (Fixman and Kovac, 1973), this modulus is expected to increase as $d\sigma/d\gamma \sim \sigma^{3/2}$, which is consistent with the experiments by Gardel et al. (2004). This provides a strong test of the underlying mechanism of network elasticity.

In addition to good agreement between theory and experiment for densely cross-linked networks, these experiments have also shown evidence of a lack of strain-stiffening behavior of these networks at lower concentrations (of polymer or cross-links), which may provide evidence for a non-affine regime of network response described above.

Discussion

Cytoskeletal filaments play key mechanical roles in the cell, either individually (for example, as paths for motor proteins) or in collective structures such as networks. The latter may involve many associated proteins for cross-linking, bundling, or coupling the cytoskeleton to other cellular structures like membranes. Our knowledge of the cytoskeleton has improved in recent years through the development of new experimental techniques, such as in visualization and micromechanical probes in living cells. At the same time, combined experimental and theoretical progress on *in vitro* model systems has provided fundamental insights into the possible mechanical mechanisms of cellular response.

In addition to their role in cells, cytoskeletal filaments have also proven remarkable model systems for the study of semiflexible polymers. Their size alone makes it possible to visualize individual filaments directly. They are also unique in the extreme separation of two important lengths, the persistence length ℓ_p and the size of a single monomer. In the case of F-actin, ℓ_p is more than a thousand times the size of a single monomer. This makes for not only quantitative but also qualitative differences from most synthetic polymers. We have seen, for instance, that the way in which semiflexible polymers entangle is very different. This makes for a surprising variation of the stiffness of these networks with only changes in the density of cross-links, even at the same concentration.

In spite of the molecular complexity of filamentous proteins as compared with conventional polymers, a quantitative understanding of the properties of single filaments provides a quantitative basis for modeling solutions and networks of filaments. In fact, the macroscopic response of cytoskeletal networks quite directly reflects, for example, the underlying dynamics of an individual semiflexible chain fluctuating in its Brownian environment. This can be seen, for instance in the measured dynamics of microtubules in cells (Caspi et al., 2000).

In developing our current understanding of cytoskeletal networks, a crucial role has been played by *in vitro* model systems, such as the one in Fig. 8-1. Major challenges, however, remain for understanding the cytoskeleton of living cells. In the cell, the cytoskeleton is hardly a passive network. Among other differences from the model systems studied to date is the presence of active contractile or force-generating elements such as motors that work in concert with filamentous proteins. Nevertheless, *in vitro* models may soon permit a systematic and quantitative study of various actin-associated proteins for cross-linking and bundling (Gardel et al., 2004), and even contractile elements such as motors.

References

Alberts B, et al., 1994, *Molecular Biology of the Cell*, 3rd edn. (Garland Press, New York).

Amblard F, Maggs A C, Yurke B, Pargellis A N, and Leibler S, 1996, Subdiffusion and anomalous local visoelasticity in actin networks, *Phys. Rev. Lett.*, **77** 4470–4473.

Boal D, 2002, *Mechanics of the Cell* (Cambridge University Press, Cambridge).

Bustamante C, Marko J F, Siggia E D, and Smith S, 1994, Entropic elasticity of lambda phage DNA, *Science*, **265** 1599–1600.

Caspi A, Granek R, and Elbaum M, 2000, Enhanced diffusion in active intracellular transport, *Phys. Rev. Lett.*, **85** 5655.

Farge E and Maggs A C, 1993, Dynamic scattering from semiflexible polymers. *Macromolecules*, **26** 5041–5044.

Fixman M and Kovac J, 1973, Polymer conformational statistics. III. Modified Gaussian models of stiff chains, *J. Chem. Phys.*, **58** 1564.

Gardel M L, Shin J H, MacKintosh F C, Mahadevan L, Matsudaira P, and Weitz D A, 2004, Elastic behavior of cross-linked and bundled actin networks, *Science*, **304** 1301.

Gisler T and Weitz D A, 1999, Scaling of the microrheology of semidilute F-actin solutions, *Phys. Rev. Lett.*, **82** 1606.

Gittes F, Mickey B, Nettleton J, and Howard J, 1993, Flexural rigidity of microtubules and actin filaments measured from thermal fluctuations in shape. *J. Cell Biol.*, **120** 923–934.

Gittes F, Schnurr B, Olmsted, P D, MacKintosh, F C, and Schmidt, C F, 1997, Microscopic viscoelasticity: Shear moduli of soft materials determined from thermal fluctuations, *Phys. Rev. Lett.*, **79** 3286–3289.

Gittes F and MacKintosh F C, 1998, Dynamic shear modulus of a semiflexible polymer network, *Phys. Rev. E.*, **58** R1241–1244.

Granek R, 1997, From semi-flexible polymers to membranes: Anomalous diffusion and reptation, *J. Phys. II (France)*, **7** 1761.

Grosberg A Y and Khokhlov A R, 1994, *Statistical Physics of Macromolecules* (American Institute of Physics Press, New York).

Head D A, Levine A J, and MacKintosh F C, 2003a, Deformation of cross-linked semiflexible polymer networks, *Phys. Rev. Lett.*, **91** 108102.

Head D A, Levine A J, and MacKintosh F C, 2003b, Distinct regimes of elastic response and deformation modes of cross-linked cytoskeletal and semiflexible polymer networks, *Phys. Rev. E.*, **68** 061907.

Hinner B, Tempel M, Sackmann E, Kroy K, and Frey E, 1998, Entanglement, elasticity, and viscous relaxation of action solutions *Phys. Rev. Lett.*, **81** 2614.

Howard J, 2001, *Mechanics of Motor Proteins and the Cytoskeleton* (Sinauer Press, Sunderland, MA).

Isambert H, and Maggs A C, 1996, Dynamics and Rheology of actin solutions, *Macromolecules* **29** 1036–1040.

Janmey P A, Hvidt S, Käs J, Lerche D, Maggs, A C, Sackmann E, Schliwa M, and Stossel T P, 1994, The mechanical properties of actin gels. Elastic modulus and filament motions, *J. Biol. Chem.*, **269**, 32503–32513.

Kroy K and Frey E, 1996, Force-extension relation and plateau modulus for wormlike chains, *Phys. Rev. Lett.*, **77** 306–309.

Landau L D, and Lifshitz E M, 1986, *Theory of Elasticity* (Pergamon Press, Oxford).

MacKintosh F C, Käs J, and Janmey P A, 1995, Elasticity of semiflexible biopolymer networks, *Phys. Rev. Lett.*, **75** 4425.

Morse D C, 1998a, Viscoelasticity of tightly entangled solutions of semiflexible polymers, *Phys. Rev. E.*, **58** R1237.

Morse D C, 1998b, Viscoelasticity of concentrated isotropic solutions of semiflexible polymers. 2. Linear response, *Macromolecules*, **31** 7044.

Odijk T, 1983, The statistic and dynamics of confined or entangled stiff polymers, *Macromolecules*, **16** 1340.

Pasquali M, Shankar V, and Morse D C, 2001, Viscoelasticity of dilute solutions of semiflexible polymers, *Phys. Rev. E.*, **64** 020802(R).

Pollard T D and Cooper J A, 1986, Actin and actin-binding proteins. A critical evaluation of mechanisms and functions, *Ann. Rev. Biochem.*, **55** 987.

Satcher R L and Dewey C F, 1996, Theoretical estimates of mechanical properties of the endothelial cell cytoskeleton, *Biophys. J.*, **71** 109–11.

Schmidt C F, Baermann M, Isenberg G, and Sackmann E, 1989, Chain dynamics, mesh size, and diffusive transport in networks of polymerized actin - a quasielastic light-scattering and microfluorescence study, *Macromolecules*, **22**, 3638–3649.

Schnurr B, Gittes F, Olmsted P D, MacKintosh F C, and Schmidt C F, 1997, Determining microscopic viscoelasticity in flexible and semiflexible polymer networks from thermal fluctuations, *Macromolecules*, **30** 7781–7792.

Storm C, Pastore J J, MacKintosh F C, Lubensky T C, and Janmey P J, 2005, Nonlinear elasticity in biological gels, *Nature*, **435** 191.

Wilhelm J and Frey E, 2003, *Phys. Rev. Lett.*, **91** 108103.

9 Cell dynamics and the actin cytoskeleton

James L. McGrath and C. Forbes Dewey, Jr.

ABSTRACT: This chapter focuses on the mechanical structure of the cell and how it is affected by the dynamic events that shape the cytoskeleton. We pay particular attention to actin because the actin structure turns over rapidly (on the order of tens of seconds to tens of minutes) and is strongly correlated with dynamic events such as cell crawling. The chapter discusses the way in which actin and its associated binding proteins provide the dominant structure within the cell, and how the actin is organized. Models of the internal structure that attempt to provide a quantitative picture of the stiffness of the cell are given, followed by an in-depth discussion of the actin polymerization and depolymerization mechanics. The chapter provides a tour of the experiments and models used to determine the specific effects of associated proteins on the actin cycle and contains an in-depth exposition of how actin dynamics play a pivotal role in cell crawling. Some conclusions and thoughts for the future close the chapter.

Introduction: The role of actin in the cell

Eukaryotic cells are wonderful living engines. They sustain themselves by bringing in nutrients across their membrane shells, manufacturing thousands of individual protein species that are needed to sustain the cell's function, and communicating with the surrounding environment using a complex set of receptor molecules that span the membrane and turn external chemical and mechanical signals into changes in cell function and composition.

The cell membrane is flexible and allows the cell to move – and to be moved – by changes in the internal cytoskeletal structure. The cytoskeleton is a spatially sparse tangled matrix of rods and rod-like elements held together by smaller proteins. One of the characteristics of this structure is that it is continuously changing. These dynamics are driven by thermal energy and by phosphorylation of the major proteins that are constituents of the cell. Even though the bond strengths in most cases are many folds larger than the mean thermal energy kT, on rare occasions the bonds will be broken and a new mechanical arrangement will replace the old. There are additional mechanisms that grow the long filaments from monomers in the cell cytoplasm. Complimentary reactions remove molecules from the ends of the filaments and also cleave the filaments, creating additional free ends. If these growth and turnover mechanisms have a bias direction, the cell will crawl.

The many different cellular states that occur can be examined, with varying degrees of difficulty. The most difficult state is when the cells of interest are embedded in a tissue matrix in a living animal, such as chondrocytes within the matrix of joint cartilage. Some cells, such as the endothelial cells that line the cardiovascular tree of mammals, are more accessible because they lie closely packed along the surface of blood vessels. This configuration is called the endothelial monolayer because the equilibrium condition for these cells is to form a single-cell-thick layer in which all cells are in contact with their neighbors. Much study has been given to endothelial cells because of their putative role in controlling the events that can cause atherosclerosis and subsequent heart disease. An additional attractiveness of studying endothelial cells is the ease with which they can be grown in culture (Gimbrone, 1976). One can study the influence of many physical and chemical properties in a setting having strong similarities to the *in vivo* conditions, while varying the environment one parameter at a time. Many of the data in this chapter were derived from experiments using endothelial cells in culture.

The interior of the cell between the cytoskeletal members is filled with cytoplasm, a water-based slurry composed primarily of electrolytes and small monomeric proteins. The small proteins diffuse through the viscous cytoplasm with a diffusion coefficient that can be measured with modern fluorescence techniques. For actin monomer in the cytoplasm of a vascular endothelial cell, the diffusion coefficient D is about $3 - 6 \times 10^{-8}$ cm^2/s (Giuliano and Taylor, 1994; McGrath et al., 1998b). By contrast, water molecules in the liquid have a self-diffusion coefficient three orders of magnitude larger.

Of the three types of cytoskeletal polymers – actin filaments, intermediate filaments, and microtubules – that determine endothelial cell shape, actin filaments are the most abundant and are located in closest proximity to the cell membrane. Confluent endothelial cells assemble ~70 percent of their 100 μM total actin into a rich meshwork of just over 50,000 actin filaments that are on average ~3 μm long (McGrath et al., 2000b). Cross-linking proteins organize actin filaments into viscoelastic gels that connect to transmembrane proteins and signaling complexes located at intercellular attachment sites and extracellular matrix adhesion sites. Of particular importance are the direct connections of actin filaments to β integrin tails by talin (Calderwood and Ginsberg, 2003) and filamins (Stossel et al., 2001), and to cadherins by vinculin and catenins (Wheelock and Knudsen, 1991). During cell locomotion and shape change events, the actin cytoskeleton is extensively remodeled (Satcher et al., 1997; Theriot and Mitchison, 1991), primarily by adding and subtracting subunits at free filament ends.

Monomeric actin has a molecular weight of 42.8 KDaltons and is found in all eukaryotic cells. It is the major contributor to the mechanical structure of the cell, and the degree to which it is polymerized into long filaments changes dramatically depending on the conditions in which the cell finds itself. The properties of endothelial cells, those that line blood vessels, are illustrated by the typical values for the actin polymerization and other mechanical properties summarized in Table 9-1. Similar numbers have been found for many eukaryotic cells (see Stossel et al. (2001)).

Table 9-1. *Mechanical properties of vascular endothelium*

Property	Typical value	Reference
Cell size – subconfluent	~50 μm dia.	Gimbrone, 1976
– confluent	~40 μm dia.	"
Cell size – aligned, confluent	20 × 60 μm	Dewey et al., 1981
Crawling speed – subconfluent	0.5 μm/min.	Tardy et al., 1997
– confluent	0.15 μm/min.	Osborn et al., 2004
– confluent + flow	0.05 to 0.15 μm/min.	"
Total polymerized actin (F-actin)	10–20 mg/ml	Hartwig et al., 1992
Fraction of actin polymerized		
– subconfluent	35–40%	McGrath et al., 2000b
– confluent	65–80%	"
Total actin content (calculated)	20–40 mg/ml	
Young's modulus for actin filaments	2.3 GPa	Gittes et al., 1993

Interaction of the cell cytoskeleton with the outside environment

Eukaryotic cells are composed of a semistructured interior and an enclosing membrane that separates the interior from the environment. In Fig. 9-1A, it can be seen how the external membrane covers the cell. Removing the membrane and the interior cytoplasm with a suitable solvent reveals the actin cytoskeleton as seen in Fig. 9-1B. These views are from the apical side of cultured endothelial cells. One can see remnants of the membrane at junctions where the bounding bilayer membrane shell was attached to the underlying structure with large protein complexes. A diagram of the current view of these complicated attachments is given in Fig. 9-2. These complexes figure prominently in transducing mechanical signals from the outside, responding to external forces on the membrane such as the shear stress produced by flow.

Endothelial cells are found to be strongly attached to a substrate. *In vivo*, the cells attach to the artery wall, which is covered with a basement membrane of collagen and other proteins. The attachment consists of large protein complexes that connect the substrate to the internal actin cytoskeleton through the intermediary of transmembrane proteins. The transmembrane proteins attach to ligands in the substrate. In equilibrium, the cell is pulled into a flat configuration varying in thickness from about 2–3 μm in the cytoplasmic periphery to around 4–6 μm over the nucleus.

In vitro cell attachment in culture medium is similar, except that there is no basement membrane; the substrate is normally covered with fibrinogen or collagen and the transmembrane proteins attach directly to this layer. Within twenty-four hours in culture, the cells begin to excrete their own substrate proteins, and this forms a surface to which the cells stick tenaciously. Fluid shear stresses up to 40 dynes/cm^2 have no ability to detach the cells. On the other hand, the individual attachment complexes turn over continuously, as shown by exquisite confocal microscopy experiments, with a time scale on the order of fifteen minutes (Davies et al., 1993; Davies et al., 1994).

Although the mechanosensitive molecular mechanisms that determine shear-stress-mediated endothelial shape change are poorly defined, a growing body of evidence supports a decentralized, integrated signaling network in which force-bearing cytoskeletal polymers attach to transmembrane proteins where conformational

Cell dynamics and the actin cytoskeleton

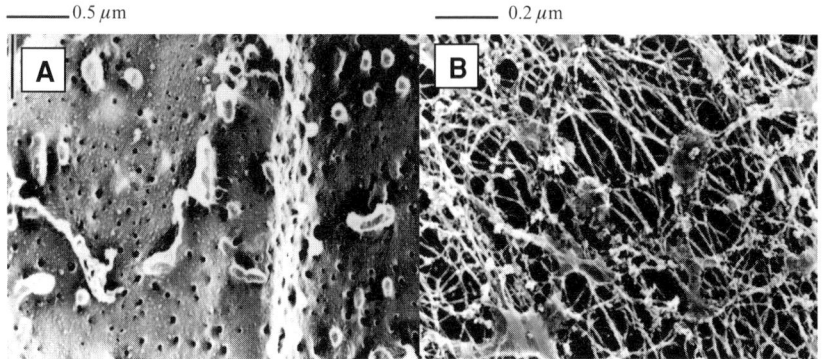

Fig. 9-1. Endothelial cells in culture, showing intact membrane (A) and the underlying actin cytoskeleton (B) after solubilization to remove most of the membrane. Note that the two pictures are at different magnifications. From Satcher et al., 1997.

changes in connected proteins initiate signaling events (Helmke et al., 2003; Kamm and Kaazempur-Mofrad, 2004). Recently, heterogeneous µm-scale displacements of cytoskeletal structures have been described in endothelial cells that, when converted to strain maps, reveal forces applied at the lumen being transmitted through the cell to the basal attachments (Helmke et al., 2003). Changes in the number, type, and structure of cytoskeletal connections alter the location and magnitude of transmitted forces and may modify the specific endothelial phenotype, depending on the spatial and temporal microstimuli that each endothelial cell senses (Davies et al., 2003).

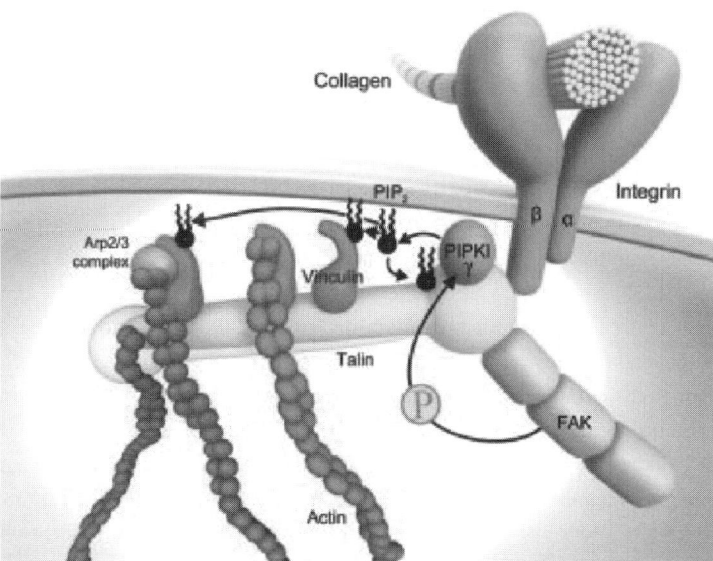

Fig. 9-2. A schematic representation of the focal adhesion complex joining the extracellular matrix to the cytoskeleton across the membrane. This diagram would represent the basal focal adhesion sites. Apical sites would have integrin receptor pairs without attachment to collagen or other materials. The cell could be activated by ligands binding to the integrin pair and causing the cytoplasmic tails to produce new biological reactions in the cytoplasm. From Brakebush and Fässler, 2003.

Properties of actin filaments

Eukaryotic cells exhibit three distinct types of internal polymerized actin structures. The first and arguably most important is the distributed actin lattice that supports most cells and appears to be very well distributed throughout the cell. That configuration is shown in Fig. 9-1B. The individual filaments are too small to be resolved with optical microscopy, and staining for fluorescence of actin can at most show a dull glow throughout the cell when staining this component.

The second type of polymerized actin is found in filament bundles called stress fibers. These stress fibers are often seen to have ends coincident with the location of attachment complexes at the cell-substrate boundary, and also at cell-cell junction complexes. These stress fibers are collections of five to twenty individual actin filaments tightly bound together by other proteins. Their large size and high density makes them prominently visible in actin fluorescence experiments. Because the filaments are visible and also change with the stress conditions of the cell, it is often believed that these stress fibers reflect the main role of actin in a cell. Careful estimates of the fraction of total cellular actin associated with stress fibers, however, suggest that they only play a small role in maintaining cell shape and providing the cytoskeleton of the cell (Satcher and Dewey, 1996). The distributed actin lattice is much more important to cell shape and mechanical properties.

A third type of actin structure is a lattice similar in appearance to Fig. 9-1B that is confined to a small region of the cell just under the surface membrane. This is termed cortical actin, and is found for example in red blood cells. Red blood cells are devoid of any distributed lattice within the rest of the cell interior, and their reliance on cortical structure is not typical of most other cells, including endothelium (Hartwig and DeSisto, 1991).

Single actin filaments have been studied intensively for over twenty-five years. Reviews of their properties can be found in the excellent treatises by Boal (2002) and Preston et al. (1990). An actin filament in a thermally active bath of surrounding molecules will randomly deform from a straight rod into a curved shape. The characteristic length, ξ_p, over which the curvature of an isolated actin filament can become significant is very long, about 10–20 μm at 37 C (Boal, 2002; Janmey et al., 1994; MacKintosh et al., 1995). This is comparable to the dimensions of the cell, whereas the distance between points where the filaments are in contact is much shorter. Typical actin filament lengths are less than 1 μm, so that the cytoskeleton appears to be a tangle of fairly stiff rods. A most important feature is the fact that the small actin filaments are bound to one another by special actin-binding proteins. The most prominent of this class is Filamin A, and its characteristics are described in the next two paragraphs. For the purposes of classifying the mechanical properties of this mixture, one can visualize a relatively dense packing of filaments bound together at relatively large angles by attached protein bridges. An examination of Fig. 9-1B suggests that the filament spacing is typically 100 nm, the intersection angles of the filaments vary

Cell dynamics and the actin cytoskeleton

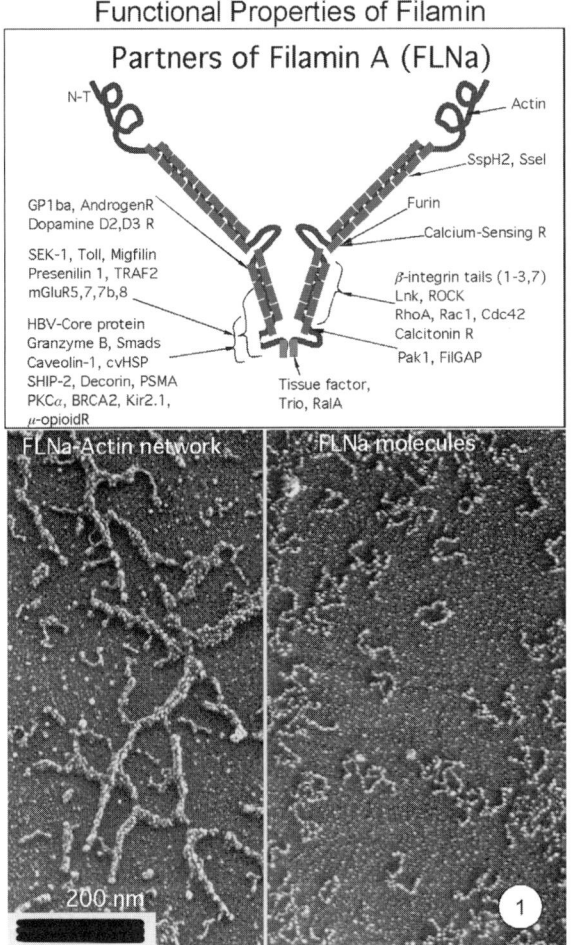

Fig. 9-3. Functional properties of filamin. The bar is 200 nm. Text and drawing courtesy of J.H. Hartwig; micrographs courtesy of C.A. Hartemink, unpublished, 2004.

significantly but are most often closer to 90 degrees than to 45 degrees, and that the spacing will vary with the density of filaments.

The role of filamin A (FLNa)

Cellular actin structure is controlled at different levels. Of particular importance are proteins that regulate actin filament assembly/disassembly reactions and those that regulate the architecture of F-actin and/or attach it to the plasma membrane. One such protein is filamin A (FLNa), a product of the X-chromosome (Gorlin et al., 1993; Gorlin et al., 1990; Stossel et al., 2001). Filamin A was initially isolated and characterized in 1975 (Stossel and Hartwig, 1976) and subsequent research has continued to find important new functions for the protein (Nakamura et al., 2002). As shown in Fig. 9-3, this large protein binds actin filaments, thus defining the cytoskeletal architecture, and attaches them to membrane by also binding to a number of membrane adhesive receptors including $\beta 1$ and $\beta 7$ integrin and GP1bα (Andrews and Fox, 1991; Andrews and Fox, 1992; Fox, 1985; Sprandio et al., 1988; Takafuta et al., 1998).

With many binding partners now described, FLNa participates in signaling cascades by spatially collecting and concentrating signaling proteins at the plasma membrane–cytoskeletal junction and may possibly function as an organizing center for actin network rearrangements (see Fig. 9-3). Important partner interactions that may be dependent on filamin include GTPase targeting and charging and linkage of the actin cytoskeleton to membrane glycoproteins such as GP1bα and β-integrins. FLNa is part of a larger family of proteins that include FLNb and FLNc, whose genes are on chromosomes 3 and 7, respectively (Bröcker and al, 1999; Krakow et al., 2004; Sheen et al., 2002; Thompson et al., 2000).

FLNa is an elongated homodimer (Hartwig and Stossel, 1981). Each subunit has an N-terminal actin-binding site joined to twenty-four repeat motifs, each ~100 residues in length. Repeats are $\beta\beta$-barrel structures that are believed to interconnect like beads on a string. Subunits self-associate into dimers using only the most C-T repeat motif. The location of known binding partner proteins along each FLNa subunit is indicated in Fig. 9-3. Molecules are 160 nm in length in the electron microscope (Fig. 9-3, bottom right) but can organize actin filaments into branching networks (Fig. 9-3, bottom left).

The FLNa concentrations in endothelial and other cells is normally such that there are many times more FLNa molecules than junctions in the cell cytoskeleton. This can be ascertained by measuring the amount of FLNa in the soluble portion of the cell, computing the molecular concentration per unit cell volume, and then comparing that to the concentration of filament junctions per unit volume of cytoskeleton observable in electron microscopy (see Fig. 9-1).

The role of cytoskeletal structure

The internal structure of the cell has several functions. One is to provide a sufficient amount of rigidity so that the cell can withstand external forces. Figure 9-4 illustrates the functions that the cytoskeleton performs when the cell is subjected to fluid shear stress. A balance of forces requires that the cytoskeleton transmit the entire applied force to the substrate.

The second function to be served is that the cytoskeleton must be malleable enough to allow the cell to accommodate new environmental parameters such as imposed mechanical forces from fluid shear stress and mechanical deformation of the artery. There are two separate time scales to be considered; the first is short time behavior, where fluctuations such as systolic and diastolic changes in flow must be accommodated, and longer times, where the actin matrix can completely disassociate and reform. This latter scale is typically on the order of tens of minutes. More discussion of these mechanisms is given later in this chapter.

The following section will draw upon the ultrastructure presented above and describe how simple mechanical models of the lattice can be used to predict the mechanical properties of the cell.

Cell dynamics and the actin cytoskeleton

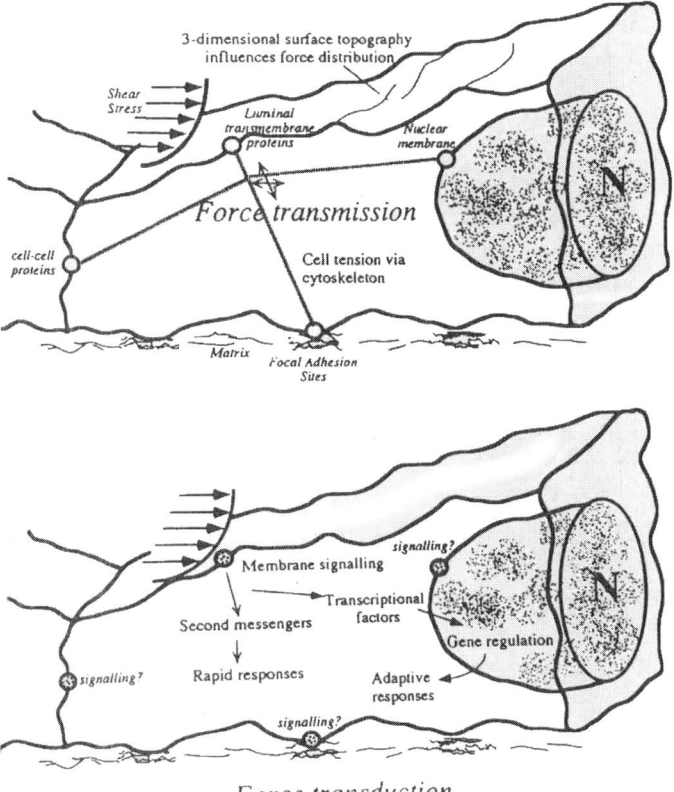

Fig. 9-4. Schematic representation of an endothelial cell subjected to fluid shear stress. The force exerted on the surface must be transmitted through the cell by the cytoskeleton to the substrate to which the cell is attached. Some structural filaments attach to the cell nucleus, so that the force is transmitted through the nucleus as well as the extranuclear part of the cell. The lower portion of the figure suggests some of the biochemical cascades triggered by the force. From Davies, 1995.

Actin mechanics

The complicated combination of semiflexible filaments and binding proteins that form the structural matrix within a cell presents a formidable challenge to the scientific investigator. Many of the key interactions, such as the details of the FLNa binding to actin polymers, are still subjects of active research. The fraction of actin molecules polymerized into structural fibers varies from cell to cell, and the mean can range from 35 percent to 80 percent depending on the state of stress, the degree of cell-cell attachment, and time. As will be described in the section on actin dynamics, the actin filaments polymerize and depolymerize continuously, so that the whole internal structure is replaced within a time scale that is tens of minutes to hours. To make the situation even more interesting, thermal fluctuations of the filaments could contribute substantially to the apparent cell stiffness. Yet just the simplicity of the randomness tempts one to find models that can at least scale the behavior of the mixture and arrive at working conclusions regarding the structural rigidity of the cell when exposed to external forces.

It is attractive to look at two bodies of literature for examples from other fields. The first is polymer physics, where a body of literature is presented on various thermally driven models. These works are well summarized in the book by Boal (2002). Included in these models are floppy chains, semiflexible chains, and welded chains. A second field is the study of porous solids, which range from structured anisotropic systems such as honeycombs to foams with random isotropic bubble inclusions. This field is reviewed in the classic treatise by Gibson and Ashby in 1988, with a second edition in 1997 (Gibson and Ashby, 1997; Gibson and Ashby, 1988).

It is possible to derive a simple model for the elastic modulus of the actin cytoskeleton by picturing it as a collection of relatively stiff elements attached to one another with stiff joints. In quantitative terms, following nomenclature from the polymer literature (Gittes and MacKintosh, 1998),[1] this means that the so-called persistence length over which the filaments will bend because of thermal agitation, l_p, is long compared to the distance between attachment points (or mesh size), l, and l is in turn large compared to the characteristic thickness of the structural elements, t. Then the intracellular structure can be compressed, stretched, and sheared by applying forces to the filament ensemble through its attachments to the membrane, to the substrate, and to the adjacent cells.

This approach was taken by Satcher and Dewey (1996) who used the analogy of a porous solid to obtain numerical values for the stiffness of cellular actin networks. A simple representative model originally proposed by Gibson and Ashby (1988) considers a three-dimensional rectangular meshwork of short rods connected to the lateral sides of adjacent elements as shown in Fig. 9-5. The key feature of the model is that the structural elements are placed into bending by applied forces. Although the bonds between the filaments are shown in the model as being rigid and the prototypical geometry is taken to be cubic cells, similar scaling of the rigidity of the matrix would be expected if the joints were simply pinned and the prototypical geometry were triangular. In that case, individual elements would buckle with applied stress. In the matrix described by Fig. 9-5, the ratio of the apparent density of the matrix ρ^* to the density of the solid filament material, ρ_s, varies as $(\rho^*/\rho_s) \sim (t/l)^2$.

Comparing the idealization of Fig. 9-5 with the actual cytoskeletal configuration of Fig. 9-1B, one can see many simplifications. First, the angles with which the filaments come together to make the matrix vary considerably. One should recognize that Fig. 9-1B is a view looking down into a three-dimensional structure, and the actual lattice is much more open than is apparent from the micrograph. The vast majority of junctions have an angle between the intersecting filaments that is greater than 45 degrees, and many approach right angles. What is important to the representation is that it puts individual structural elements into bending, thereby causing a deflection δ that can be simply computed from beam theory. The scaling of the geometry then allows the overall elastic modulus to be computed as a function of the density of filaments and their bending stiffness. Because the characteristic length l of the cellular structures is on the order of 100 nm and the individual structural elements have a typical dimension t of about 7 nm, one would expect the beams to be reasonably stiff in bending.

---

[1] Gittes and MacKintosh use the symbols ξ for the characteristic dimension of the lattice and a for the characteristic thickness of the structural element.

Cell dynamics and the actin cytoskeleton

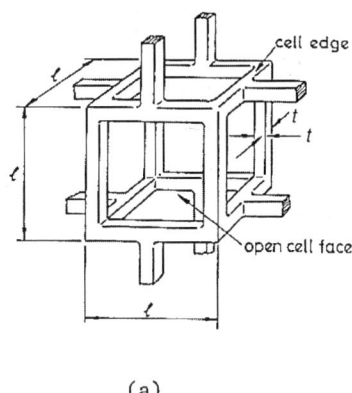

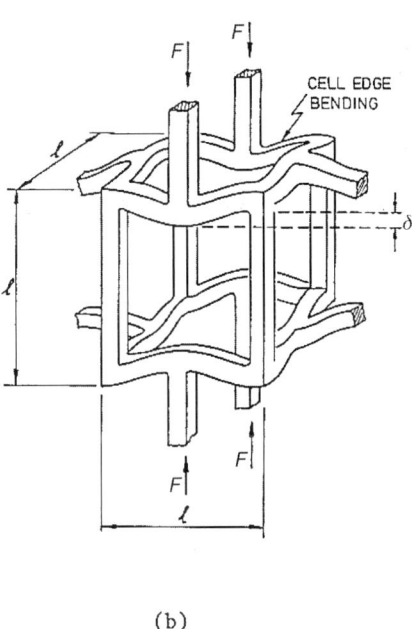

Fig. 9-5. The porous solid model of Gibson and Ashby applied to represent the structure of actin filaments within the cell. (a) is the undistorted lattice and (b) shows the action of an applied force. The vertices of the filaments are assumed to be at right angles and tightly bound for purposes of the calculation. In reality, the angles between filaments vary and the lattice spacing is not uniform; further, the bonds between individual filaments are not necessarily rigid, but may simply pin the joint. From Satcher and Dewey, 1996.

The results of the calculation show that strain ε in the lattice, $\varepsilon = \delta/l$, is proportional to the stress per unit area, σ, of the matrix, and the proportionality constant is the effective Young's modulus, E^*, for the material. Using bending theory for the strain ε, it is found (Satcher and Dewey, 1996) that

$$(E^*/E_s) \approx C_1(\rho^*/\rho_s)^2$$

where C_1 is a constant and E^* is the Young's modulus of solid actin, which is taken to be 2.3 GPa (Gittes et al., 1993). In order to complete the quantitative calculation, C_1 must be determined. That was done using empirical tabulations for polymer foams ranging over many orders of magnitude in density and structural properties as represented in Gibson and Ashby (1988). The result is that $C_1 \approx 1$. An identical scaling law is obtained if the lattice element is put into shear rather than compression, with the front-running constant of 3/8 instead of one.

This scaling can be compared to the results derived from semiflexible polymers with crosslinking. The latest results in this field are discussed in detail in Chapter 8. Published theories by MacKintosh and colleagues (Gittes and MacKintosh, 1998; MacKintosh et al., 1995; Chapter 8) show that the shear modulus G' for densely cross-linked gels scales as $G' \sim (\rho^*/\rho_s)^{5/2}$. Although the power law is similar, the value of G' depends on the "entanglement length," which is measured by thermal fluctuations of the filaments. It seems plausible that for open structures for which the typical dimensions are $l_p \gg l \gg t$ the thermal agitation would not play a strong role and the so-called enthalpic contribution would be negligible. From the point of view of existing experimental data taken in dilute solutions (Janmey et al., 1991), it is very difficult to decide between the effective modulus scaling as $(\rho^*/\rho_s)^2$ or $(\rho^*/\rho_s)^{5/2}$.

Two common features between the models are important. First, the elastic modulus is independent of frequency, at least for times over which the cross-links do not turn over and for which the frequencies are sufficiently low so as to not change the basic mechanisms of deformation of the filaments. Second, the existence of cross-linking is crucial; without cross-linking, the effective modulus would be substantially lower than the observed values.

A practical consequence of the predictions of the theory is that a drop in the fraction of actin polymerized can have a very profound effect on the rigidity of the cell. In separate experiments, we have observed a drop by factors of two to three in the fraction of actin polymerized following changes from a packed monolayer to freely crawling cells (McGrath et al., 2000b; McGrath et al., 1998b). We therefore find that the freely crawling cells have a much lower effective modulus and a higher motility. This has potential implications for wound healing, endothelialization of graft materials, and the integrity of monolayers subjected to fluid-flow forces.

We have examined the theory and observations relative to the internal cytoskeleton of a static collection of cells. Cells are in a dynamic state, with the internal cytoskeleton changing continuously. The following section explores the interesting dynamics that occur within the cell and presents quantitative models for the processes that are at work.

Actin dynamics

Abundant, essential, and discovered more than a half-century ago, actin is one of the most studied of all the proteins. Many investigations have focused on the dynamic character of actin, leading to a rich quantitative understanding of actin assembly and disassembly. In this section we briefly review this history and summarize the modern understanding of actin dynamics and its regulation by key binding proteins.

The emergence of actin dynamics

The appreciation that actin has both dynamic and mechanical properties can be traced to the work of its discoverer, F.B. Straub (Mommaerts, 1992). Trying to understand the difference between a highly viscous mixture of 'myosin B' and a less viscous

Cell dynamics and the actin cytoskeleton

mixture of 'myosin A,' Straub discovered these were not different myosins at all, but that the myosin B preparations were 'contaminated' by another protein that ACTivated myosIN to make the viscous solution (Feuer et al., 1948). Straub's laboratory later revealed that the contaminating actin could itself convert between low- and high-viscosity solutions with the introduction of physiological salts and/or ATP (Straub and Feuer, 1950). Further data from Straub demonstrated that the phase change occurred because a globular protein ('G-actin') polymerized into long filaments ('F-actin') and that the filamentous form could hydrolyze ATP. With the discovery of sarcomere structure (Huxley and Hanson, 1954) and the sliding filament model of muscle contraction in the 1950's (Huxley, 1957), actin's role as the structural thin filament of muscle was in place. The significance of actin's dynamic properties, however, remained unclear.

Over the next several decades actin was identified in every eukaryotic cell investigated. Extracts formed from macrophages (Stossel and Hartwig, 1976b) or ananthamoeba (Pollard and Ito, 1970) could be made to 'gel' in a manner that involved actin polymerization. Actin filaments were found to be concentrated near ruffling membranes and in fibroblast 'stress fibers' (Goldman et al., 1975; Lazarides and Weber, 1974), and cell movements could be halted with the actin-specific cytochalasins (Carter, 1967). Cytochalasins were found to block actin polymerization (Brenner and Korn, 1979), to inhibit the gelation of extracts (Hartwig and Stossel, 1976; Stossel and Hartwig, 1976; Weihing, 1976), and to reduce the strengths of reconstituted actin networks (Hartwig and Stossel, 1979). With these findings, actin's role as a structural protein appeared universal. However, unlike in muscle cells, actin's dynamic properties allow nonmuscle cells to tailor diverse and dynamic mechanical structures. Elucidating the intrinsic and regulated dynamics of actin was rightly viewed as a fundamental question in cell physiology, and became a lifelong pursuit for many talented biologists and physicists.

The intrinsic dynamics of actin

The pioneering work on protein self-assembly by Fumio Oosawa (see Fig. 9-6) led to some of the earliest insights into the process of actin polymerization. Using light scattering to follow the progress of polymerization, Oosawa noticed that actin polymerization typically began several minutes after the addition of polymer-inducing salts. He concluded that during the early 'lag' phase of polymerization, monomers form nuclei in an unfavorable reaction, but that once formed, nuclei are rapidly stabilized by monomer addition (Oosawa and Kasai, 1962). Because the addition of monomers to filaments is a highly favorable reaction, the lag phase could be overcome by adding preformed filaments as nuclei. Oosawa's studies indicated that the nucleus was formed of three monomers, a number that has stood further scrutiny (Wegner and Engel, 1975).

Oosawa described the mechanism of actin assembly by constructing some of the first kinetic models of the process (Oosawa and Asakura, 1975). Because experiments revealed actin filaments to have biochemically distinct 'barbed' and 'pointed' ends (Woodrum et al., 1975), and because ATP- and ADP-bound monomers are capable of assembling at these ends, Oozawa considered four assembly and four disassembly

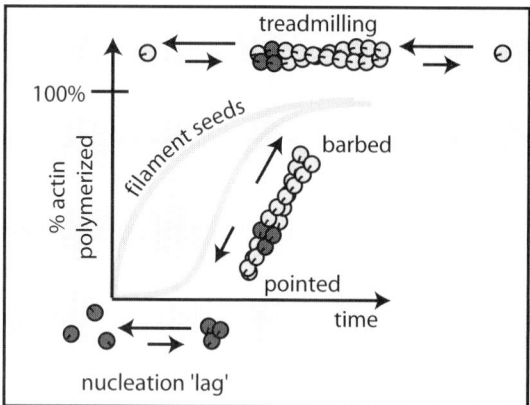

Fig. 9-6. Intrinsic actin dynamics. Oosawa and colleagues found that actin polymerizes in two phases: (a) an unfavorable nucleation phase involving an actin trimer and (b) a rapid assembly phase in which actin monomers add to both barbed and pointed ends. In the presence of excess ATP, actin does not polymerize to equilibrium but to a 'treadmilling' steady-state in which monomers continuously add to the barbed end and fall off of the pointed end.

reactions. Oozawa also assumed that spontaneous filament fragmentation and annealing ultimately determined the number of filaments in solution; this was later validated with detailed modeling (Murphy et al., 1988; Sept et al., 1999).

Obtaining reliable constants for the assembly and disassembly rates in the Oosawa model was a challenge that would occupy many years. Indirect assays gave conflicting results because of their inability to control or accurately measure the number of filaments in solutions. In 1986, Pollard's direct electron microscopic measurements of growth on individual nuclei produced numbers that are now the most widely cited (Pollard, 1986).

A decade before Pollard's experiments, Albert Wegner proposed that actin assembly does not proceed to equilibrium, but to a steady-state in which assembled subunits continuously traverse filaments from the barbed to the pointed end (Wegner and Engel, 1975). His now-classic experiments revealed that even after bulk polymerization had halted, filaments continued to hydrolyze ATP. Wegner decided that the results could only be explained if ATP-carrying monomers had a higher affinity for barbed ends than for pointed ends, and that the energy of ATP hydrolysis must be used to sustain the imbalance. Wegner suggested that, on average, ATP-G-actin assembles at barbed ends, hydrolyzes ATP in the filament interior, and disassembles bound to ADP at pointed ends. Thus Wegner was the first to introduce the concept of actin filament 'treadmilling,' although the term itself must be credited to Kirschner some years later (Kirschner, 1980). While Pollard's rate constants were not measured at steady-state, they clearly suggested barbed ends had a higher affinity for monomer in the presence of ATP and thus supported the existence of treadmilling.

A difficulty with both the Pollard and the Wegner studies was the lack of an accurate measure of the ATP hydrolysis rate on F-actin. Within the same year of Pollard's paper, Carlier and colleagues reported that ATP hydrolysis occurred less than a second after subunit addition (Carlier et al., 1987). Because this time scale for ATP hydrolysis

Cell dynamics and the actin cytoskeleton

is comparable to the time between monomer assembly events in most experiments, Pollard's assumption that ATP-actin was the disassembling species in his experiments must be reevaluated. Also significant were data revealing that inorganic phosphate was released from the cleaved nucleotide several minutes after hydrolysis (Carlier and Pantaloni, 1986). The latter discovery meant that three species needed to be considered for actin dynamics: ATP, ADP, and a long-lived intermediate ADP·Pi. Furthermore, an ADP·Pi monomer species should be generated by disassembly to either reassemble or release inorganic phosphate (Pi) and become a source of ADP-G-actin. Unfortunately, there has been no focused effort to determine all twelve assembly/disassembly rate constants and the two rates of Pi release (G-actin and F-actin) needed to properly update Oosawa's model. One reason for the missing effort is that both biochemical and structural data indicate that ADP·Pi and ATP-F-actin are similar (Otterbein et al., 2001; Rickard and Sheterline, 1986; Wanger and Wegner, 1987), so that distinguishing between the two species may be unnecessary in many contexts. Assuming equivalence between ADP·Pi and ATP-actin species, we have recently published a broad mathematical model of the steady-state actin cycle that predicts a broad range of experimentally observed behaviors (Bindschadler et al., 2004). While this agreement is encouraging, it does not replace the need for newly designed experiments that definitively establish rates.

In Bindschadler et al. (2004) we carefully tabulated consensus rate constants for intrinsic actin dynamics. While controversies persist concerning the mechanism of ATP hydrolysis and the rate of nucleotide exchange on G-actin, a growing consensus on these topics can be inferred from agreements by independent laboratories. As mentioned, the rates for the ADP·Pi-actin species remain unmeasured. Also unaddressed is the fact that the rate constants must depend on the nucleotide content of filaments and so the single values reported by Pollard cannot constitute the complete story.

Regulation of dynamics by actin-binding proteins

Many efforts over the past twenty years have focused on elucidating the mechanisms by which actin-associated proteins modulate actin dynamics. These efforts are essential because intracellular signaling pathways do not modify actin itself, and so the regulation of binding proteins provides an indirect route for changing cytoskeletal structure and cell shape. Unlike solutions of pure actin, cells contain short, dynamic filaments in highly structured networks and often a large fraction of unpolymerized actin. Here we review properties of actin-binding proteins thought to account for the major differences between the dynamics of cellular and purified actin (see Fig. 9-7).

ADF/cofilin: targeting the rate-limiting step in the actin cycle

Named for their activity as "actin depolymerizing factors" and their ability to form cofilamentous structures with F-actin, the ADFs and cofilins form two subgroups of a family of proteins (ADF/cofilins or ACs) expressed in most mammalian cells. Efforts to understand AC function have been challenged by a multiplicity of AC functions that differ slightly between the ADFs and cofilins and also among species (Bamburg, 1999). In general, ACs bind to both monomeric and filamentous actin with rather

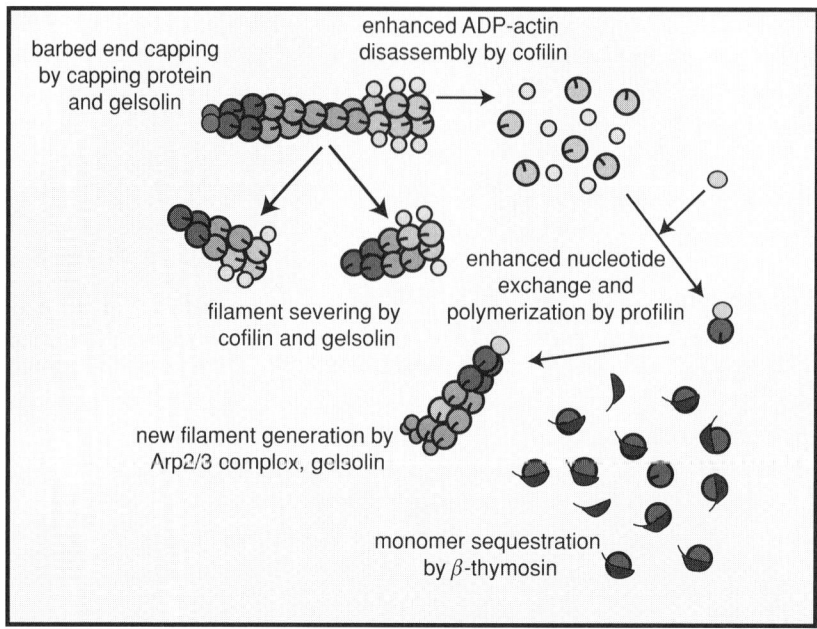

Fig. 9-7. Regulation of actin dynamics. Actin-binding proteins modulate every phase of the actin cycle including assembly and disassembly kinetics, nucleotide exchange, and filament number and length.

exclusive affinity for the ADP-bound conformations (Carlier et al., 1997; Maciver and Weeds, 1994). On filaments, structural data indicate that ACs bind the sides of filaments and destablize the most interior interactions between subunits (Bobkov et al., 2004; Galkin et al., 2003; McGough et al., 1997). AC-decorated filaments break along their length (Maciver et al., 1991) and rapidly disassemble at their ends (Carlier et al., 1997). Despite debate over whether ACs should be thought of primarily as filament-severing proteins or catalysts of ADP subunit disassociation (Blanchoin and Pollard, 1999; Carlier et al., 1997), both effects may occur as manifestations of the same structural instability on filaments. ACs have been shown to increase the rate of Pi release on filaments, and to slow the rate at which ADP monomers recharge with ATP on monomer (Blanchoin and Pollard, 1999). Thus ACs both hasten the production of ADP-actin and stabilize the ADP form. Some AC proteins bind ADP-G-actin with a much higher affinity than ADP-F-actin to create a thermodynamic drive toward the ADP-G-actin state (Blanchoin and Pollard, 1999). Like their multiple functions, ACs have multiple avenues for regulation including pH sensitivity, inactivation by PIP_2 binding, and serine phosphorylation (Bamburg, 1999).

Given this seemingly perfect arsenal of disassembling functions *in vitro*, observations that ACs trigger polymerization and generate new barbed ends in cells (Ghosh et al., 2004) certainly appear contradictory. There are at least two likely explanations for the paradox. First, with a large pool of sequestered ATP-actin available to assemble at free barbed ends (see the discussion of thymosins), the conditions inside a cell are primed for assembly (Condeelis, 2001). Thus filaments generated by AC severing may not have sufficient time to disassemble before they become nuclei for new filament

Cell dynamics and the actin cytoskeleton

growth. Supporting this, enhanced assembly occurs at early time points *in vitro* when ACs are added to solutions containing an excess of ATP-G-actin. (Blanchoin and Pollard, 1999; Du and Frieden, 1998; Ghosh et al., 2004).

The cellular data on cofilin-mediated growth should not be interpreted to mean that ACs are not involved in filament dissolution *in vivo*. Indeed, theoretical calculations indicate that severing alone cannot explain how filament turnover in cells occurs orders of magnitude faster than unregulated actin (Carlier et al., 1997). Thus, the second explanation for the paradox is that AC-mediated disassembly is directly linked to filament assembly. If pointed-end disassembly is rate-limiting for the actin cycle *in vivo* as it is *in vitro*, the enhanced production of ADP-G-actin should lead to a larger supply of ATP-monomer and enhanced polymerization elsewhere. Thus both of the 'destructive' activities of ACs – severing and enhanced disassembly – can lead to filament renewal and rapid turnover in the cellular environment.

Profilin: a multifunctional protein to close the loop

Profilin was the first monomer-binding protein discovered and originally thought to sequester G-actin in a nonpolymerizable form (Tobacman and Korn, 1982; Tseng and Pollard, 1982). However, later data made clear that profilins do not prevent actin assembly, but actually drive the assembly phase of the actin cycle (Pollard and Cooper, 1984). Today profilins are known to catalyze the rate of nucleotide exchange on G-actin by as much as an order of magnitude (Goldschmidt-Clermont et al., 1991; Selden et al., 1999). In cells this means that newly released ADP-G-actin is recharged to the ATP state shortly after binding to profilin. Significantly, the profilin-G-actin complex is capable of associating with filament barbed ends, but not with pointed ends (Pollard and Cooper, 1984). Profilin binds to a structural hinge on the barbed end of an actin filament and slightly opens the hinge to expose the nucleotide-binding pocket and promote nucleotide exchange (dos Remedios et al., 2003; Schutt et al., 1993). Profilin binding in this region also sterically blocks association of G-actin with pointed ends. Because there is no evidence that profilin blocks actin assembly at barbed ends, profilin presumably instantly disassociates from monomer after assembly. Further, the profilin-actin complex assembles at barbed ends at the same rates as ATP-G-actin alone (Kang et al., 1999; Pantaloni and Carlier, 1993). With these properties, profilin-bound actin becomes a subpopulation of barbed-end-specific monomer (Kang et al., 1999). In our published analysis of the actin cycle (Bindschadler et al., 2004), we found that profilin's functions provide the perfect complement to cofilin's disassembly functions. Together the two proteins appear to overcome every major barrier to increasing the rate of filament treadmilling (Bindschadler et al., 2004).

Arp2/3 complex and formins: making filaments anew

A perplexing and important question for cell biologists in the 1980s and 1990s was "how are new filaments created in cells?" One answer was that new filaments are generated when existing filaments are first severed and then elongate; however there was no reasonable mechanism for the *de novo* generation of filaments in cells. In the late 1990s it became clear that the Arp2/3 complex was dedicated to this task (Mullins et al., 1998; Pollard and Beltzner, 2002). The two largest members of this seven-protein complex are the actin-related proteins Arp2 and Arp3 (Machesky et al., 1994). Like the other members of the Arp family, Arp2 and Arp3 share a strong structural similarity to actin. In the Arp2/3 complex, these similarities are used to create a pocket that recruits an actin monomer to form a pseudotrimer that nucleates a new filament (Robinson et al., 2001). The complex holds the growing filament at its pointed end, leaving the barbed end free for rapid assembly (Mullins et al., 1998). For robust nucleation of filaments, the Arp2/3 complex requires activation, first by WASp/Scar family proteins (Machesky et al., 1999), and secondarily by binding to preexisting F-actin (Machesky et al., 1999). The arrangement gives autocatalytic growth of branched filament networks *in vitro*: new filaments grow from the sides of old ones to create a $\sim 70°$ included angle (Mullins et al., 1998). This same network geometry is found at the leading edge of cells and the Arp2/3 complex localizes to branch points in the cellular networks (Svitkina and Borisy, 1999).

While it is now clear that the Arp2/3 complex is an essential ingredient of the actin cytoskeleton, its discovery is new enough that many details of its mechanism are clouded in controversy. The most visible controversy has been over the nature of the Arp2/3 complex/F-actin interaction. Carlier and colleagues argue that the complex incorporates at the barbed ends of actin filaments to create a bifurcation in filament growth (Pantaloni et al., 2000); however, using direct visualization of fluorescently labeled filaments, several laboratories have demonstrated that new filaments can grow from the sides of preexisting filaments (Amann and Pollard, 2001a; Amann and Pollard, 2001b; Fujiwara et al., 2002; Ichetovkin et al., 2002). One paper appears to resolve the confusion with data indicating that branching occurs primarily from the sides of ATP-bound regions of the mother filament very near the barbed end (Ichetovkin et al., 2002). However, others believe that branching can occur on any subunit but filaments release or *debranch* rapidly from ADP segments of the mother filaments. This idea is supported by a correlation between the kinetics of debranching and Pi release on the mother filament (Dayel and Mullins, 2004), but contradicted by a report indicating that ATP hydrolysis on Arp2 is the trigger for debranching (Le Clainche et al., 2003). A disheartened reader looking for clearer understandings should consult the most recent reviews on the Arp2/3 complex.

Very recently, it has become clear that the Arp2/3 complex is not the only molecule capable of *de novo* filament generation in cells. In yeast, members of the formin family of proteins generate actin filament bundles needed for polarized growth (Evangelista et al., 2002), and in mammalian cells formin family members help generate stress fibers (Watanabe et al., 1999) and actin bundles involved in cytokinesis (Wasserman, 1998). Dimerized FH2 domains of formins directly nucleate filaments in a most remarkable manner (Pruyne et al., 2002; Zigmond et al., 2003). The FH2 dimer remains attached to the barbed ends of actin filaments even as it allows insertion of new subunits at that same end (Kovar and Pollard, 2004; Pruyne et al., 2002). By tracking the barbed ends of growing filaments, formins block associations with capping protein (Kovar et al., 2003; Zigmond et al., 2003). In cells where capping protein and cross-linking proteins are abundant, formin-based nucleation should naturally lead to bundles of long filaments, while Arp2/3-complex-generated filaments should naturally arrange into branched networks of short filaments.

Cell dynamics and the actin cytoskeleton

Capping protein: 'decommissioning' the old

Capping protein is an abundant heterodimeric protein that binds with high affinity to the barbed ends of actin filaments to block both assembly and disassembly at these ends (Cooper and Pollard, 1985; Isenberg et al., 1980). Vertebrates express multiple isoforms of both the α and β subunits (Hart et al., 1997; Schafer et al., 1994). With the conditions of cells favoring polymerization at free barbed ends, capping protein is essential to control the degree of polymerization. The association rates of capping protein with barbed ends in combination with high cellular concentrations of capping protein (~ 2 μM) should only allow a newly crated, unprotected barbed end to grow for ~ 1 s (Schafer et al., 1996). On the other hand, because the residency time of capping proteins on barbed ends is ~ 30 minutes (Schafer et al., 1996), short capped filaments will depolymerize from their pointed ends in cells. Capping proteins are thought to be integral to the recycling of monomers in dendritically arranged filaments at the leading edge of cells (Pollard et al., 2000). Consistent with this idea are findings that perturbations of capping activity dramatically alter the geometry of Arp2/3-complex-induced networks in reconstitution studies (Pantaloni et al., 2000; Vignjevic et al., 2003).

In addition to blocking barbed-end dynamics, capping protein diminishes the lag phase of actin polymerization (Pollard and Cooper, 1984). In this 'nucleating' activity, capping protein is probably stabilizing small oligomers rather than generating filaments *de novo* (Schafer and Cooper, 1995). Because the growing filaments are capped at their barbed end, this function is probably not active in cells with abundant sequestering proteins that can prevent assembly at pointed ends. The only known regulation of capping protein activity is by phospholipids. Phospholipids can both inactivate free capping protein (Heiss and Cooper, 1991) and remove bound capping protein from barbed ends (Schafer et al., 1996).

Gelsolin: rapid remodeling in one or two steps

If the job of actin-binding proteins is to remodel the actin cytoskeleton, then gelsolin has exceptional qualifications. Activated by micromolar Ca^{2+} (Yin and Stossel, 1979), gelsolin binds to the sides of actin filaments and severs them (Yin et al., 1980). However, unlike cofilin, gelsolin remains attached to the new barbed end created by severing to block further polymerization (Yin et al., 1981; Yin et al., 1980). Because gelsolin has nM affinity for barbed ends, it functions as a permanent cap that can only be removed through subsequent binding by phospholipids (Janmey and Stossel, 1987). In platelets and neutrophils, activated gelsolin remodels actin in two steps (Barkalow et al., 1996; Glogauer et al., 2000). Because the majority of filaments in resting cells are capped, cellular activation first leads to gelsolin severing to create a large number of dynamically stable filaments. Shortly thereafter, these filaments become nuclei for new growth as phospholipid levels increase to result in massive uncapping.

While gelsolin seems built for acute remodeling, expression studies clearly indicate a role for gelsolin at steady-state. Gelsolin null fibroblasts have impaired motility, reduced membrane ruffling, slow filament turnover, and abundant stress fibers (Azuma et al., 1998; McGrath et al., 2000a; Witke et al., 1995), and gelsolin overexpression produces the opposite trends (Cunningham et al., 1991). With its high

Ca^{2+} requirements, it is unclear if these steady-state effects are mediated by intermittent and localized gelsolin activity, or constant but low levels of activity. Gelsolin also has a fascinating role in apoptosis where caspases cleave the protein to create an unregulated severing peptide (Kothakota et al., 1997). Continuous severing by this peptide helps create a mechanically compromised cell that eventually detaches from its substrate (Kothakota et al., 1997).

β4-thymosin: accounting (sometimes) for the other half

The 15 kD protein β4-thymosin is present in mammalian cells at levels that equal or exceed actin itself (Safer and Nachmias, 1994). Its discovery appeared to resolve the critical question of how mammalian cells maintained nearly half of their actin in an unpolymerized form despite intracellular conditions that favored polymerization. β4-thymosins are unstructured in solution and partially wrap around the G actin monomer to control its associations (Safer et al., 1997). β4-thymosins bind to ATP-G-actin (but not ADP-G-actin) with an affinity comparable to the pointed end ATP critical concentration, but less than the barbed end ATP critical concentration (Carlier et al., 1993). In this way when barbed ends are mostly capped, β4-thymosin functions as a 'sequestering' protein that maintains a pool of nonfilamentous actin, but when free barbed ends are abundant the pool diminishes. Profilin can compete for the ATP-G-actin pool maintained by β4-thymosin (Carlier et al., 1993), possibly by emerging with the charged monomer after forming a complex that includes all three proteins (Yarmola et al., 2001). Thus with or without profilin, β4-thymosin helps to reserve ATP-G-actin for future assembly at barbed ends. The high concentration of β4-thymosin can support extensive and sudden conversions from G-actin to filaments. This conversion is likely occurring in the dramatic polymerization-induced shape change of both platelets (Safer et al., 1990) and neutrophils (Cassimeris et al., 1992), both of which contain abundant β4-thymosin/G-actin complex at rest. However β4-thymosin apparently is not required for the continuous shape change during crawling, because motile amoebae are thought to be void of thymosin-family proteins (Pollard et al., 2000).

Dynamic actin in crawling cells

In this section we explore a most conspicuous and well-studied function of the actin cytoskeleton: its ability to serve as the engine for cell crawling. By driving the expansion of the plasma membrane in the direction of cell advancement, actin polymerization initiates the crawling cycle (Fig. 9-8). The networks formed by polymerization evolve to structures that provide mechanical support for cell extensions; that link the cell to its substrate; and that support the myosin-based contractions needed for cell translation. The network must also disassemble to recycle its constituents for further rounds of assembly. Thus the actin network at the leading edge of motile cells provides both the structure and the forces needed for crawling (see Fig. 9-9). Here we review the current understanding of the geometry and dynamics of these networks, and address the important question of how polymerization might lead to pushing forces.

Cell dynamics and the actin cytoskeleton

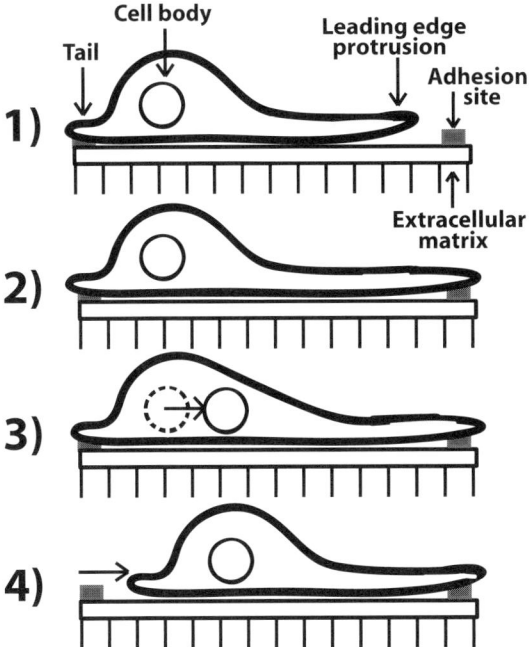

Fig. 9-8. The four steps in cell migration. The classic schematic of crawling breaks the process down into a four-step cycle. The cycle begins with the protrusion of the leading edge driven by actin polymerization. The extended cell forms new attachments in advance of its body and then contracts against this attachment to break tail adhesions and translate forward. From Mitchison and Cramer, 1996.

Actin in the leading edge

The extension of the plasma membrane that interrogates new regions of substrate can come in several forms. Mammalian cells crawling in culture environments extend both finger-like projections, called fillopodia, and broad, thin, veil-like projections called lamellipodia. Which structure occurs more frequently is a strong function of cell type and substrate conditions (Pelham and Wang, 1997). Cells that crawl in amoeboid fashion – a class that includes the leukocytes of the mammalian immune system – use bulkier protrusions known as pseudopodia. By all accounts, the initiator of filament assembly in each of these cellular protrusions is the Arp2/3 complex activated by membrane-bound WASp/Scar family proteins. Pollard and Borisy have offered the most detailed proposal for how actin networks evolve in lamellipodia (Pollard and Borisy, 2003), and filopodia appear to be triggered from rearrangements of a lamellipodial network (Svitkina et al., 2003). Because the three-dimensional character of pseudopodia makes them less amenable to ultrastructural and fluorescence studies, far less is known about the geometry and dynamics of pseudopodial networks.

Synthesizing data from electron micrographs of cytoskeletal structure in the lamellipodia of fast-moving keratocytes (Svitkina and Borisy, 1999), immunochemical analysis of Arp2/3 complex, ADF/cofilin, and capping protein location in these same samples (Svitkina and Borisy, 1999), and live cell fluorescence data revealing regions of actin assembly and disassembly in fibroblasts (Watanabe and Mitchison, 2002),

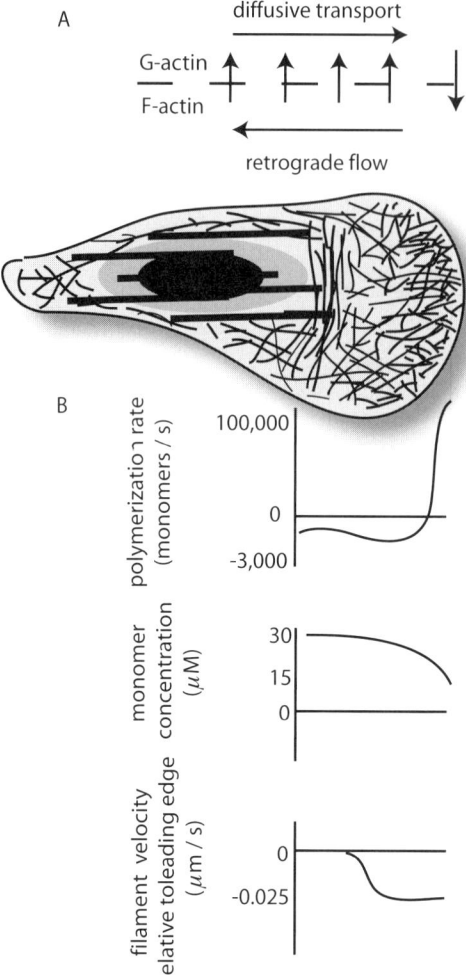

Fig. 9-9. **Actin dynamics in crawling cells.** (A) For steady crawling, actin polymerization and depolymerization must complete a balanced cycle. However, the demand for assembling monomers at the leading edge causes a spatial segregation of these processes and flows. Actin that assembles at the leading edge but does not incorporate into a protrusion flows in retrograde fashion as it is disassembled. The emerging G-actin population is presumably returned to the leading edge by diffusion. (B) The graphs show possible profiles for polymerization, monomer concentration, and retrograde flow across the front of a crawling cell. The calculated numbers combine measurements of actin dynamics in fibroblasts (Vallotton et al., 2004; Watanabe and Mitchison, 2002) and the leading edge monomer demand for these cells (Abraham et al., 1999), and assumes monomer is returned by diffusion with a diffusivity of 6×10^{-8} cm^2/s. From McGrath et al., 1998a.

Pollard and Borisy propose the lamellipodia are filled with a neatly segregated, highly dynamic network (Pollard and Borisy, 2003). In their model, filaments are first generated by activated Arp2/3 complex to form brush-like networks within one micron of the advancing plasma membrane. The newly born filaments remain short because their growth is rapidly terminated by abundant capping protein. The continuous assembly of ATP-actin at the leading edge explains why the ADP-actin specific ADF/cofilins are excluded from this region (Svitkina and Borisy, 1999). ADF/cofilin-family proteins

Cell dynamics and the actin cytoskeleton

take up residence at distances one micron and more from the leading edge where they bind to aged ADP-actin, disassembling the network into monomers for rapid recycling and further polymerization (Svitkina and Borisy, 1999). More recent and sophisticated analysis of actin filament dynamics are consistent with a segregation of the lamellipod into a ∼1 micron membrane-proximal region dominated by polymerization and an immediately adjacent region where significant depolymerization occurs (Vallotton et al., 2004).

A leading theory for filopodia generation proposes that these structures emerge from a rearrangement of the dendritic networks of the lamellipod (Svitkina et al., 2003). In the convergence/elongation theory, convergent filaments with barbed ends that abut the plasma and are protected against capping are zippered together by fascin as they grow to several microns in length. Consistent with this model, recent knockdown studies reveal that filopodia-rich phenotypes occur in capping-protein-depleted cells and point to an essential role for the anticapping activies of the Ena/VASP family in filopodia formation (Mejillano et al., 2004).

Nearer the cell body, filaments surviving the destructive actions of ADF/cofilins mature into a contractile network. In fast-moving cells like fish keratocytes, the surviving network remains fixed with respect to the substrate as the cell crawls past (Theriot and Mitchison, 1991). In slower-moving cells like fibroblasts, polymerization exceeds the rate of cell advancement, and much of the network flows toward the cell center (Theriot and Mitchison, 1992). Filament survival is facilitated by association with the long, side-binding protein tropomyosin, which blocks the association of ADF/cofilins (Bernstein and Bamburg, 1982; Cooper, 2002; DesMarais et al., 2002), and growth to several microns is likely facilitated by the anticapping activities of formins (Higashida et al., 2004). In fibroblasts, the polarity of filaments is graded such that all filaments at the leading edge are oriented with their barbed ends facing the periphery, but the interior bundles have a well-mixed polarity (Cramer et al., 1997). The gradation appears to facilitate both pushing at the edge of cells and myosin-based contractions by muscle-like filaments sliding within the cell interior. Just as some of the filaments of the lamellipod mature into contractile stress fibers, the focal contacts that transmit these stresses to surfaces also begin life in lamellipodia as nascent focal complexes and mature into focal contacts as they become part of the more central structures of the cell (DeMali et al., 2002).

Monomer recycling: the other 'actin dynamics'

For steady migration a cell must have a constant supply of monomer delivered to its leading edge. This monomer certainly derives from filament disassembly at more interior regions, and so the rates of assembly and migration are tied to the rate of monomer supply. If monomer is provided by diffusion, the supply rate is equal to the product of the diffusion coefficient and the gradient of the monomer concentration. The diffusion coefficient of the fastest of two kinetically distinguishable populations of actin tracers is $\sim 6 \times 10^{-8}$ cm^2/s (Giuliano and Taylor, 1994; Luby-Phelps et al., 1985; McGrath et al., 1998a). Assigning this diffusion coefficient to actin monomer has caused difficulties for modelers of the lamellipod (Abraham et al., 1999; Mogilner and Edelstein-Keshet, 2002). If the assembly of monomers at the leading edge of a crawling cell is driven by mass action, then the concentration of monomer at the leading edge must exceed 15 µM for a single filament to keep pace with the plasma membrane in keratocytes. Fick's law, in combination with a monomer diffusion coefficient of 6×10^{-8} cm^2/s, requires a gradient of 20 µM/µm moving toward the interior of the cell. If this gradient persists over 10 microns, as modelers have assumed, then a maximum concentration of ~ 200 µM G-actin near the cell body is prohibitively high because cells typically carry less than 100 µM of actin total.

Assuming that the more diffusive population of actin is strictly monomer may be wrong, as short, diffusing filaments from recent severing and growth events are likely. Indeed, investigators have proposed that the diffusion of small oligomers may explain why the diffusion coefficient for the mobile actin in cytoplasm is ~ 6 times slower than the value for monomer-sized ficoll (Luby-Phelps et al., 1987). Through comparisons with the diffusion of sugar particles of various sizes, one concludes that actin diffuses as a molecule five times bigger than its hydrodynamic radius (Luby-Phelps et al., 1987), and so the discrepancy cannot be explained by the fact that G-actin is complexed with smaller molecules of thymosin or profilin. The possibility of filament diffusion appears to justify the use of the high G-actin diffusion coefficient of 3×10^{-7} cm^2/s in models (Abraham et al., 1999; Mogilner and Edelstein-Keshet, 2002). The value is that for the diffusion of monomer-sized ficoll and leads to the derivation of G-actin profiles with more gradual gradients of ~ 4 µM/µm. However it must be noted that all proteins studied diffuse slower in cytoplasm than size-matched 'inert' sugars, even those that do not oligomerize (Luby-Phelps et al., 1985), and so the most reasonable explanation for the discrepancy between the diffusion of actin and sugar particles is that the sugars are not good models for protein diffusivity.

In support of the interpretation of the fast-diffusing species as monomer, roughly half of the tracer actin is in this population consistent with biochemical fractionation (McGrath et al., 2000c). Further, when cells are treated with jasplankinolide, a membrane-permeating toxin that blocks filament depolymerization (Bubb et al., 2000), all actin becomes immobile, indicating that the fast-diffusing species is assembly competent (McGrath et al., 1998a; Zicha et al., 2003). As noted, recent speckle microscopy experiments indicate that the major zone of depolymerization in lamellipodia is found 2 microns behind the leading edge (Ponti et al., 2004; Watanabe and Mitchison, 2002), and so the gradient may in fact be steep but confined to a dynamic region considerably smaller than the 10 microns assumed in the models. Over this small distance the experimentally determined values for monomer diffusion should be sufficient for the steady resupply of monomer to the leading edge.

Recent data also suggest that G-actin is returned to the leading edge of cells by active transport mechanisms. This prospect is raised by the recent results of Grahm Dunn and colleagues (Zicha et al., 2003). Using a modification of the photobleaching technique, these experimentalists marked a population of actin several microns behind the leading edge of a protruding cell and found that some of the marked actin incorporates into new protrusions at rates that defy simple diffusion. They provide some evidence that the phenomenon is halted by myosin inhibitors and does not depend on microtubules. This suggests that the privileged monomer does not ride as cargo on a motor complex, but that it happens to be near a convective channel. The difficulty with the prospect of motors moving G-actin as cargo is that the association between

Cell dynamics and the actin cytoskeleton

actin and motor proteins would likely be specific. The prospect of a return mechanism that does not discriminate among molecules is more attractive because the problem of recycling cytoskeletal pieces is not limited to G-actin; whatever mechanism is at work must be able to carry all the building blocks to the leading edge.

The biophysics of actin-based pushing

While compelling support for actin polymerization forces have existed for decades (Tilney et al., 1973), the mechanism by which polymerization leads to pushing remains unclear. Today there are two leading theories: (1) a series of related 'ratchet' models that explain pushing as a natural consequence of the polymerization of semiflexible filaments against a membrane (Dickinson and Purich, 2002; Mogilner and Oster, 1996; Mogilner and Oster, 2003; Peskin et al., 1993) and (2) a mesoscopic model that explains how pushing forces derive from the formation of actin networks on curved surfaces.

Understanding of the biophysics of actin-based pushing in the *Listeria* system has progressed through a steadily tightening cycle of theory and experiment that continues to this day. In large part, these efforts have centered around the actin-based motility of the bacterium *Listeria monocytogenes*. This intracellular pathogen invades host cytoplasm and hi-jacks the same force-producing mechanisms that drive leading-edge motility. Riding a wave of actin polymerization, the bacterium becomes motile so that it can eventually exit the dying infected cell for an uninfected neighbor.

The first ratchet model proposed that *Listeria* was a 1-D Brownian particle blocked from rearward diffusion by the presence of the growing actin tail (Mogilner and Oster, 1996). Observations that *Shigella* move at speeds similar to *Listeria* despite being more than twice as large violated a prediction of the Brownian Ratchet and motivated a new theory. The "Elastic Ratchet" proposed that filaments, rather than *Listeria*, fluctuate due to thermal excitation (Mogilner and Oster, 1996). Filaments fluctuate away from the *Listeria* surface to allow space for polymerization. Lengthened filaments apply propulsive pressure as they relax to unstrained configurations.

More recent biophysical measurements established that *Listeria* are tightly bound to their tails (Gerbal et al., 2000a; Kuo and McGrath, 2000), rigorously eliminating the Brownian Ratchet theory and demanding a revision of the Elastic Ratchet theory. One study also established that *Listeria* motion is discontinuous, with frequent pauses and nanometer-sized steps (Kuo and McGrath, 2000). The developments led to the first proposal for how elastic filaments could push *Listeria* while attached. In the 'Actoclampin' model (Dickinson and Purich, 2002), filaments diffuse axially due to bending fluctuations within a surface-bound complex. The complex binds ATP-bound subunits and releases the filament upon hydrolysis. In this scheme, flexed filaments push the *Listeria* and lagging filaments act as tethers (Fig. 9-10A).

More than molecular stepping, the force-velocity curve of the polymerization engine constrains theoretical models, but current data appear to be in disagreement. Recently, we published a curve for *Listeria monocytogenes* (McGrath et al., 2003), using methylcellulose to manipulate the viscoelasticity of extracts, particle tracking to determine viscoelastic parameters near motile *Listeria*, and a modified Stokes equation to infer forces. In 2003, Mogilner and Oster published an evolution of the

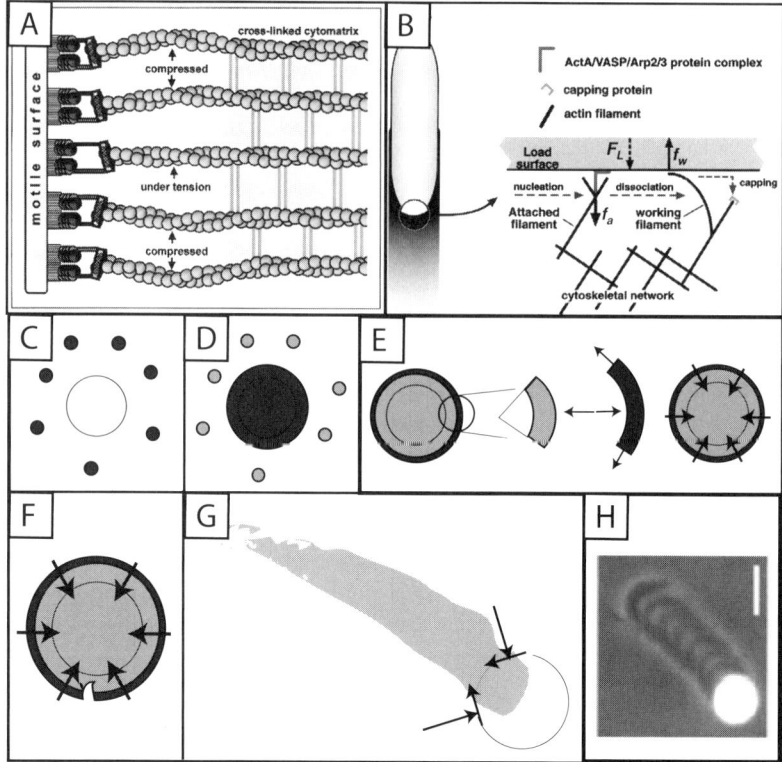

Fig. 9-10. **Theories of force production in actin-based motility.** *(A–B) In Molecular Ratchet models, filament fluctuations and growth near the surface combine to create motility.* (A) In the Actoclampin model (from Dickinson and Purich, 2002) all filaments are similarly attached to the motile surface, but some are compressed and others are stretched taut. (B) The Tethered Ratchet (from Mogilner and Oster, 2003), considers distinct attached and pushing (working) filaments. *(C–G) In the Elastic Propulsion theory, elastic stresses lead to symmetry breaking and motion.* (C) The first layer of actin polymerized at the surface is pushed outward (D) by the next layer, creating hoop stresses in the gel and normal pressure on the sphere (E). Fluctuations in stress levels and strengths cause a local fracturing event (F) and unraveling of the gel that leads to a motile state (G) in which stresses build and relax periodically as the particle moves forward. (H) *Large particles create tails with periodic actin density ('hopping'), suggesting stress building and relaxation.* Bar is 10 microns. From Bernheim-Groswasser, 2003.

Elastic Ratchet model (Mogilner and Oster, 2003) that quantitatively predicts the force-velocity curve of McGrath et al. (2003). The "Tethered Ratchet" features "working" filaments that push as Elastic Ratchets and "tethered" filaments anchored at complexes that also nucleate dendritic branches (Fig. 9-10B).

While Molecular Ratchets appear to account for the force-velocity data in McGrath et al. (2003), shallower force-velocity curves obtained by Wiesner et al. (2003) are interpreted in terms of a very different theory termed Elastic Propulsion (Gerbal et al., 2000b) (Fig. 9-10C–H). The theory describes stress build-up in continuum, elastic actin networks that grow on curved surfaces. During nucleation, older layers are displaced radially by newer polymerization at the nucleating surface. Because the displaced layers must stretch, they generate 'hoop' or 'squeezing' stresses around the curved particle or bacterium. Propulsive tails emerge from symmetric actin 'clouds' because the stresses eventually exceed gel strengths and cause the network to partially unravel. In the motile configuration, hoop stresses continue to build due to surface curvature, and frictional tractions hold the tail on the particle (Fig. 9-10G). Once the propulsive stresses exceed frictional resistances, the object slips forward and tail stresses relax. The process repeats as polymerization continues and stresses rebuild.

Further evidence supporting the Elastic Propulsion theory is a report by Bernheim-Groswasser et al. (2002) that large (>4 micron) VCA-coated particles advance in a micron-scale 'hopping' pattern (Bernheim-Groswasser et al., 2002). In this phenomenon, the density of actin in tails varies in a periodic fashion to give tails a banded pattern (Fig. 9-10H), periods of high actin intensity are also periods of low velocities and vice versa (Bernheim-Groswasser et al., 2002). The pattern is interpreted as the build up and release of squeezing stresses in Elastic Propulsion. The pattern is prominent on large particles because it takes longer to build the critical stresses for slipping on surfaces with lower curvature.

Despite evidence supporting Elastic Propulsion, we found that it cannot be the only mechanism for generating pushing forces in reconstitution experiments. Recognizing that the nucleating surface must be curved for the Elastic Propulsion mechanism but not for Molecular Ratchets, we tested whether actin-based motility could occur on flat surfaces. In Schwartz et al. (2004) we created flat particles by compressing heated polystyrene spheres. Not only did we find that flat surfaces could be substrates for pushing forces, we found that disks pushed on flat faces moved faster than did the coated versions of the spheres from which they were manufactured.

Conclusion

This chapter on cell dynamics and the role of the actin cytoskeleton should be considered as a snapshot of a rapidly evolving field of research. In the last three years, PubMed lists 4,000 articles with the two key words actin and cytoskeleton. Indeed, many of the fascinating topics we work on today – such as actin filament branching and severing, connection of the actin cytoskeleton to membrane-associated protein complexes, and the behavior of actin bundles commonly termed stress fibers – receive negligible mention here. The field is very rich, and the inquiries are diverse. It is a fruitful field of research with much fundamental biology and biophysics to be discovered.

Can we point to specific therapies that could be influenced by research in the areas described in this chapter? One possibility is understanding how the endothelium acts to realign and create small-vessel proliferation in cancerous tissue. Finding means to defeat the invagination of tumors would create the possibility of starving growing tumors and killing cancerous tissue selectively. The motility of the endothelium depends intimately on the organization and turnover of the actin cytoskeleton.

A second grand challenge to which this research points is understanding the mechanisms of atherosclerosis proliferation. A key step is the trans-endothelial migration of leukocytes and an inflammatory response cycle that leads to intimal smooth muscle cell proliferation. In order for the leukocytes to cross an intact layer of endothelial cells, the cells must first stick to the endothelium and then induce the underlying endothelial cells to retract, paving a path through which the leukocyte can enter the arterial wall. Understanding the cytoskeleton dynamics associated with this process could lead to therapies preventing intimal proliferation and subsequent plaque buildup in the arteries.

The chapter makes clear the tremendous detail with which science now understands both the dynamics and biophysics of the actin machinery that controls cell shape. Applying this knowledge to disease will require not only continued discovery of mechanisms and rates, but also the organization of the vast information into a predictive computational model. Thanks to decades of investigation by scientists emphasizing quantitative experiments, modeling of the actin cytoskeleton is advanced compared to the hundreds of other subcellular systems required to quantitatively describe cellular life. Models of other systems will inevitably join models of the actin cytoskeleton over the next few years to begin the broad integration of knowledge. The possibility exists to use actin models as examples, and to begin today to design information architectures that can handle such massive amounts of information. Only with a quantitative means of describing the highly nonlinear interaction between the many important cellular systems can we hope to represent the complex and highly nonlinear behavior within cells, and eventually tissues and organs.

Acknowledgements

The authors wish to thank John H. Hartwig for many years of fruitful collaboration on problems related to cell mechanics and the cytoskeleton. His insights with regard to the role of actin-binding proteins has been especially valuable. JLM acknowledges support from Whitaker Research Grant #RG-01-033 and thanks Mr. Ian Schwartz for Fig. 9-8. CFD is grateful for many years of support by the NHLBI, National Institutes of Health.

References

Abraham, V.C., V. Krishnamurthi, D.L. Taylor, and F. Lanni. 1999. The actin-based nanomachine at the leading edge of migrating cells. *Biophys. J.*, 77:1721–32.

Amann, K.J., and T.D. Pollard. 2001a. The Arp2/3 complex nucleates actin filament branches from the sides of pre-existing filaments. *Nat. Cell Biol.*, 3:306–10.

Amann, K.J., and T.D. Pollard. 2001b. Direct real-time observation of actin filament branching mediated by Arp2/3 complex using total internal reflection fluorescence microscopy. *Proc. Natl. Acad. Sci. USA*, 98:15009–13.

Andrews, R., and J. Fox. 1991. Interaction of purified actin-binding protein with the platelet membrane glycoprotein Ib-IX complex. *J. Biol. Chem.*, 266:7144–47.

Andrews, R., and J. Fox. 1992. Identification of a region in the cytoplasmic domain of the platelet membrane glycoprotein Ib-IX complex that binds to purified actin-binding protein. *J. Cell Biol.*, 267:18605–11.

Azuma, T., W. Witke, T.P. Stossel, J.H. Hartwig, and D.J. Kwiatkowski. 1998. Gelsolin is a downstream effector of rac for fibroblast motility. *Embo J.*, 17:1362–70.

Bamburg, J.R. 1999. Proteins of the ADF/cofilin family: essential regulators of actin dynamics. *Annu. Rev. Cell Dev. Biol.*, 15:185–230.

Barkalow, K., W. Witke, D.J. Kwiatkowski, and J.H. Hartwig. 1996. Coordinated regulation of platelet actin filament barbed ends by gelsolin and capping protein. *J. Cell Biol.*, 134:389–99.

Bernheim-Groswasser, A., S. Wiesner, R.M. Golsteyn, M.F. Carlier, and C. Sykes. 2002. The dynamics of actin-based motility depend on surface parameters. *Nature*, 417:308–11.

Bernstein, B.W., and J.R. Bamburg. 1982. Tropomyosin binding to F-actin protects the F-actin from disassembly by brain actin-depolymerizing factor (ADF). *Cell Motil.*, 2:1–8.

Bindschadler, M., E.A. Osborn, C.F. Dewey, Jr., and J.L. McGrath. 2004. A mechanistic model of the actin cycle. *Biophys. J.*, 86:2720–39.

Blanchoin, L., and T.D. Pollard. 1999. Mechanism of interaction of Acanthamoeba actophorin (ADF/Cofilin) with actin filaments. *J .Biol. Chem.*, 274:15538–46.

Boal, D. 2002. *Mechanics of the Cell*. Cambridge, UK: Cambridge University Press.

Bobkov, A.A., A. Muhlrad, A. Shvetsov, S. Benchaar, D. Scoville, S.C. Almo, and E. Reisler. 2004. Cofilin (ADF) affects lateral contacts in F-actin. *J. Mol. Biol.*, 337:93–104.

Brakebush, C., and R. Fässler. 2003. The integrin-actin connection, an eternal love affair. *EMBO J.*, 22:2324–2333.

Brenner, S.L., and E.D. Korn. 1979. Substoichiometric concentrations of cytochalasin D inhibit actin polymerization. Additional evidence for an F-actin treadmill. *J. Biol. Chem.*, 254:9982–5.

Bröcker, F., and et al. 1999. Assignment of human filamin gene FLNB to human chromosome band 3p14.3 and identification of YACs containing the complete FLNB transcribed region. *Cytogenet. Cell Genet.*, 85:267–8.

Bubb, M.R., I. Spector, B.B. Beyer, and K.M. Fosen. 2000. Effects of jasplakinolide on the kinetics of actin polymerization. An explanation for certain in vivo observations. *J. Biol. Chem.*, 275:5163–70.

Calderwood, D., and M. Ginsberg. 2003. Talin forges the links between integrins and actin. *Nat. Cell Biol.*, 5:694–7.

Carlier, M.B., A. Zenebergh, and P.M. Tulkens. 1987. Cellular uptake and subcellular distribution of roxithromycin and erythromycin in phagocytic cells. *J. Antimicrob Chemother*, 20 Suppl B:47–56.

Carlier, M.F., C. Jean, K.J. Rieger, M. Lenfant, and D. Pantaloni. 1993. Modulation of the interaction between G-actin and thymosin beta 4 by the ATP/ADP ratio: possible implication in the regulation of actin dynamics. *Proc. Natl. Acad. Sci. USA*, 90:5034–8.

Carlier, M.F., V. Laurent, J. Santolini, R. Melki, D. Didry, G.X. Xia, Y. Hong, N.H. Chua, and D. Pantaloni. 1997. Actin depolymerizing factor (ADF/cofilin) enhances the rate of filament turnover: implication in actin-based motility. *J. Cell Biol.*, 136:1307–22.

Carlier, M.F., and D. Pantaloni. 1986. Direct evidence for ADP-Pi-F-actin as the major intermediate in ATP-actin polymerization. Rate of dissociation of Pi from actin filaments. *Biochemistry*, 25:7789–92.

Carter, S.B. 1967. Effects of cytochalasins on mammalian cells. *Nature*, 213:261–4.

Cassimeris, L., D. Safer, V.T. Nachmias, and S.H. Zigmond. 1992. Thymosin beta 4 sequesters the majority of G-actin in resting human polymorphonuclear leukocytes. *J. Cell Biol.*, 119:1261–70.

Condeelis, J. 2001. How is actin polymerization nucleated in vivo? *Trends Cell Biol.*, 11:288–93.

Cooper, J.A. 2002. Actin dynamics: tropomyosin provides stability. *Curr. Biol.*, 12:R523–5.

Cooper, J.A., and T.D. Pollard. 1985. Effect of capping protein on the kinetics of actin polymerization. *Biochemistry*, 24:793–9.

Cramer, L.P., M. Siebert, and T.J. Mitchison. 1997. Identification of novel graded polarity actin filament bundles in locomoting heart fibroblasts: implications for the generation of motile force. *J. Cell Biol.*, 136:1287–305.

Cunningham, C.C., T.P. Stossel, and D.J. Kwiatkowski. 1991. Enhanced motility in NIH 3T3 fibroblasts that overexpress gelsolin. *Science*, 251:1233–6.

Davies, P. 1995. Flow-mediated endothelial mechanotransduction. *Physiol. Rev.*, 75:519–560.

Davies, P., A. Robotewskyj, and M. Griem. 1993. Endothelial cell adhesion in real time. *J. Clin. Invest*, 9:2640–52.

Davies, P., J. Zilberberg, and B. Helmke. 2003. Spatial microstimuli in endothelial mechanosignaling. *Circ. Res.*, 92:359–70.

Davies, P.F., A. Robotewskyj, and M.L. Griem. 1994. Quantitative studies of endothelial cell adhesion: directional remodeling of focal adhesion in response to flow forces. *J. Clinical Invest*, 93:2031–8.

Dayel, M.J., and R.D. Mullins. 2004. Activation of Arp2/3 complex: Addition of the first subunit of the new filament by a WASP protein triggers rapid ATP hydrolysis on Arp2. *PLoS Biol.*, 2:E91.

DeMali, K.A., C.A. Barlow, and K. Burridge. 2002. Recruitment of the Arp2/3 complex to vinculin: coupling membrane protrusion to matrix adhesion. *J. Cell Biol.*, 159:881–91.

DesMarais, V., I. Ichetovkin, J. Condeelis, and S.E. Hitchcock-DeGregori. 2002. Spatial regulation of actin dynamics: a tropomyosin-free, actin-rich compartment at the leading edge. *J. Cell Sci.*, 115:4649–60.

Dewey, C., Jr., S. Bussolari, M. Gimbrone, Jr., and P. Davies. 1981. The dynamic response of vascular endothelial cells to fluid shear stress. *J. Biomech. Eng.*, 103:177–185.

Dickinson, R.B., and D.L. Purich. 2002. Clamped-filament elongation model for actin-based motors. *Biophys. J.*, 82:605–17.

dos Remedios, C.G., D. Chhabra, M. Kekic, I.V. Dedova, M. Tsubakihara, D.A. Berry, and N.J. Nosworthy. 2003. Actin binding proteins: regulation of cytoskeletal microfilaments. *Physiol. Rev.*, 83:433–73.

Du, J., and C. Frieden. 1998. Kinetic studies on the effect of yeast cofilin on yeast actin polymerization. *Biochemistry*, 37:13276–84.

Evangelista, M., D. Pruyne, D.C. Amberg, C. Boone, and A. Bretscher. 2002. Formins direct Arp2/3-independent actin filament assembly to polarize cell growth in yeast. *Nat. Cell Biol.*, 4:32–41.

Feuer, G., F. Molnar, E. Pettko, and F. Straub. 1948. Studies on the composition and polymerization of actin. *Hungarica Acta Physiologica*, 1:150–163.

Fox, J. 1985. Identification of actin-binding protein as the protein linking the membrane skeleton to glycoproteins on platelet plasma membrane. *J. Biol. Chem.*, 260:11970–11977.

Fujiwara, I., S. Suetsugu, S. Uemura, T. Takenawa, and S. Ishiwata. 2002. Visualization and force measurement of branching by Arp2/3 complex and N-WASP in actin filament. *Biochem. Biophys. Res. Commun.*, 293:1550–5.

Galkin, V.E., A. Orlova, M.S. VanLoock, A. Shvetsov, E. Reisler, and E.H. Egelman. 2003. ADF/cofilin use an intrinsic mode of F-actin instability to disrupt actin filaments. *J. Cell Biol.*, 163:1057–66.

Gerbal, F., P. Chaikin, Y. Rabin, and J. Prost. 2000a. An elastic analysis of Listeria monocytogenes propulsion. *Biophys. J.*, 79:2259–75.

Gerbal, F., V. Laurent, A. Ott, M.F. Carlier, P. Chaikin, and J. Prost. 2000b. Measurement of the elasticity of the actin tail of Listeria monocytogenes. *Eur. Biophys. J.*, 29:134–40.

Ghosh, M., X. Song, G. Mouneimne, M. Sidani, D.S. Lawrence, and J.S. Condeelis. 2004. Cofilin promotes actin polymerization and defines the direction of cell motility. *Science*, 304:743–6.

Gibson, L., and M. Ashby. 1997. *Cellular solids: structure and properties*. Oxford: Pergamon Press.

Gibson, L.J., and M.F. Ashby. 1988. *Cellular solids: structure and properties*. Oxford, New York: Pergamon Press. ix, 357 pp.

Gimbrone, M.A. 1976. Culture of vascular endothelium. *Prog. Hemostasis Thromb.*, 3:1–28.

Gittes, F., and F.C. MacKintosh. 1998. Dynamic shear modulus of a semiflexible polymer network. *Phys. Rev. E.*, 58:R1241–R1244.

Gittes, F., B. Michey, J. Nettleton, and J. Howard. 1993. Flexural rigidity of microtubules and actin filaments measured from thermal fluctuations in shape. *J. Cell Biol.*, 120:923–34.

Giuliano, K.A., and D.L. Taylor. 1994. Fluorescent actin analogs with a high affinity for profilin in vitro exhibit an enhanced gradient of assembly in living cells. *J. Cell Biol.*, 124:971–83.

Glogauer, M., J. Hartwig, and T. Stossel. 2000. Two pathways through Cdc42 couple the N-formyl receptor to actin nucleation in permeabilized human neutrophils. *J. Cell Biol.*, 150:785–96.

Goldman, R.D., E. Lazarides, R. Pollack, and K. Weber. 1975. The distribution of actin in non-muscle cells. The use of actin antibody in the localization of actin within the microfilament bundles of mouse 3T3 cells. *Exp. Cell Res.*, 90:333–44.

Goldschmidt-Clermont, P.J., L.M. Machesky, S.K. Doberstein, and T.D. Pollard. 1991. Mechanism of the interaction of human platelet profilin with actin. *J. Cell Biol.*, 113:1081–9.

Gorlin, J., E. Henske, S. Warren, C. Kunst, M. D'Urso, G. Palmieri, J. Hartwig, G. Bruns, and D. Kwiatkowski. 1993. Actin-binding protein (ABP-280) filamin gene (FLN) maps telomeric to the colar vision locus (R/GCP) and centromeric to G6PD in Xq28. *Genomics*, 17:496–8.

Gorlin, J., R. Yamin, S. Egan, M. Stewart, T. Stossel, D. Kwiatkowski, and J. Hartwig. 1990. Human endothelial actin-binding protein (ABP-280, non-muscle filamin): a molecular leaf spring. *J. Cell Biol.*, 111:1089–1105.

Hart, M.C., Y.O. Korshunova, and J.A. Cooper. 1997. Vertebrates have conserved capping protein alpha isoforms with specific expression patterns. *Cell Motil. Cytoskeleton*, 38:120–32.

Hartwig, J. 1992. Mechanism of actin rearrangements mediating platelet activation. *J. Cell Biol.*, 118:1421–42.

Hartwig, J., and M. DeSisto. 1991. The cytoskeleton of the resting human blood platelet: Structure of the membrane skeleton and its attachment to actin filaments. *J. Cell Biol.*, 112:407–425.

Hartwig, J., and T. Stossel. 1981. The structure of actin-binding protein molecules in solution and interacting with actin filaments. *J. Mol. Biol.*, 145:563–81.

Hartwig, J.H., and T.P. Stossel. 1976. Interactions of actin, myosin, and an actin-binding protein of rabbit pulmonary macrophages. III. Effects of cytochalasin B. *J. Cell Biol.*, 71:295–303.

Hartwig, J.H., and T.P. Stossel. 1979. Cytochalasin B and the structure of actin gels. *J. Mol. Biol.*, 134:539–53.

Heiss, S., and J. Cooper. 1991. Regulation of CapZ, an actin capping protein of chicken muscle, by anionic phsopholipids. *Biochemistry*, 30:8753–58.

Helmke, B., A. Rosen, and P. Davies. 2003. Mapping mechanical strain of an endogenous cytoskeletal network in living endothelial cells. *Biophys. J.*, 84:2691–99.

Higashida, C., T. Miyoshi, A. Fujita, F. Oceguera-Yanez, J. Monypenny, Y. Andou, S. Narumiya, and N. Watanabe. 2004. Actin polymerization-driven molecular movement of mDia1 in living cells. *Science*, 303:2007–10.

Huxley, A.F. 1957. Muscle structure and theories of contraction. *Prog. Biophys. Biophys. Chem.*, 7:255–318.

Huxley, H., and J. Hanson. 1954. Changes in the cross-striations of muscle during contraction and stretch and their structural interpretation. *Nature*, 173:973–6.

Ichetovkin, I., W. Grant, and J. Condeelis. 2002. Cofilin produces newly polymerized actin filaments that are preferred for dendritic nucleation by the Arp2/3 complex. *Curr. Biol.*, 12:79–84.

Isenberg, G., U. Aebi, and T.D. Pollard. 1980. An actin-binding protein from Acanthamoeba regulates actin filament polymerization and interactions. *Nature*, 288:455–9.

Janmey, P.A., U. Euteneuer, P. Traub, and M. Schliwa. 1991. Viscoelastic properties of vimentin compared with other filamentous biopolymer networks. *J. Cell Biol.*, 113:155–60.

Janmey, P.A., S. Hvidt, J. Kas, D. Lerche, A. Maggs, E. Sackmann, M. Schliwa, and T.P. Stossel. 1994. The mechanical properties of actin gels. Elastic modulus and filament motions. *J. Biol. Chemistry*, 269:32 503–32.

Janmey, P.A., and T.P. Stossel. 1987. Modulation of gelsolin function by phosphatidylinositol 4,5-bisphosphate. *Nature*, 325:362–4.

Kamm, R.D., and M.R. Kaazempur-Mofrad. 2004. On the molecular basis for mechanotransduction. *Mech. Chem. Biosystems*, 1:201–209.

Kang, F., D.L. Purich, and F.S. Southwick. 1999. Profilin promotes barbed-end actin filament assembly without lowering the critical concentration. *J. Biol. Chem.*, 274:36963–72.

Kirschner, M.W. 1980. Implications of treadmilling for the stability and polarity of actin and tubulin polymers in vivo. *J. Cell Biol.*, 86:330–4.

Kothakota, S., T. Azuma, C. Reinhard, A. Klippel, J. Tang, K. Chu, T.J. McGarry, M.W. Kirschner, K. Koths, D.J. Kwiatkowski, and L.T. Williams. 1997. Caspase-3-generated fragment of gelsolin: effector of morphological change in apoptosis. *Science*, 278:294–8.

Kovar, D.R., J.R. Kuhn, A.L. Tichy, and T.D. Pollard. 2003. The fission yeast cytokinesis formin Cdc12p is a barbed end actin filament capping protein gated by profilin. *J. Cell Biol.*, 161:875–87.

Kovar, D.R., and T.D. Pollard. 2004. Insertional assembly of actin filament barbed ends in association with formins produces piconewton forces. *Proc. Natl. Acad. Sci. USA*, 101:14725–30.

Krakow, D., S. Robertson, L. King, T. Morgan, E. Sebald, C. Bertolotto, S. Wachsmann-Hagiu, D. Acuna, S. Shapiro, T. Takafuta, S. Aftimos, C. Kim, H. Firth, C. Steiner, V. Cormier-Daire, A. Superti-Furga, L. Bonafe, J. Graham Jr, A. Grix, C. Bacino, J. Allanson, M. Bialer, R. Lachman,

D. Rimoin, and D. Cohn. 2004. Mutations in the gene encoding filamin B disrupt vertebral segmentation, joint formation and skeletogenesis. *Nat. Genet.*, online.

Kuo, S., and J. McGrath. 2000. Steps and fluctuations of Listeria monocytogenes during actin-based motility. *Nature*, 407:1026–29.

Lazarides, E., and K. Weber. 1974. Actin antibody: the specific visualization of actin filaments in non-muscle cells. *Proc. Natl. Acad. Sci. USA*, 71:2268–72.

Le Clainche, C., D. Pantaloni, and M. F. Carlier. 2003. ATP hydrolysis on actin-related protein 2/3 complex causes debranching of dendritic actin arrays. *Proc. Natl. Acad. Sci. USA*, 100:6337–42.

Luby-Phelps, K., P.E. Castle, D.L. Taylor, and F. Lanni. 1987. Hindered diffusion of inert tracer particles in the cytoplasm of mouse 3T3 cells. *Proc. Natl. Acad. Sci. USA*, 84:4910–3.

Luby-Phelps, K., F. Lanni, and D.L. Taylor. 1985. Behavior of a fluorescent analogue of calmodulin in living 3T3 cells. *J. Cell Biol.*, 101:1245–56.

Machesky, L.M., S.J. Atkinson, C. Ampe, J. Vandekerckhove, and T.D. Pollard. 1994. Purification of a cortical complex containing two unconventional actins from Acanthamoeba by affinity chromatography on profilin-agarose. *J. Cell Biol.*, 127:107–15.

Machesky, L.M., R.D. Mullins, H.N. Higgs, D.A. Kaiser, L. Blanchoin, R.C. May, M.E. Hall, and T.D. Pollard. 1999. Scar, a WASp-related protein, activates nucleation of actin filaments by the Arp2/3 complex. *Proc. Natl. Acad. Sci. USA*, 96:3739–44.

Maciver, S.K., and A.G. Weeds. 1994. Actophorin preferentially binds monomeric ADP-actin over ATP-bound actin: consequences for cell locomotion. *FEBS Lett.*, 347:251–6.

Maciver, S.K., H.G. Zot, and T.D. Pollard. 1991. Characterization of actin filament severing by actophorin from Acanthamoeba castellanii. *J. Cell Biol.*, 115:1611–20.

MacKintosh, F.C., J. Kass, and P.A. Janmey. 1995. Elasticity of semiflexible biopolymer networks. *Phys. Rev. Lett.*, 75:4425–28.

McGough, A., B. Pope, W. Chiu, and A. Weeds. 1997. Cofilin changes the twist of F-actin: implications for actin filament dynamics and cellular function. *J. Cell Biol.*, 138:771–81.

McGrath, J., J. Hartwig, and S. Kuo. 2000a. The Mechanics of F-actin microenvironments depend on the chemistry of probing surfaces. *Biophys. J.*, 79:3258–66.

McGrath, J., E. Osborn, Y. Tardy, C. Dewey Jr., and J. Hartwig. 2000. Regulation of the actin cycle in vivo by actin filament severing. *Proc. Natl. Acad. Sci. USA*, 97:6532–37.

McGrath, J., Y. Tardy, C. Dewey Jr., J. Meister, and J. Hartwig. 1998a. Simultaneous measurements of actin filament turnover, filament fraction, and monomer diffusion in endothelial cells. *Biophys. J.*, 75:2070–78.

McGrath, J.L., N.J. Eungdamrong, C.I. Fisher, F. Peng, L. Mahadevan, T.J. Mitchison, and S.C. Kuo. 2003. The force-velocity relationship for the actin-based motility of Listeria monocytogenes. *Curr. Biol.*, 13:329–32.

McGrath, J.L., Y. Tardy, C.F. Dewey, Jr., J.J. Meister, and J.H. Hartwig. 1998b. Simultaneous measurements of actin filament turnover, filament fraction, and monomer diffusion in endothelial cells. *Biophys. J.*, 75:2070–8.

Mejillano, M.R., S. Kojima, D.A. Applewhite, F.B. Gertler, T.M. Svitkina, and G.G. Borisy. 2004. Lamellipodial versus filopodial mode of the actin nanomachinery: pivotal role of the filament barbed end. *Cell*, 118:363–73.

Mitchison, T.J., and L.P. Cramer. 1996. Actin-based cell motility and cell locomotion. *Cell*, 84:371–9.

Mogilner, A., and L. Edelstein-Keshet. 2002. Regulation of actin dynamics in rapidly moving cells: a quantitative analysis. *Biophys. J.*, 83:1237–58.

Mogilner, A., and G. Oster. 1996. Cell motility driven by actin polymerization. *Biophys. J.*, 71:3030–45.

Mogilner, A., and G. Oster. 2003. Force Generation by Actin Polymerization II: The Elastic Ratchet and Tethered Filaments. *Biophys. J.*, 84:1591–605.

Mommaerts, W.F. 1992. Who discovered actin? *Bioessays*, 14:57–9.

Mullins, R.D., J.A. Heuser, and T.D. Pollard. 1998. The interaction of Arp2/3 complex with actin: nucleation, high affinity pointed end capping, and formation of branching networks of filaments. *Proc. Natl. Acad. Sci. USA*, 95:6181–6.

Murphy, D.B., R.O. Gray, W.A. Grasser, and T.D. Pollard. 1988. Direct demonstration of actin filament annealing in vitro. *J. Cell Biol.*, 106:1947–54.

Nakamura, F., E. Osborn, P.A. Janmey, and T.P. Stossel. 2002. Comparison of filamin A-induced cross-linking and Arp2/3 complex-mediated branching on the mechanics of actin filaments. *J. Biol. Chem.*, 277:9148–54.

Oosawa, F., and S. Asakura. 1975. Thermodynamics of the polymerization of protein. New York: Academic Press.

Oosawa, F., and M. Kasai. 1962. A theory of linear and helical aggregations of macromolecules. *J. Mol. Biol.*, 4:10–21.

Osborn, E.A., A. Rabodzey, C.F. Dewey, Jr., and J.H. Hartwig. 2006. Endothelial actin cytoskeleton remodeling during mechanostimulation with fluid shear stress. *Am. J. Physiol. Cell Physiol.*, 290(2): C444–52.

Otterbein, L.R., P. Graceffa, and R. Dominguez. 2001. The crystal structure of uncomplexed actin in the ADP state. *Science*, 293:708–11.

Pantaloni, D., R. Boujemaa, D. Didry, P. Gounon, and M.F. Carlier. 2000. The Arp2/3 complex branches filament barbed ends: functional antagonism with capping proteins. *Nat. Cell Biol.*, 2:385–91.

Pantaloni, D., and M.F. Carlier. 1993. How profilin promotes actin filament assembly in the presence of thymosin beta 4. *Cell*, 75:1007–14.

Pelham, R.J., Jr., and Y. Wang. 1997. Cell locomotion and focal adhesions are regulated by substrate flexibility. *Proc. Natl. Acad. Sci. USA*, 94:13661–5.

Peskin, C.S., G.M. Odell, and G.F. Oster. 1993. Cellular motions and thermal fluctuations: the Brownian ratchet. *Biophys. J.*, 65:316–24.

Pollard, T.D. 1986. Rate constants for the reactions of ATP- and ADP-actin with the ends of actin filaments. *J. Cell Biol.*, 103:2747–54.

Pollard, T.D., and C.C. Beltzner. 2002. Structure and function of the Arp2/3 complex. *Curr. Opin. Struct. Biol.*, 12:768–74.

Pollard, T.D., L. Blanchoin, and R.D. Mullins. 2000. Molecular mechanisms controlling actin filament dynamics in nonmuscle cells. *Annu. Rev. Biophys. Biomol. Struct.*, 29:545–76.

Pollard, T.D., and G.G. Borisy. 2003. Cellular motility driven by assembly and disassembly of actin filaments. *Cell*, 112:453–65.

Pollard, T.D., and J.A. Cooper. 1984. Quantitative analysis of the effect of Acanthamoeba profilin on actin filament nucleation and elongation. *Biochemistry*, 23:6631–41.

Pollard, T.D., and S. Ito. 1970. Cytoplasmic filaments of Amoeba proteus. I. The role of filaments in consistency changes and movement. *J. Cell Biol.*, 46:267–89.

Ponti, A., M. Machacek, S.L. Gupton, C.M. Waterman-Storer, and G. Danuser. 2004. Two distinct actin networks drive the protrusion of migrating cells. *Science*, 305:1782–6.

Preston, T. M., C.A. King, and J.S. Hyams. 1990. The cytoskeleton and cell motility. Glasgow and London: Blackie 202 pp.

Pruyne, D., M. Evangelista, C. Yang, E. Bi, S. Zigmond, A. Bretscher, and C. Boone. 2002. Role of formins in actin assembly: nucleation and barbed-end association. *Science*, 297: 612–5.

Rickard, J.E., and P. Sheterline. 1986. Cytoplasmic concentrations of inorganic phosphate affect the critical concentration for assembly of actin in the presence of cytochalasin D or ADP. *J. Mol. Biol.*, 191:273–80.

Robinson, R.C., K. Turbedsky, D.A. Kaiser, J.B. Marchand, H.N. Higgs, S. Choe, and T.D. Pollard. 2001. Crystal structure of Arp2/3 complex. *Science*, 294:1679–84.

Safer, D., R. Golla, and V.T. Nachmias. 1990. Isolation of a 5-kilodalton actin-sequestering peptide from human blood platelets. *Proc. Natl. Acad. Sci. USA*, 87:2536–40.

Safer, D., and V.T. Nachmias. 1994. Beta thymosins as actin binding peptides. *Bioessays*, 16:590.

Safer, D., T.R. Sosnick, and M. Elzinga. 1997. Thymosin beta 4 binds actin in an extended conformation and contacts both the barbed and pointed ends. *Biochemistry*, 36:5806–16.

Satcher, R., C.F. Dewey Jr, and J. Hartwig. 1997. Mechanical remodeling of the endothelial surface and actin cytoskeleton induced by fluid flow. *Microcirculation*, 4:439–53.

Satcher, R.L., Jr., and C. F. Dewey, Jr. 1996. Theoretical estimates of mechanical properties of the endothelial cell cytoskeleton. *Biophys. J.*, 71:109–18.

Schafer, D.A., and J. A. Cooper. 1995. Control of actin assembly at filament ends. *Annu. Rev. Cell Dev. Biol.*, 11:497–518.

Schafer, D.A., P.B. Jennings, and J.A. Cooper. 1996. Dynamics of capping protein and actin assembly in vitro: uncapping barbed ends by polyphosphoinositides. *J. Cell Biol.*, 135:169–79.

Schafer, D.A., Y.O. Korshunova, T.A. Schroer, and J.A. Cooper. 1994. Differential localization and sequence analysis of capping protein beta-subunit isoforms of vertebrates. *J. Cell Biol.*, 127:453–65.

Schwartz, I.M., M. Ehrenberg, M. Bindschadler, and J.L. McGrath. 2004. The role of substrate curvature in actin pushing forces. *Cur. Biol.*, 14: 1094–8.

Schutt, C.E., J.C. Myslik, M.D. Rozycki, N.C. Goonesekere, and U. Lindberg. 1993. The structure of crystalline profilin-beta-actin. *Nature*, 365:810–6.

Selden, L.A., H.J. Kinosian, J.E. Estes, and L.C. Gershman. 1999. Impact of profilin on actin-bound nucleotide exchange and actin polymerization dynamics. *Biochemistry*, 38:2769–78.

Sept, D., J. Xu, T.D. Pollard, and J.A. McCammon. 1999. Annealing accounts for the length of actin filaments formed by spontaneous polymerization. *Biophys. J.*, 77:2911–9.

Sheen, V., Y. Feng, D. Graham, T. Takfuta, S. Shapiro, and C. Walsh. 2002. Filamin A and filamin B are co-expressed within neurons during periods of neuronal migration and can physically interact. *Human Molec. Genetics*, 11:2845–54.

Sprandio, J., S. Shapiro, P. Thiagarajan, and S. McCord. 1988. Cultured human umbilical vein endothelial cells contain a membrane glycoprotein immunologically related to platelet glycoprotein Ib. *Blood*, 71:234.

Stossel, T., J. Condeelis, L. Cooley, J. Hartwig, A. Noegel, M. Schleicher, and S. Shapiro. 2001. Filamins as integrators of cell mechanics and signalling. *Nature Reviews*, 2:138–145.

Stossel, T.P., and J.H. Hartwig. 1976. Interactions of actin, myosin, and a new actin-binding protein of rabbit pulmonary macrophages. II. Role in cytoplasmic movement and phagocytosis. *J. Cell Biol.*, 68:602–19.

Straub, F., and G. Feuer. 1950. Adenosinetriphosphate the functional group of actin. *Biochim. Biophys. Acta.*, 4:455–70.

Svitkina, T.M., and G.G. Borisy. 1999. Arp2/3 complex and actin depolymerizing factor/cofilin in dendritic organization and treadmilling of actin filament array in lamellipodia. *J. Cell Biol.*, 145:1009–26.

Svitkina, T.M., E.A. Bulanova, O.Y. Chaga, D.M. Vignjevic, S. Kojima, J.M. Vasiliev, and G.G. Borisy. 2003. Mechanism of filopodia initiation by reorganization of a dendritic network. *J. Cell Biol.*, 160:409–21.

Takafuta, T., G. Wu, G. Murphy, and S. Shapiro. 1998. Human β-filamin is a new protein that interacts with the cytoplasmic tail of glycoprotein 1bα. *J. Biol. Chem.*, 273:17531–538.

Tardy, Y., N. Resnick, T. Nagel, M.A. Gimbrone, Jr., and C.F. Dewey, Jr. 1997. Shear stress gradients remodel endothelial monolayers in vitro via a cell proliferation-migration-loss cycle. *Arterioscler Thomb. Vasc. Biol.*, 17(11):3102–6.

Theriot, J.A., and T.J. Mitchison. 1991. Actin microfilament dynamics in locomoting cells. *Nature*, 352:126–31.

Theriot, J.A., and T.J. Mitchison. 1992. Comparison of actin and cell surface dynamics in motile fibroblasts. *J. Cell Biol.*, 119:367–77.

Thompson, T., Y.-M. Chan, A. Hack, M. Brosius, M. Rajala, H. Lidov, E. McNally, S. Watkins, and L. Kunkel. 2000. Filamin 2 (FLN2): a muscle-specific sacroglycan interacting protein. *J. Cell Biol.*, 148:115–126.

Tilney, L.G., S. Hatano, H. Ishikawa, and M.S. Mooseker. 1973. The polymerization of actin: its role in the generation of the acrosomal process of certain echinoderm sperm. *J. Cell Biol.*, 59:109–26.

Tobacman, L.S., and E.D. Korn. 1982. The regulation of actin polymerization and the inhibition of monomeric actin ATPase activity by Acanthamoeba profilin. *J. Biol. Chem.*, 257:4166–70.

Tseng, P.C., and T.D. Pollard. 1982. Mechanism of action of Acanthamoeba profilin: demonstration of actin species specificity and regulation by micromolar concentrations of MgCl2. *J. Cell Biol.*, 94:213–8.

Vallotton, P., S.L. Gupton, C.M. Waterman-Storer, and G. Danuser. 2004. Simultaneous mapping of filamentous actin flow and turnover in migrating cells by quantitative fluorescent speckle microscopy. *Proc. Natl. Acad. Sci. USA*, 101:9660–5.

Vignjevic, D., D. Yarar, M.D. Welch, J. Peloquin, T. Svitkina, and G.G. Borisy. 2003. Formation of filopodia-like bundles in vitro from a dendritic network. *J. Cell Biol.*, 160:951–62.

Wanger, M., and A. Wegner. 1987. Binding of phosphate ions to actin. *Biochim. Biophys. Acta.*, 914:105–13.

Wasserman, S. 1998. FH proteins as cytoskeletal organizers. *Trends Cell Biol.*, 8:111–5.

Watanabe, N., T. Kato, A. Fujita, T. Ishizaki, and S. Narumiya. 1999. Cooperation between mDia1 and ROCK in Rho-induced actin reorganization. *Nat. Cell Biol.*, 1:136–43.

Watanabe, N., and T.J. Mitchison. 2002. Single-molecule speckle analysis of actin filament turnover in lamellipodia. *Science*, 295:1083–6.

Wegner, A., and J. Engel. 1975. Kinetics of the cooperative association of actin to actin filaments. *Biophys. Chem.*, 3:215–25.

Weihing, R.R. 1976. Cytochalasin B inhibits actin-related gelation of HeLa cell extracts. *J. Cell Biol.*, 71:303–7.

Wiesner, S., E. Helfer, D. Didry, G. Ducouret, F. Lafuma, M.F. Carlier, and D. Pantaloni. 2003. A biorimetic motility assay provides insight into the mechanism of actin-based motility. *J. Cell Biology*, 160:307–98

Wheelock, M.J., and K.A. Knudsen. 1991. Cadherins and associated proteins. *In Vivo*, 5:505–13.

Witke, W., A.H. Sharpe, J.H. Hartwig, T. Azuma, T.P. Stossel, and D.J. Kwiatkowski. 1995. Hemostatic, inflammatory, and fibroblast responses are blunted in mice lacking gelsolin. *Cell*, 81:41–51.

Woodrum, D.T., S.A. Rich, and T.D. Pollard. 1975. Evidence for biased bidirectional polymerization of actin filaments using heavy meromyosin prepared by an improved method. *J. Cell Biol.*, 67:231–7.

Yarmola, E.G., S. Parikh, and M.R. Bubb. 2001. Formation and implications of a ternary complex of profilin, thymosin beta 4, and actin. *J. Biol. Chem.*, 276:45555–63.

Yin, H.L., J.H. Hartwig, K. Maruyama, and T. P. Stossel. 1981. Ca^{2+} control of actin filament length. Effects of macrophage gelsolin on actin polymerization. *J. Biol. Chem.*, 256:9693–7.

Yin, H.L., and T.P. Stossel. 1979. Control of cytoplasmic actin gel-sol transformation by gelsolin, a calcium-dependent regulatory protein. *Nature*, 281:583–6.

Yin, H.L., K.S. Zaner, and T.P. Stossel. 1980. Ca^{2+} control of actin gelation. Interaction of gelsolin with actin filaments and regulation of actin gelation. *J. Biol. Chem.*, 255:9494–500.

Zicha, D., I.M. Dobbie, M.R. Holt, J. Monypenny, D.Y. Soong, C. Gray, and G.A. Dunn. 2003. Rapid actin transport during cell protrusion. *Science*, 300:142–5.

Zigmond, S.H., M. Evangelista, C. Boone, C. Yang, A.C. Dar, F. Sicheri, J. Forkey, and M. Pring. 2003. Formin leaky cap allows elongation in the presence of tight capping proteins. *Curr. Biol.*, 13:1820–3.

10 Active cellular protrusion: continuum theories and models

Marc Herant and Micah Dembo

ABSTRACT: This chapter attempts to develop a general perspective on the phenomenon of active protrusion by ameboid cells. Except in rare special cases, principles of mass and momentum conservation require that one consider at least two phases to explain the process of cellular protrusion. These phases are best identified with the cytosolic and cytoskeletal components of the cytoplasm. A continuum mechanical formalism of Reactive Interpenetrative Flows (RIF) is contructed. It is general enough to encompass within its framework a large range of theories of active cell deformation and movement. Most physically plausible theories of protrusion fall into one of two classes: protrusion driven by cytoskeletal self-interactions; or protrusion driven by cytoskeletal–membrane interactions. These are described within the RIF formalism. The RIF formalism is cast in a form that is amenable to computer simulations through standard numerical algorithms. An example is next given of the numerical study of the most elementary protrusive event possible: the formation of a single pseudopod by an isolated round cell. Some final thoughts are offered on the role of modeling in understanding cellular mechanical activity.

Cellular protrusion: the standard cartoon

Over the past decades, experiments have consistently demonstrated that active protrusion in animal cells is accompanied by a local increase in cytoskeletal density through active polymerization. A review of the evidence for this is beyond the scope of this chapter, but key points include the high concentration of filamentous actin observed by fluorescence or EM at the leading edge of growing protrusions, as well as the abrogation of protrusion by nearly any disruption of actin polymerization, such as that caused by cytochalasin. In addition, in most cases of ameboid free protrusion, it does not appear that molecular motors such as myosins are mandatory; instead, simple polymerizing activity seems sufficient.

This has led to a "standard cartoon" model of cellular protrusion that figures prominently in many reviews or textbooks of cell biology (see Fig. 10-1). In this picture, cytoskeletal monomers are added by polymerization at the leading edge and removed by depolymerization at the base of the protrusion. The free monomers then diffuse back to the front to be reincorporated in the cytoskeleton in a process that has been called "treadmilling."

Active cellular protrusion: continuum theories and models

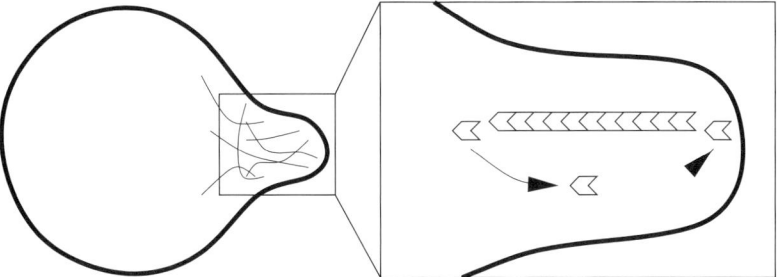

Fig. 10-1. The standard cartoon of cellular protrusion

As a general rule such cartoons are powerful but dangerous instruments of knowledge. On the one hand, this cartoon is a means of communicating a complex mechanism in a compact, readily intelligible way. It makes clear the following important points.

1. The cytoplasm is an *inhomogeneous* medium; its properties at the leading edge of a protrusion are different than they are elsewhere in the cell.
2. Cytoplasmic dynamics require that there exist a simultaneous forward flow of material in the form of water and cytoskeletal monomers and backward flow of cytoskeletal material in the form of filaments (this is the treadmilling).
3. There is a net flow of cytoplasmic volume into the protrusion (otherwise, the protrusion would not grow!).

On the other hand, the apparent simplicity of the cartoon glosses over important qualitative and quantitative issues that need to be addressed with rigor for the whole scheme to stand scrutiny as follows. First, momentum conservation (or force balance) is not evident; when extending a pseudopod, a cell has to exert an outward-directed force, if only against the cortical tension that tends to minimize the surface area (but there may be other opposing forces). By the principle of action and reaction, this requires some sort of bracing. Without a thrust plate supporting the outward-directed force, the pseudopod would collapse back into the main body of the cell. Second, the cartoon is inevitably silent on the mode of force production. The intuitive picture of a growing scaffolding pushing out a "tent" of membrane can be misleading and in any case gives little quantitative information on the protrusive force.

Even before going into details, it is clear from these issues that a theory of cellular protrusion must embody certain attributes: cytoplasmic inhomogeneity; differential flow of cytoskeleton and cytosol; and volume and momentum conservation. The Reactive Interpenetrative Flow (RIF) formalism described next is a natural choice to address these constraints in a rigorous, quantitative manner.

The RIF formalism

Consider the three principal structural components of animal cells:

- *The cortical membrane* defines the boundary of the cell by controlling (and often preventing) volume fluxes with the external world. It is furthermore highly

flexible, fluid, and virtually inextensible. Together, these three properties make it a good conductor of stress.
- *The cytoskeleton* resists deformation through viscoelastic properties and is able to generate active forces through molecular motors (for example myosin) or other interactions (such as electrostatic).
- *The cytosol* flows passively through the cytoskeleton; it is a medium for the propagation of signals. Furthermore, it can be converted to cytoskeleton via the polymerization of dissolved monomers (for example G-actin → F-actin) and vice versa.

Note that in general, many chemical entities will simultaneously contribute to the cytoskeletal and cytosolic phases, the best example being actin in filamentous or globular form. However, given biomolecules can be classified as being either part of the cytoskeleton, where they are able to transmit stresses, or part of the cytosol, where they are able to diffuse, but not both. Note also that this classification ignores membrane-enclosed organelles. In particular, the contribution to the mechanical properties of the cell of the largest of those, the nucleus, can occasionally be important.

If one makes the key assumption that at the mesoscopic scale, that is, at a scale small compared to the whole cell but large compared to individual molecules, the properties of the cell can be represented by continuous fields. Then the general framework of continuum mechanics can be applied to animal cells just as it is done with any other material. More specifically, one can write down a closed set of equations to compute the evolution of $\theta_n(\mathbf{x}, t)$ the network phase (cytoskeleton) volume fraction, $\theta_s(\mathbf{x}, t)$ the solvent phase (cytosol) volume fraction, $\mathbf{v}_n(\mathbf{x}, t)$ the network velocity field, $\mathbf{v}_s(\mathbf{x}, t)$ the solvent velocity field, where $\mathbf{x}$ is the position vector and t is the time. The scalar fields θ_n and θ_s and the vector fields $\mathbf{v}_n$ and $\mathbf{v}_s$ are thus defined on a simply connected, compact domain in Euclidian space that defines the physical extent of the cell. The boundary of this domain and constraints associated with it through boundary conditions are then to be a representation of the physical cortical membrane. This is the method of Reactive Interpenetrative Flows (see Dembo et al. 1986), in other words 'reactive' because it allows conversion of one phase into another, and 'interpenetrative' because it allows for different velocity fields for each phase.

The evolution equations for the quantities θ_n, θ_s, $\mathbf{v}_n$, and $\mathbf{v}_s$ are determined by the laws of mass and momentum conservation.

Mass conservation

The fact that we have only two phases (cytoskeleton and cytosol) mandates that the sum of their volume fractions is unity:

$$\theta_n + \theta_s = 1. \tag{10.1}$$

Net cytoplasmic volume flow is given by the sum of the flow of cytosolic volume and cytoskeletal volume, that is, $\mathbf{v} = \theta_n \mathbf{v}_n + \theta_s \mathbf{v}_s$. Because the cytoplasm is in a condensed phase, it is to an excellent approximation incompressible ($\nabla \cdot \mathbf{v} = 0$), so

Active cellular protrusion: continuum theories and models

that the incompressibility condition yields:

$$\nabla \cdot (\theta_n \mathbf{v}_n + \theta_s \mathbf{v}_s) = 0. \tag{10.2}$$

Finally, conservation of cytoskeleton implies that the rate of change of network concentration at a given point in space (Eulerian derivative) is the sum of an advective transport term describing the net inflow of network, and a source term $\mathcal{J}$, which represents the net rate of *in situ* cytoskeletal production by polymerization:

$$\frac{\partial \theta_n}{\partial t} = -\nabla \cdot (\theta_n \mathbf{v}_n) + \mathcal{J}. \tag{10.3}$$

Obviously $\mathcal{J}$ depends on a prescription for local chemical activity that needs to be provided separately. Eq. 10.3 naturally has a counterpart for the solvent

$$\frac{\partial \theta_s}{\partial t} = -\nabla \cdot (\theta_s \mathbf{v}_s) - \mathcal{J}, \tag{10.4}$$

which, when taken together with Eq. 10.1, unsurprisingly reduces to Eq. 10.2. As a result, only Eqs. 10.1, 10.2, and 10.3 are needed and Eq. 10.4 is redundant.

Momentum conservation

The momentum equations for the solvent and network phases are simplified by two observations. First, due to the small dimensions and velocities involved, the inertial terms are negligible in comparison with typical cellular forces. Second, the essentially aqueous nature of the cytosol implies that its characteristic viscosity is not very different from that of water (0.02 poise). Because this is much less than typical cytoplasmic viscosities (of order 1000 poise) we shall assume that the entire viscous stress is carried by the cytoskeletal (network) phase, while the cytosolic (solvent) phase remains approximately inviscid.

Within such an approximation, the only two forces that act on the solvent are pressure gradients and solvent-network drag – that is the drag force that occurs when the solvent moves through the network because of mismatched velocities. In the spirit of Darcy's law, the solvent momentum equation can then be written

$$-\theta_s \nabla P + \mathcal{H} \theta_s \theta_n (\mathbf{v}_n - \mathbf{v}_s) = 0. \tag{10.5}$$

P is the cytoplasmic pressure, and it is assumed that, as for the partial pressures of a mixture of gases, it is shared by the cytosolic and cytoskeletal phases according to concentrations (volume fractions). $\mathcal{H}$ is the solvent-network drag coefficient more familiar as the product $\theta_n \mathcal{H}$, which represents the hydraulic conductivity that appears in the usual form of Darcy's equation. Theoretical considerations (for example see Scheidegger 1960) as well as experiments on polymer networks (Tokita and Tanaka 1991) give estimates of $\mathcal{H}$ that lead to small drag forces compared to other forces acting within the cytoplasm, chief among them the cytoskeletal vicosity. This is not surprising, because $\mathcal{H}$ should be approximately proportional to the solvent viscosity, which is small compared to network viscosity.

The smallness of $\mathcal{H}$ in turn implies that pressure gradients will be small, or that the pressure is close to uniform inside the cell. Thus from the point of view of overall cell shape and motion that is determined by cytoskeletal dynamics (Eq. 10.6), the precise value $\mathcal{H}$ does not matter as long as it is sufficiently small. However, from the point of view of internal cytosolic flow, which can play an important transport role, the value of $\mathcal{H}$ does matter and pressure gradients, even though small, are not negligible.

It is in the network (cytoskeleton) momentum equation that the rich complexity of cell mechanics becomes evident. Aside from pressure gradients and solvent-network drag, the network is also subject to viscous, elastic, and interaction forces and the network momentum equation can therefore be written:

$$-\theta_n \nabla P - \mathcal{H}\theta_s\theta_n(\mathbf{v}_n - \mathbf{v}_s) + \nabla \cdot \left[\nu\left(\nabla \mathbf{v}_n + (\nabla \mathbf{v}_n)^T\right)\right] - \nabla \cdot \Psi = 0, \qquad (10.6)$$

Here, ν is the network viscosity and Ψ is the part of the network stress tensor remaining under static conditions. The latter can include interfilament interactions (such as contractility due to actin myosin assembly), filament-membrane interactions (such as Brownian ratchets), elastic forces due to deformations, and so forth.

Boundary conditions

These partial differential equations must of course be complemented by boundary conditions and this is where the characteristics of the plasma membrane come into play. From a mass conservation point of view, the key issue is that of permeability. In most circumstances, it seems reasonable that the membrane remains impermeable to the cytoskeleton (which may even be anchored to the membrane) so that therefore:

$$\mathbf{v}_M \cdot \mathbf{n} = \mathbf{v}_n \cdot \mathbf{n} \qquad (10.7)$$

where $\mathbf{v}_M$ is the velocity of the boundary and $\mathbf{n}$ is the outward normal unit vector. If we also assume that the membrane is impermeable to the cytosol (which appears to be true in some cases and not in others) we also have

$$\mathbf{v}_M \cdot \mathbf{n} = \mathbf{v}_s \cdot \mathbf{n}, \qquad (10.8)$$

but this condition can certainly be relaxed to allow a net volume flux through the boundary.

From a momentum conservation point of view, there are two main possibilities: either the boundary is constrained by interaction with a solid surface, as in the case of a cell/dish or cell/pipette interface; or it is free membrane bathed by an inviscid external medium. In the former case, the boundary condition boils down to constraints on the normal (and in the case of no slip, tangential) components of the velocities. In the latter case, the boundary condition amounts to a stress continuity requirement:

$$\nu\left(\nabla \mathbf{v}_n + (\nabla \mathbf{v}_n)^T\right) \cdot \mathbf{n} - \Psi \cdot \mathbf{n} - P\mathbf{n} = -2\gamma\kappa\mathbf{n} - P_{\text{ext}}\mathbf{n}, \qquad (10.9)$$

where γ is the surface tension, κ is the mean curvature of the membrane, and $\mathbf{n}$ is the outward normal to the membrane. The surface tension (and sometimes the permeability to the cytosol) is thus the main contribution of the cortical membrane to the governing evolution equations.

Active cellular protrusion: continuum theories and models

Constitutive equations

The mass and momentum conservation equations are cast in a very general framework that needs to be further constrained to provide closure of the system. These additional prescriptions (the constitutive equations) embody the biological specifications of the cell. For instance, it is likely that the viscosity ν will depend on the cytoskeletal density: a law such as

$$\nu = \nu_0 \theta_n \qquad (10.10)$$

prescribing a well-defined linear relation between viscosity and network concentration is such a constitutive relation. In principle, this could be verified empirically by investigating the rheology of the cytoplasm at various cytoskeletal concentrations. However, in general such experimental evidence is sparse and often difficult to interpret. One is thus usually reduced to educated guesses for the constitutive laws that govern $\mathcal{J}$ (the network formation or cytoskeletal polymerization rate), $\mathcal{H}$ (the resistance to solvent flow through the network), Ψ (the network stress due to elasticity and static interactions), ν (the network viscosity), and γ (the tension of the cortical membrane). Conversely, the main advantage of this formalism is that it is sufficiently general to accommodate most theories of protrusions: as we shall see below, it all depends on the proper adjustment of the constitutive equations.

Cytoskeletal theories of cellular protrusion

As has been touched on, it appears that polymerization of large amounts of actin in the vicinity of a membrane causes outward force and protrusion. It also appears that this phenomenon is probably not directly dependent on molecular motors such as myosins, especially as their contractile activity tends to 'pull' rather than 'push' the cytoskeleton. This has led to theories of protrusion such as the Brownian ratchet model, in which the free energy released by the addition of monomers to a filament is transduced to generate a pressure against a membrane that sterically interferes with the reaction. Without going into the specifics, however, it is clear that such cytoskeletal theories of protrusion can be categorized into two classes:

- Network–membrane interaction theories in which the cytoskeleton and the membrane repel one another through a force field. The classic Brownian ratchet model belongs to this class, as it relies on the hard-core potential of actin monomers pushing on the membrane (Peskin et al. 1993).
- Network–network interaction theories in which the cytoskeleton interacts with itself, resulting in a repulsive force. This could be due to electrostatic interactions (actin is negatively charged) or thermal agitation.

In what follows we shall formalize these classes of theory in a way that enables linkage to the RIF approach.

We wish to emphasize that we are only discussing *free* protrusions – that is, protrusions that emerge from the cell body without adhesion to an external substrate. When adhesion occurs, additional classes of theory become tenable, but these are not considered here.

Network–membrane interactions

The basic idea behind a network–membrane disjoining stress is that there exists a repulsive force between actin monomers (polymerized or unpolymerized) and the cortical membrane. For free (G-actin) subunits this has no dynamical consequences, as redistribution occurs freely in the cytosol. However, once subunits are sequestered into the cytoskeleton by polymerization, the repulsive force has dynamical consequences because it endows the cytoskeleton with a macroscopic stress. In other words, while free monomers *cannot* push back against the membrane, filaments *can* because they are braced by the entire inner cytoskeletal scaffolding of the cell (these concepts first came to the fore with the work of Hill and Kirschner, 1982).

To put these notions on a more formal footing, let us postulate a mean-field repulsive force between monomers (free or in a filament) and the cortical membrane that derives from the interaction potential $\psi^M(r)$ where r is the distance from the membrane and where we set the constant by requiring $\lim_{r \to \infty} \psi^M(r) = 0$. From equilibrium thermodynamics, we can relate the ratio of free monomer concentration far away ($[M^{\text{free}}(\infty)]$) and at distance r with the interaction potential as a Boltzmann factor:

$$\frac{[M^{\text{free}}(r)]}{[M^{\text{free}}(\infty)]} = \exp\left(-\frac{\psi^M(r)}{k_B T}\right). \tag{10.11}$$

Here we can think of $\psi^M(r)$ as the average work required to bring a monomer from infinity to distance r from the membrane.

At the same time we have the chemical reaction between the polymerized and unpolymerized state of the monomer.

$$M^{\text{free}} \rightleftharpoons M^{\text{bound}}. \tag{10.12}$$

Taking actin as an example where addition of free monomers into polymers occurs at the barbed ends of filaments, there exists a critical free-monomer density $[M^{\text{free}}_{\text{crit}}]$ above which reaction 10.12 is driven to the right and below which it is driven to the left.

We need to examine 10.12 in the light of 10.11. Let us assume that the membrane–monomer interaction potential $\psi^M(r)$ decreases monotonously with distance from the membrane, and further that it is infinite (or very large) at zero distance from the membrane (see Fig. 10-2; in other words, the monomer is excluded from the membrane by a hard-core potential). Then it should be clear that in a region near the membrane, $[M^{\text{free}}(r)]$ is smaller than $[M^{\text{free}}_{\text{crit}}]$, and hence that there cannot be any monomers added into polymers in this region. This region is labeled the 'gap' in Fig. 10-2. Advected polymers may appear in the gap, but unless they are stabilized by capping, they will tend to disassemble.

Further away from the membrane, $\psi^M(r)$ decreases to the point that $[M^{\text{free}}(r)]$ is smaller than $[M^{\text{free}}_{\text{crit}}]$ and this allows the elongation of polymers by driving free monomers from the free to the bound state. Assuming that the polymerization of free monomers is rapid, we then have a region where $[M^{\text{free}}(r)] \simeq [M^{\text{free}}_{\text{crit}}]$. This means that for the region of interest, that is, where polymerization is allowed to take place (outside the gap) but within the region where membrane–monomer interaction is still

Active cellular protrusion: continuum theories and models

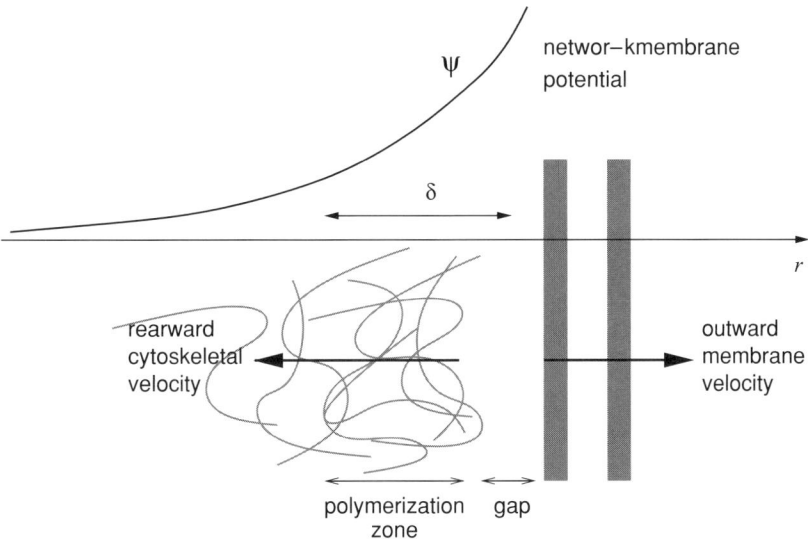

Fig. 10-2. Cartoon of network-membrane interactions.

significant, we have

$$\psi^M(r) = k_B T \ln \frac{[M^{\text{free}}(\infty)]}{[M^{\text{free}}_{\text{crit}}]}. \tag{10.13}$$

This region is labeled 'polymerization zone' in Fig. 10-2; within a network–membrane interaction model, it is that region that determines the dynamics of protrusion. Of note is that polymerization could take place further back, but that regions further to the rear do not have a dynamical impact because there, $\psi^M(r) \sim 0$.

The force exerted by the membrane on a monomer (free *or* bound) is given by $-\nabla \cdot \psi^M$, and by action and reaction, this is the opposite of the force $\mathbf{f}_M$ exterted by the monomer on the membrane. If δ is the range of the potential (we assume the gap region to be small), then we approximately have

$$\mathbf{f}_M = \frac{k_B T}{\delta} \ln \frac{[M^{\text{free}}(\infty)]}{[M^{\text{free}}_{\text{crit}}]} \mathbf{n}. \tag{10.14}$$

If we further assume that most of the monomers in the vicinity of the membrane are sequestered in the cytoskeletal phase, the total force-per-unit area of the membrane is given by the number of monomers within range times the force per unit monomer:

$$\Psi^M = \frac{\theta_n}{V_M} \delta \times \frac{k_B T}{\delta} \ln \frac{[M^{\text{free}}(\infty)]}{[M^{\text{free}}_{\text{crit}}]} \mathbf{n} = \frac{\theta_n}{V_M} k_B T \ln \frac{[M^{\text{free}}(\infty)]}{[M^{\text{free}}_{\text{crit}}]} \mathbf{n}, \tag{10.15}$$

where V_M is the volume of a monomer. For actin networks, $V_M = (4\pi/3)(2.7 \times 10^{-7} \text{cm})^3$, so that for normal temperature conditions

$$\Psi^M = 5 \times 10^5 \theta_n \ln \frac{[M^{\text{free}}(\infty)]}{[M^{\text{free}}_{\text{crit}}]} \text{ dyn cm}^{-2}. \tag{10.16}$$

Natural logarithms of even very large numbers are seldom more than 10, and maximal volume fractions of cytoskeleton are ~2% (see Hartwig and Shevlin, 1986). The maximum disjoining stress at the membrane/cytoskeleton interface is thus of order 10^5 dyn cm^{-2} (10^4 Pa, 100 cm H$_2$O).

Network dynamics near the membrane
One should be cautious not to assimilate the force-per-unit area of membrane or stress given in Eq. 10.16 to a protrusive force; in general the net outward force at the membrane as could theoretically be measured with a constraining spring would be considerably less. There are two reasons for this.

1. Imperfect cytoskeletal bracing against backflow: unless there exists some sort of mechanism to brace the cytoskeleton near the membrane, it will simply slide back, negating any outward force. In general, such bracing is expected to be provided by the viscoelastic properties of the cytoskeleton interior to the boundary layer, which transports stress to whatever is bracing the cell (in other words, the substratum).
2. Imperfect cytoskeletal decoupling from the membrane: if the cytoskeleton is somehow anchored to the membrane and cannot flow back, the stress F_M is simply carried by the anchors and no outward force results.

We will assume here that the second condition is appropriately fulfilled, although experimental evidence is sometimes contradictory (for example, *Listeria* actin tails are attached to the *Listeria*, Gerbal et al., 2000). Let us further assume that the counteracting force to rear flow is provided by interior cytoskeletal viscosity. In that case, simple dimensional analysis (which can be made more rigorous, see Herant et al., 2003) gives

$$\Delta v \simeq \frac{\Psi^M \delta}{\nu} \qquad (10.17)$$

where ν is the viscosity and Δv is the velocity change near the membrane (see Fig. 10-2). In the case of perfect bracing and no external opposing force, the protrusive velocity is then Δv, but in the general case it will be less. In the limit of a stalled protrusion, the protrusive velocity is zero while the retrograde flow of cytoskeleton is $-\Delta v$. Note that within this picture, the important parameter is not the magnitude of Ψ^M but rather its product with the range of the membrane–network interaction $\Psi^M \delta$.

Finally, an interesting finding is that, if one assumes that the constitutive laws for the viscosity and for the network–membrane force have the same dependence on the cytoskeletal concentration, for example $\nu = \nu_0 \theta_n$, $\Psi^M \propto \Psi_0^M \theta_n$, then one notices that Eq. 10.17 implies that Δv is independent of the network concentration near the membrane. Using numerical values for ν_0 (6×10^6 poise, see Herant et al., 2003) and Ψ_0^M (upper limit 5×10^6 dyn cm^{-2}, see Eq. 10.16), one gets $\Delta v < \delta$ s^{-1}. Typical velocities Δv at the leading edge are at least 10 nm s^{-1} and may range to as high as 0.5 μm s^{-1}, (see Theriot and Mitchison 1992), which means that δ, the characteristic range of interaction, must be greater than 10 nm and even reach 0.5 μm.

Active cellular protrusion: continuum theories and models

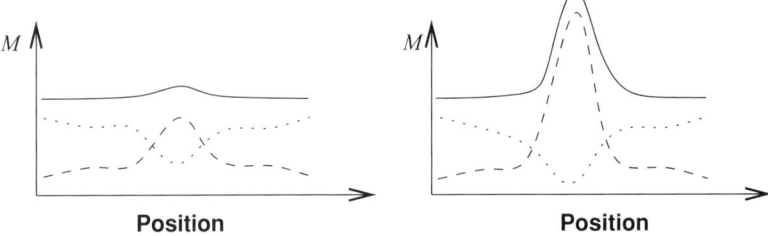

Fig. 10-3. Monomer concentration vs. position (dots – free monomers; dashes – bound monomers; solid – total monomers).

Special cases of network–membrane interaction: polymerization force, brownian and motor ratchets

In recent years, the concept of "polymerization force" has gained traction as a putative explanation for cellular protrusion. While the concept is often used in a vague manner, it has been formalized on more solid physical grounds in two interesting models: the Brownian ratchet model (see Mogilner and Oster, 1996) and the clamped filament elongation model (Dickinson and Purich, 2002). In essence, these models are special cases of network–membrane interactions in that they rely on the hard-core repulsion between monomers and membrane as an interaction potential. For instance, Eq. 10.14 is identical to that commonly given for the force produced by a Brownian ratchet (see for example Howard, 2001), where δ is taken to be the incremental lengthening of the polymer by addition of a monomer.

Network–network interactions

The basic idea behind a network swelling stress is similar to that of a membrane–network disjoining stress and we shall follow the approach of the previous section. One begins with the assumption that there exists a repulsive force between actin monomers, free or bound. Again, for free (G-actin) subunits, this has no dynamical consequences, as redistribution occurs freely in the cytosol. However, once subunits are sequestered into the cytoskeleton by polymerization, the repulsive force has dynamical consequences because it endows the cytoskeleton with a macroscopic stress. Under these conditions, one can intuitively perceive how the energy of the chemical process of polymerization can be transformed into expansion work.

Fig. 10-3 illustrates this principle. On the left, polymerizing activity is moderate; wherever there is a mild excess of monomers bound in filaments, free monomers are driven out by the repulsive interaction. The end result is that the total concentration of monomers (bound and unbound) varies little.

The amount of variation is determined by the relative magnitudes of the time scale of thermal-driven diffusion of monomer into regions of excess polymer $\tau_{\text{diff}} = l^2/D$ and the time scale of force-driven diffusion out of regions of excess polymer $\tau_{\text{force}} = l^2/[D(\psi/k_B T)]$ (where l is the length scale of the region, D the monomeric diffusion coefficient, and ψ the repulsive potential, and where we have used Einstein's relation between viscosity and diffusion coefficients).

If, however, polymerization becomes intense – for instance due to uncapping of filaments – it can drive the number of monomers sequestered in filaments above that of the background, and we have the case depicted on the right of Fig. 10-3. The local free monomer concentration goes near zero but cannot be negative, so that the total monomer concentration has a significant bump. A monomer moving around therefore sees a repulsive potential force in the bump that is higher than the baseline away from the bump.

Our formal development here will approximately parallel that of the network–membrane interaction problem in the previous section. We assume a pairwise repulsive potential force (potential ϕ) between actin monomers either free or part of a filament. The total force exerted on a monomer M is therefore the result of a sum on all other monomers M_i:

$$\sum_i \mathbf{F}_{MM_i} = -\sum_i \frac{\partial \phi_{MM_i}}{\partial \mathbf{r}_{MM_i}} \simeq -\nabla \psi^n. \tag{10.18}$$

Here ψ^n is the part of the potential that comes from fixed monomers sequestered in filaments. Generally, this will be the dominant contribution wherever network is highly concentrated, as free monomers will naturally diffuse away and lower the free monomer concentration (see Fig. 10-3). Just like the case of membrane–network interactions, we have a Boltzmann factor,

$$\frac{[M^{\text{free}}(\text{bump})]}{[M^{\text{free}}(\text{baseline})]} = \exp\left(-\frac{\psi^n(\text{bump})}{k_B T}\right) \tag{10.19}$$

where we have assumed that the network concentration outside of the "bump" is so low as to make $\psi^n(\text{bump}) \gg \psi^n(\text{baseline}) \simeq 0$. In this picture, $\psi^n(\text{bump})$ is the work of bringing one free monomer from baseline concentration into bump (Fig. 10-3).

Again, we have the chemical reaction

$$M^{\text{free}} \rightleftharpoons M^{\text{bound}}, \tag{10.20}$$

which goes to the right if $[M^{\text{free}}(\mathbf{x})] > [M^{\text{free}}_{\text{crit}}]$ and to the left if $[M^{\text{free}}(\mathbf{x})] < [M^{\text{free}}_{\text{crit}}]$ where, as before, $[M^{\text{free}}_{\text{crit}}]$ is the critical free monomeric concentration above which free ends of polymers are lengthened by monomer addition.

In regions of very high network density, ψ^n is large, and by Eq. 10.19 this leads to $[M^{\text{free}}] < [M^{\text{free}}_{\text{crit}}]$. This drives Eq. 10.20 to the left (depolymerization) so that one can say that such a region cannot be created by polymerization (although external or contractile forces could compress network above such a threshold). It is therefore clear that the highest network concentration achievable by chemical network polymerization is that for which $[M^{\text{free}}(\mathbf{x})] \simeq [M^{\text{free}}_{\text{crit}}]$, and that therefore, the highest achievable network repulsive stress per monomer is:

$$\psi^n = k_B T \ln \frac{[M^{\text{free}}(\text{baseline})]}{[M^{\text{free}}_{\text{crit}}]}. \tag{10.21}$$

Eq. 10.21 makes it clear that in general, the stress contribution by polymerizing a single monomer into the bump is at most of order $10 \, k_B T$. Let V_M be the volume of

Active cellular protrusion: continuum theories and models

a monomer, we have

$$\Psi^n(\text{bump}) \simeq 10 \, k_B T \frac{\theta_n(\text{bump})}{V_M}, \qquad (10.22)$$

where Ψ^n is the spatially averaged stress density that will contribute in the network momentum equation (Eq. 10.6). Note here that a proper treatment would integrate Ψ^n from a network concentration of 0 to θ_n, including a threshold effect (see Fig. 10-3) and would also yield a θ_n^2 term instead of θ_n. Considering the other uncertainties of the problem, we have preferred to sacrifice accuracy to simplicity.

If we use typical biological numbers, for example actin, $V_M = (4\pi/3)(2.7 \times 10^{-7} \text{cm})^3$, and maximal volume fractions of cytoskeleton $\theta_n \sim 2\%$, we obtain a maximum swelling stress of the cytoskeleton of order 10^5 dyn cm^{-2} (10^4 Pa, 100 cm H_2O), the same as the maximum stress for the network–membrane interaction.

Network dynamics with swelling

It is obvious that network–network repulsion will tend to smooth out nonuniform cytoskletal distributions by expansion of overdense regions into underdense regions. It is thus of interest to look at the dynamics of a clump of network in a low-density environment. If the principal retarding force is taken to be viscosity of the network itself, simple dimensional analysis shows that the time scale of expansion is:

$$\tau = \frac{\nu}{\Psi^n}. \qquad (10.23)$$

Of note is that there is no intrinsic scale to the problem; instead it is set by the length scale of the clump of overdense network d, and so the characteristic velocity is $v = d/\tau$. In addition, if both ν and Ψ^n have the same functional dependence (for example, linear) on the network volume fraction θ_n, then τ becomes independent of θ_n. We have used and provided experimental support for $\nu = \nu_0 \theta_n$ where $\nu_0 = 6 \times 10^6$ poise (Herant et al. 2003) and from the calculation above, the swelling stress $\Psi^n = \Psi_0^n \theta_n$ is such that at most $\Psi_0^n \simeq 5 \times 10^6$ dyn cm^{-2} (there is no reason it cannot be much less), which gives a minimal expansion time scale of order one second.

Other theories of protrusion

For the sake of completeness, we would like to briefly touch upon alternative theories of protrusion that are not currently in vogue because they do not fit in the standard cartoon (Fig. 10-1). Of note is that all these theories are also amenable to modeling within the RIF formalism and that it is out of concern for keeping this survey within a reasonable length that we do not pursue a quantitative analysis for each of these models (see Fig. 10-4).

Hydrostatic pressure protrusion. Hydrostatic pressure-driven protrusion is a venerable model (Mast, 1926) that probably has applicability in a limited number of circumstances. The basic idea is that an increase in internal pressure (presumably due to contractile activity somewhere in the cell) leads to bulging and protrusion of the membrane. Hidden behind the apparent simplicity of the concept are a number of

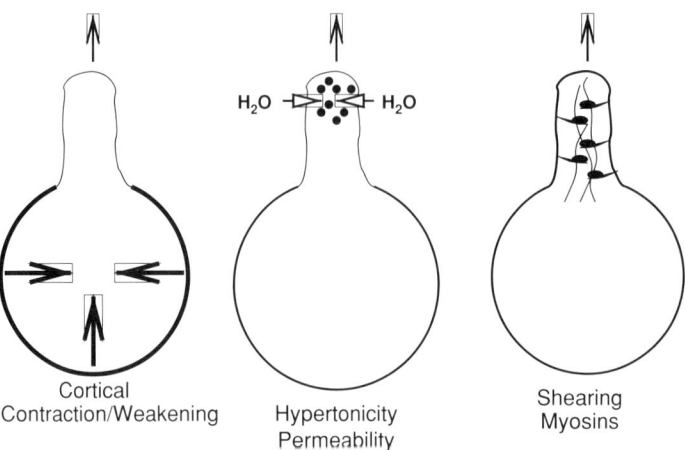

Fig. 10-4. Three alternative theories of cellular protrusion.

factors that merit consideration. In order for there to be a *local* protrusion of membrane, either the increase of hydrostatic pressure has to be local to the region, or the compliance of the cortical membrane must increase locally. Except in very large cells (for example, *Amoeba proteus*), or in compartmentalized organisms, the former condition is difficult to realize. This is because the hydraulic resistance to cytosolic flow through the cytoskeleton is typically extremely small. As a result, any local pressure excess tends to be quickly erased by solvent flow. The only way around this constraint is to have a large distance leading to a large hydraulic resistance, or an isolated compartment in which pressure can be locally increased without driving cytosolic flow to other parts of the cell.

The alternative is that of a local weakening of the cortical tension driving a local Marangoni-type of flow. Again there are some difficulties with such a mechanism. Recall that cell cortical tension is the result of contributions from the tension of the plasma membrane and from the cytoskeletal cortex underlying the membrane. It is unlikely that membrane tension can be lowered locally, because the massless, fluid-mosaic nature of the plasma membrane should make it a good conductor of stress that equilibrates surface tension rapidly around the cell. (Experimental and modeling evidence hint at the membrane tension being a global property of the cell, Raucher and Sheetz, 1999; Herant et al., 2003.) It is, however, possible for the cytoskeletal cortex to be locally weakened (see Lee et al., 1997). In a regime in which it carries substantial (tensile) stresses, such a weakening may result in pseudopod formation.

Hypertonic protrusion. Although as far as we are aware they have not been the subject of much recent study, models based on osmotic swelling accompanied by modified membrane permeability were once actively pursued, especially in the context of the extension of the acrosomal process of the *Thyone* sperm, a setting that may not have general applicability to ameboid protrusion (Oster and Perelson, 1987). Here the basic idea is that through the action of locally activated severing enzymes, osmotic tension

Active cellular protrusion: continuum theories and models

near the tip of a protrusion is increased, and that the permeability of the plasma membrane is sufficient to allow significant inflow from the extracellular environment. It is then possible for the volume of the protrusion to grow, and the filling with polymerized cytoskeleton is considered to take place after the fact for structural reasons. There are many problems with such a model – possibly explaining why it has lain fallow for a while, now. To mention just one problem, to the extent we are able to ascertain it, it appears that cellular volume does not change appreciably during the extension of protrusions, even big ones.

Shearing motor protrusion. At a most elementary level, myosin motors are shearing motors in the sense that they actively slide filaments parallel to one another. If one imagines a reasonably stiff assembly of cytoskeletal filaments perpendicular to the plasma membrane, it is conceivable that this structure could be driven out by a shearing motor mechanism as shown in Fig. 10-4 (Condeelis, 1993). Once of some popularity, this model seems more or less abandoned in the context of free protrusions, probably because there is evidence that molecular motors are not required – although we would caution that in our view, the case is far from being experimentally airtight.

Numerical implementation of the RIF formalism

A detailed discussion of the numerical strategies that can be used to solve the evolution equations is beyond the scope of this chapter. We will therefore limit ourselves to a brief outline of the methodology. Because it is well suited to free-boundary problems in the low-Reynolds-number limit, we use a Galerkin finite element scheme implemented in two spatial dimensions (for problems with cylindrical symmetry) on a mesh of quadrilateral cells. Grid and mass advection are implemented following cannonical methods that can be found in standard texts and reviews.

Briefly, the calculation is advanced over a time-step Δt determined by the Courant condition or other fast time scale of the dynamics. We evolve over Δt by means of sequential operations (this is operator splitting):

1. We advect the mesh boundary according to the network flow and then reposition mesh nodes for optimal resolution while preserving mesh topology, boundaries, and interfacial surfaces (Knupp and Steiberg, 1994).
2. We advect mass from the old mesh positions to the new mesh using a general Eulerian-Lagrangian scheme with upwind interpolation (Rash and Williamson, 1990).
3. We use constitutive laws to compute necessary quantities such as viscosities and surface tensions.
4. Finally, the momentum equations and the incompressibility condition together with the applicable boundary conditions are discretized using the Galerkin approach and the resulting system is solved for the pressure, network velocity, and solvent velocity on the advected mesh using an Uzawa style iteration (Temam, 1979).

Because of the multiphase nature of the flow (one has to solve the triplet $\mathbf{v}_n$, $\mathbf{v}_s$, and P rather than just for $\mathbf{v}$ and P), this last step requires some modifications from the usual treatment.

By adding the solvent and network momentum equation (Eq. 10.5 and 10.6) together, $\mathbf{v}_s$ can be eliminated to obtain a "bulk" cytoplasmic momentum equation:

$$-\nabla P + \nabla \cdot \left[\nu(\nabla \mathbf{v}_n + (\nabla \mathbf{v}_n)^T)\right] - \nabla \cdot \Psi = 0. \tag{10.24}$$

This is to be complemented with the appropriate boundary condition for stress across the membrane, which, in usual situations, looks like

$$\left[\nu(\nabla \mathbf{v}_n + (\nabla \mathbf{v}_n)^T)\right] \cdot \mathbf{n} - P\mathbf{n} - \Psi \cdot \mathbf{n} = -P_{\text{ext}}\mathbf{n} - 2\gamma\kappa\mathbf{n}, \tag{10.25}$$

where P_{ext} is the external pressure, γ and κ the surface tension and mean curvature.

The solvent momentum equation (Eq. 10.5) gives an expression for $\mathbf{v}_s$:

$$\mathbf{v}_s = \mathbf{v}_n - \frac{\nabla P}{\mathcal{H}\theta_n} \tag{10.26}$$

which can then be substituted in the incompressibility condition to yield:

$$\nabla \cdot \left(\mathbf{v}_n - \frac{1}{\mathcal{H}}\frac{\theta_s}{\theta_n}\nabla P\right) = 0. \tag{10.27}$$

In situations where there is zero membrane permeability (in other words, no transmembrane solvent flow) the boundary condition simplifies to:

$$\nabla P = 0. \tag{10.28}$$

Following the standard Uzawa method, an initial guess for the pressure field allows the computation of the network velocity field by Eq. 10.24. This velocity field can then be used to update the pressure field by Eq. 10.27, and so on through iterations between the two equations. Once the network velocity field $\mathbf{v}_n$ and pressure field P have converged to a self-consistent solution, the solvent velocity field $\mathbf{v}_s$ can be trivially extracted through the use of Eq. 10.26 with automatic enforcement of the incompressibility condition.

An example of cellular protrusion

Probably the simplest possible case of cellular protrusion is the emergence of a single pseudopod from an isolated, initially round nonadherent cell. This configuration has the advantages of a simple geometry with two-dimensional cylindrical symmetry and of avoiding potential confounding factors that appear when the mechanics of adhesion is involved. The formation of such pseudopods has been studied in individual neutrophils by Zhelev et al. (2004) and modeled in some detail by Herant et al. (2003). Here we present a simplified version of this process as a pedagogical introduction to the basic principles that are involved.

We will describe behaviors under both a cytoskeleton-membrane-repulsion and cytoskeleton-swelling model. In both cases, we make the following assumptions:

- Initial condition is that of a round cell of diameter 8.5 μm.
- Cortical tension is $\gamma = 0.025$ dyn cm^{-1}.

Active cellular protrusion: continuum theories and models

- Equilibrium network (cytoskeleton) volume fraction is $\theta_0 = 0.1\%$ (Watts and Howard 1993) everywhere inside the cell *except*:
  - Network fraction over a spherical patch of cell surface (diameter 1 µm) is fixed to one percent.
  - Away from the frontal patch, network fraction evolves to its equilibrium according to first order kinetics with time-scale $\tau_n = 20$ seconds, that is $d\theta_n/dt = (\theta_0 - \theta_n)/\tau_n$.
  - Network viscosity is given by $\nu_0 \theta_n$ where $\nu_0 = 6 \times 10^6$ poise so that the baseline interior viscosity of the cell is $\nu = \nu_0 \theta_0 = 6000$ poise.

These values are reasonable approximations of the characteristics of human neutrophils (see Herant et al., 2003).

The idea is that in both the repulsion and the swelling models, the excess cytoskeleton at the activated patch of cortex will drive the formation of a pseudopod. Bracing by the viscous interior of the cell allows outward protrusion against the restoring force of the cortical tension that tends to sphericize the cell.

Protrusion driven by membrane–cytoskeleton repulsion

The mechanics of membrane–cytoskeleton repulsion as a driver of protrusion is straightforward as follows: (i) a region with increased cytoskeletal density appears due to enhanced polymerization at the leading edge of the pseudopod. (ii) Due to repulsion from the cortical membrane, cytoskeleton flows into the cell while cytosol is sucked forward. (iii) Cytoskeletal flow into the cell is opposed by viscous resistance of the underlying cytoplasm. By action and reaction, this leads to bulging out of the membrane, eventually creating and lengthening a pseudopod.

As has been pointed out, the quantity that matters for the dynamics of membrane–cytoskeleton interaction is the product of the membrane–cytoskeleton interaction stress density Ψ^M with the range δ of the interaction. In numerical simulations that encompass the whole cell, it is not practical to try to accurately model the details of the cytoskeletal dynamics near the membrane as depicted in Fig. 10-2. Instead, the stress contribution to the network momentum equation (Eq. 10.6) is integrated over the range δ of network–membrane interaction. So, if we allow a generous average of $6\,k_B T$ interaction energy per actin monomer within range of the membrane, we have

$$\Psi^M_{ij} = 3 \times 10^6 \theta_n n_i n_j \, \text{dyn cm}^{-2}, \qquad (10.29)$$

where **n** is the unit normal vector to the membrane. If we would like the pseudopod to extend a few µm within about a minute, we need the flow velocity of cytoskeleton to be of order $0.2\,\mu\text{m s}^{-1}$. By Eqs. 10.17 and 10.29, this means that $\delta \sim 0.5$ µm.

Fig. 10-5 shows the outline of the cell together with the network velocity field sixty seconds into a two-dimensional cylindrical symmetry simulation starting from a round cell. The flow clearly changes direction right at the pseudopod surface with the disjoining membrane–cytoskeleton force driving out the membrane while causing a centripetal flow of cytoskeleton into the cell.

The observant reader has probably noted that in this simulation, the frontal surface of the pseudopod has a curiously uniform velocity. This is because the additional

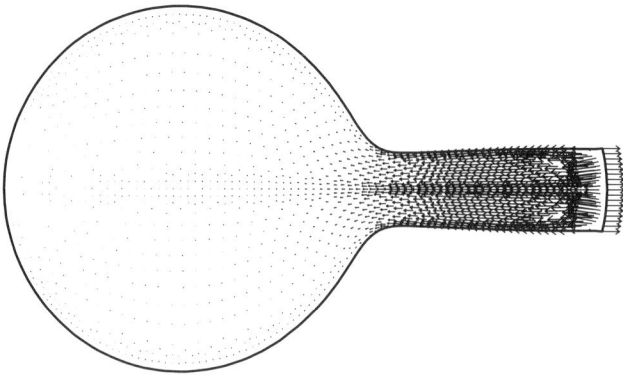

Fig. 10-5. Simulation Protrusion driven by membrane – cytoskeleton repulsion. Length of the pseudopod is ~6 μm, pseudopod velocities are ~0.1 μm s^{-1}.

velocity constraint of "no-shear" was introduced at the tip of the pseudopod. In the absence of such a prescription, membrane–network repulsion can drive severe rippling instabilities whereby the membrane grows folds that rapidly lengthen in a way reminiscent of the Rayleigh-Taylor instability (heavy over light fluid). From a biological point of view, the "no-shear" condition that we had to impose could point to rigid transverse cross-linking of filaments at the leading edge that prevent sliding. This instability also provides a potential mechanism for the formation of microvilli.

Protrusion driven by cytoskeletal swelling

The mechanics of cytoskeletal swelling as a driver of protrusion is similar to that of cytoskeletal–membrane repulsion. A region with increased cytoskeletal density appears due to enhanced polymerization in a compartment close to the leading edge of the pseudopod. Under repulsive self-interaction, the dense clump of cytoskeleton swells, drawing in cytosol like an expanding sponge draws water. This is the volume flow that accounts for the growth of the pseudopod. Finally, expansion into the cell is balanced by viscous resistance of the underlying cytoplasm. This braces outward expansion, which – while it does not have to work against viscosity of an external medium (assumed to be inviscid) – does have to work against cortical tension as new cellular area is created by the growth of the pseudopod.

Following the discussion of network swelling and in parallel with the simulation of cytoskeletal–membrane repulsion, we assume a contribution of 6 $k_B T$ per actin monomer to the cytoskeletal swelling stress density

$$\Psi^n = 3 \times 10^6 \theta_n \text{ dyn cm}^{-2}. \tag{10.30}$$

Fig. 10-6 shows the outline of the cell together with the network velocity field sixty seconds into the simulation, starting from a round cell. As in the simulation of protrusion driven by cytoskeletal–membrane repulsion, there is an obvious

Active cellular protrusion: continuum theories and models

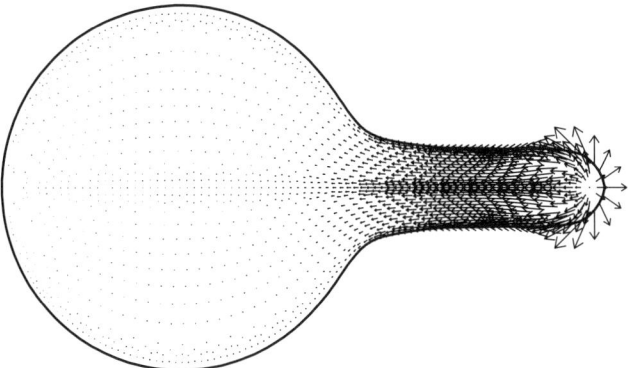

Fig. 10-6. Protrusion driven by cytoskeletal swelling. Length of the pseudopod is ~ 6 μm, pseudopod velocities are ~ 0.1 μm s^{-1}.

retrograde flow of cytoskeleton. However, one will note that the center of expansion is slightly behind the leading edge, so that there also exists a small region of forward cytoskeletal flow at the front of the pseudopod. This is in contrast with protrusion from cytoskeletal–membrane repulsion where the flow of cytoskeleton is retrograde all the way to the tip of the pseudopod and then changes to forward motion at the membrane. (In reality, this occurs at the "gap" region depicted in Fig. 10-2.)

Discussion

It is reassuring that choices of physico-biological parameters (such as viscosity or interaction energies) that seem by and large reasonable can lead to a plausible mechanism for the protrusion of a pseudopod. The morphology and extension velocity of the pseudopod is within range of the experimental data obtained for the neutrophil (Zhelev et al., 1996). In addition, retrograde flow similar to that observed in lamellipodial protrusion (for example, see Theriot and Mitchison, 1992) is evident. There are, however, some difficulties with each model. In the case of the membrane–cytoskeleton repulsion model, the range of interaction had to be set to 0.5 μm to obtain sufficient elongation velocity. This is a large distance compared to a monomer, but small compared to the persistence length of a filament (for instance, see Kovar and Pollard, 2004) which implies that the stress field would most likely have to be stored in the large scale strain energy from deformation of the cytoskeleton (Herant et al., 2006). In the case of the swelling model, it is difficult to build up clumps of cytoskeleton of significant density and size (as are observed) without prompt smoothing and dissipation by expansion. Although none of those caveats are model killers in the sense that more complicated explanations can be invoked to rescue them, they hint at more mechanical complexity than is presented in the rather simple approach followed here.

A separate but nonetheless important issue is that of discrimination between models. As Figs. 10-5 and 10-6 make clear, each model can produce a similar pseudopod.

However, the cytoskeleton velocity field is different in a way that is intrinsic to the mechanism of protrusion invoked in each case. In the swelling model, there exists a stagnation point (or center of expansion) within the cell across which the cytoskeletal flow goes from centripetal to centrifugal. The distance of the stagnation point from the leading edge depends on the force resisting protrusion (for instance in the limit of a hard wall, the stagnation point is at the membrane), but in certain conditions (see Herant et al., 2003 for an example), it can lie far back from the leading edge. This is not the case in the membrane–cytoskeleton repulsion model, where the stagnation point lies in the "gap" region (Fig. 10-2) very near the membrane. Such distinction may be a way to experimentally establish the distinction between swelling and repulsion models of protrusion.

Finally, it would have been possible to compute models of alternative protrusion theories such as those in Fig. 10.4 using the RIF formalism. We do not do so here because these other theories are not currently favored.

Conclusions

For reasons of brevity, we have not included computational examples of protrusion models that fail to produce cell-like behavior. The fact that these tend to make only rare appearances in the literature can be misleading: in reality good models with good parameters are needles in haystacks (for example, see Drury and Dembo, 2001). This is mostly due to the fact that the spaces of inputs (model specifications, such as viscosity) and outputs (model behavior, such as cellular shape) that need to be explored are both extremely large, as expected in complex biological systems.

In general, our experience has been that in hindsight, it is not difficult to discover why it is that a particular model with a particular choice of parameters does not lead to the behavior one was hoping for. On the other hand, we have also found that a priori expectations about the kind of results a given model will produce are rarely precisely matched by a numerical simulation. As a result, the process of constructing mechanical models of cell behavior is one of iteration during which one refines one's experience and intuition by running many numerical experiments. It is also our experience that this process often leads to deeper insights about the fundamental mechanisms at play in certain cellular events such as protrusions.

With this in mind, caution is in order when dealing with cartoon descriptions of the mechanics of living cells. As compelling as a mechanical diagram in the form of rods, ropes, pulleys, and motors working together may be, the actual implementation of such models within a quantitative model may not lead to what one was expecting! Thus, in a Popperian way, one of the principal benefits of a rigorous framework for cell mechanics is the ability to falsify invalid theories (Popper, 1968). This is why, although such frameworks can be difficult to develop and use, they are necessary to reach a real understanding of the mechanics of living cells.

We thank Juliet Lee as well as the editors for comments on an early version of this chapter. This work was supported by Whitaker biomedical engineering research grant RG-02-0714 to MH and NIH grant RO1-GM 61806 to MD.

References

Condeelis, J. (1993). Life at the Leading Edge: The Formation of Cell Protrusions. *Annu. Rev. Cell Biol.*, 9, 411–44.

Dembo, M., & Harlow, F. (1986). Cell Motion, Contractile Networks, and the Physics of Interpenetrating Reactive Flow. *Biophys. J.*, 50, 109–21.

Dickinson R., & Purich, D. L. (2002). Clamped-filament Elongation Model for Actin-Based Motors. *Biophys. J.*, 82, 605–17.

Drury, J. L., & Dembo M. (2001). Aspiration of Human Neutrophils: Effects of Shear Thinning and Cortical Dissipation. *Biophys. J.*, 81, 3166–77.

Gerbal, F., Laurent V., Ott A., Carlier M., Chaikin P., & Prost J. (2000). Measurement of the Elasticity of the Actin Tail of *Listeria monocytogenes*. *Eur. Biophys. J.*, 29, 134–40.

Herant, M., Marganski, W. A., & Dembo, M. (2003). The Mechanics of Neutrophils: Synthetic Modeling of Three Experiments. *Biophys. J.*, 84, 3389–413.

Herant, M., Heinrich, V., & Dembo, M. (2005). Mechanics of Neutrophil Phagocytosis: Behavior of the Cortical Tension. *J. Cell Sci.*, 118, 1789–97.

Herant, M., Heinrich, V., & Dembo, M. (2006). Mechanics of Neutrophil Phagocytosis: Experiments and Quantitative Models. *J. Cell Sci.*, 119, in press.

Hartwig, J. H., & Shevlin, P. (1986). The Architecture of Actin Filament and the Ultrastructural Location of Axtin-binding Protein in the Periphery of Lung Macrophages. *J. Cell Biol.*, 103, 1007–20.

Hill, T. L., & Kirschner, M. W. (1982). Subunit treadmilling of microtubules or actin in the presence of cellular barriers: possible conversion of chemical free energy into mechanical work. *Proc. Nat. Acad. Sci.*, 79, 490–4.

Howard, J. 2001. *Mechanics of Motor Proteins and the Cytoskeleton.* Sunderland, MA: Sinauer Associates.

Kovar, D. R., & Pollard, T. D. (2004). Insertional Assembly of Actin Filament Barbed End in Association with Foweins Produces Piconewton Forces. *Proc. Nat. Acad. Sci.*, 101, 14725–14730.

Knupp, P., & Steinberg, S. 1994. *Fundamentals of Grid Generation.* Boca Raton, FL: CRC Press.

Lee, E., Shelden, E. A., & Knecht, D. A. (1997). Changes in Actin Filament Organization during Pseudopod Formation. *Exp. Cell Res.*, 235, 295–9.

Masayuki, T., & Tanaka, T. (1991). Friction coefficient of polymer networks of gels. *J. Chem. Phys.*, 95, 4613–9.

Mast, S. O. (1926). Structure, Movement, Locomotion, and Stimulation in Amoeba. *J. Morphology and Physiology*, 41, 347–425.

Mogilner, A., & Oster, G. (1996). Cell Motility Driven by Actin Polymerization. *Biophys. J.*, 71, 3030–48.

Oster, G. F., & Perelson, A. S. (1987). The Physics of Cell Motility. *J. Cell Sci. Suppl.*, 8, 35–54.

Peskin, C. S., Odell, G. M., & Oster, G. F. (1993). Cellular Motions and Thermal Fluctuations: the Brownian Ratchet. *Biophys. J.*, 65, 316–24.

Popper, K. 1968. *The Logic of Scientific Discovery.* New York, NY: Harper & Row.

Rash, P. J., & Williamson, D. L. (1990). On Shape-Preserving Interpolation and Semi-Lagrangian Transport. *SIAM J Statist Computation*, 11, 656–87.

Raucher, D., & Sheetz, M. P. (1999). Characteristics of a Membrane Reservoir Buffering Membrane Tension. *Biophys. J.*, 77, 1992–2002.

Scheidegger, A. E. 1960. *The Physics of Flow through Porous Media.* New York: MacMillan.

Temam, R. 1979. *Navier-Stokes Equations Theory and Numerical Analysis.* Studies in Mathematics and Its Applications. Amsterdam: North-Holland.

Theriot, J. A., & Mitchison, T. J. (1992). Comparison of Actin and Cell Surface Dynamics in Motile Fibroblasts. *J. Cell Biol.*, 118, 367–77.

Tokita, M., & Tanaha, T. (1991). Friction Coefficient of Polymer Networks of Gels. *J. Chem. Phys.*, 95, 4613–19.

Watts R. G., & Howard, T. H. (1993). Mechanisms for Actin Reorganization in Chemotactic Factor-Activated Polymorphonuclear Leukocytes. *Blood*, 81, 2750–7.

Zhelev, D. V, Alteraifi, A. M., & Hochmuth, R. M. (1996). F-actin Network Formation in Tethers and in Pseudopods Stimulated by Chemoattractant. *Cell Motil. Cytoskeleton*, 35, 331–44.

Zhelev, D. V., Alteraifi, A. M., & Chodniewicz, D. (2004). Controlled Pseudopod Extension of Human Neutrophils Stimulated with Different Chemoatlractants. *Biophys. J.*, 87, 688–95.

11 Summary

Mohammad R.K. Mofrad and Roger D. Kamm

The primary objective of this book was to bring together various points of view regarding cell mechanics, contrasting and comparing these diverse perspectives. This final short chapter summarizes the various models discussed in an attempt to identify commonalities as well as any irreconcilable differences.

A wide range of computational and phenomenological models were described for cytoskeletal mechanics, ranging from continuum models for cell deformation and mechanical stress to actin-filament-based models for cell motility. A concise review was also presented (Chapter 2) of numerous experimental techniques, which typically aim to quantify cytoskeletal mechanics by exerting some sort of perturbation on the cell and examining its static and dynamic responses. These experimental observations along with computational approaches have given rise to several often contradictory theories for describing the mechanics of living cells, modeling the cytoskeleton as a simple mechanical elastic, viscoelastic, or poroviscoelastic continuum, a porous gel, a soft glassy material, or a tensegrity (tension integrity) network incorporating discrete structural elements that bear compression.

With such remarkable disparity among these models, largely due to the diversity of scales and biomechanical issues of interest, it may appear to the uninitiated that various authors are describing entirely different cells. Yet depending on the test conditions or length scales of interest, identical cells may be viewed so differently as either a continuum or as a discrete collection of structural elements.

Experimental data are accumulating, and promising methods have been proposed to describe cell rheology. While there has been some convergence toward a range of values for the cytoskeletal shear modulus, the range remains large, spanning several orders of magnitude. This suggests either disparities in the measurement methods, considerable variability between cells or between cell types, or differences in the methods employed to interpret the data. A unique aspect of cellular mechanics is that active as well as passive characteristics need to be considered.

A variety of different approaches have been described to simulate cell or cytoskeletal stiffness. Likely there is not a single "correct" model; rather, one model may prove useful under certain circumstances while another model may be better suited in others. In part, the model of choice will depend on the length scale of interest. Cells contain a microarchitecture comprised of filaments ranging down to $\sim$10 nm in diameter with separation distances of ~100 nm. When considering whole-cell deformations, a continuum description may be appropriate; when the force probe is on the scale of an AFM tip, then details of the filament organization are almost certainly critical.

As a practical matter, it is important to determine what constitutive law best fits the observed structural behavior. While a linear elastic or even linear viscoelastic material description is sufficient to mimic certain observations, other more complex descriptions will almost certainly be needed to encompass a range of excitation frequencies and large deformations. These are just now being identified. There seems to be a growing consensus that the constitutive behavior of a cell corresponds to that of a soft glassy material (see Chapter 3) even though the underlying basis for this behavior is not yet clearly understood. Albeit lacking a fundamental understanding, these measurements and the relative simplicity of the generalized form that they exhibit provide at least two critical new insights. First is that the cell responds as though the relaxation times are distributed according to a power law, suggesting many relaxation processes at low frequencies but progressively fewer as frequency is increased. Second, cytoskeletal stiffness and friction or viscosity are interrelated, in that the same underlying principles likely govern both. Both stiffness and friction appear to be governed by a single parameter, the "effective temperature," that reflects the extent to which the material is solid-like or fluid-like. Bursac et al. (2005) speculate that this might relate to a process in which the cytoskeleton is "trapped" in a collection of energy wells but can occasionally "escape" utilizing, for example, either thermal or chemical (such as, ATP-derived) energy. In this connection, the effective temperature might be a measure of molecular agitation, reflecting the relative ability to escape. As appealing as these ideas might be, however, they remain to be fully demonstrated, and so remain intriguing speculation.

As Chapter 2 points out, while there appears to be some degree of convergence regarding the values and frequency dependence of viscoelastic parameters for the cytoskeleton, the results obtained remain somewhat dependent upon the method used to probe the cell. In publications as recent as this past year, values for cytoskeletal stiffness ranging from ~20 Pa (Tseng et al., 2004) to 1.1 MPa (Marquez et al., 2005) have appeared, and the bases for these discrepancies still need to be resolved. In particular, as most (but not all) of the data on which the soft glassy material model is based are obtained from one measurement method (magnetic twisting cytometry), one still needs to exercise caution in making broad generalizations.

While some of the models appear quite disparate, there are some significant similarities. The cellular solids and biopolymer (Chapter 8) theories differ in terms of how the individual elements in the structure resist deformation, with the cellular solids model considering these to be beams subject to bending, and the biopolymer theory treating them as entropic chains that lose configurational entropy as the material is stretched. Recent studies (Gardel et al., 2004) are beginning to reconcile these differences and, perhaps not surprisingly, are finding that both descriptions might apply depending upon the concentrations of actin and cross-linkers and the state of stress in the material. Neither of these models, however, can be readily connected to the observed behavior as a soft glassy material.

Another microstructural model is based on the concept of tensegrity (Chapter 6), and is most closely related to the cellular solids model in that cytoskeletal structure

Summary

is defined by an interconnected network of elastic elements. The key distinction here, though, is that stiffness is conferred not by the stiffness of the individual elements, but rather primarily by the baseline stresses they support. These stresses are imposed either by cell adhesions to extracellular structures or by internal members such as microtubules that are in compression. In either case, the elastic properties of the elements take on secondary importance, provided they are sufficiently stiff to undergo relatively small changes in length under normal stress.

In a sense, the continuum descriptions (Chapters 4, 5 and 10) are for the most part independent of behavior at the microstructural level, and simply make use of constitutive laws that can either be based on experiments or derived directly from one of these microstructural models. Consequently, while the continuum models can be useful in describing how deformations or stresses distribute throughout the cell, they provide no information on the deformations at the microscale (that is, within the individual elements of the matrix), and are entirely dependent on information contained in the constitutive relation.

Although this one text could not possibly capture all the work being done on cell mechanics in that it represents a broad spectrum of these activities, it should immediately become clear that one fruitful direction for future research is in the modeling of dynamic processes – cell migration, phagocytosis and division. In fact, with only a few exceptions (notably the work described in Chapters 7, 9, and 10) the cell is treated as a traditional engineering material, meaning one with properties that are time invariant. Cells, on the other hand, are highly dynamic in that their cytoskeletal structures are constantly changing in response to a variety of external stimuli including, especially, external forces. Consequently, each time we probe a cell to measure its mechanical properties, we may alter those same properties. One exception to this statement is the use of the Brownian motions of intracellular structures to infer stiffness, but these measurements are still being refined; as currently implemented, they are subject to some degree of uncertainty. Still, this represents an important direction for research, and we are sure to see refinements and wider use of these nonintrusive methods in the future.

While advances in cell mechanics are considerable, many open questions still remain. Mechanotransduction, the active response of living cells to mechanical signals remains an active area of investigation. It is well known that living cells respond to mechanical stimulation in a variety of ways that affect nearly every aspect of their function. Such responses can range from changes in cell morphology to activation of signaling cascades to changes in cell phenotype. Mechanotransduction is an essential function of the cell, controlling its growth, proliferation, protein synthesis, and gene expression.

Despite the wide relevance and central importance of mechanically induced cellular response, the mechanisms for sensation and transduction of mechanical stimuli into biochemical signals are still largely unknown. What we know is that living cells can sense mechanical stimuli. Forces applied to a cell or physical cues from the extracellular environment can elicit a wide range of biochemical responses that affect the cell's phenotype in health and disease.

Various mechanisms have been proposed to explain this phenomenon. They include: changes in membrane fluidity that act to increase receptor mobility and lead to enhanced receptor clustering and signal initiation (Haidekker et al., 2000); stretch-activated ion channels (Hamil and Martinac, 2001); mechanical disruption of microtubules (Odde and co-workers, 1999); and forced deformations within the nucleus (Maniotis et al., 1997). Constrained autocrine signaling, whereby the strength of autocrine signaling is regulated by changes in the volume of extracellular compartments into which the receptor ligands are shed, is yet another mechanism (Tschumperlin et al., 2004). Changing this volume by mechanical deformation of the tissues can increase the level of autocrine signaling.

Finally, others have proposed conformational changes in intracellular proteins along the force-transmission pathway, connecting the extracellular matrix with the cytoskeleton through focal adhesions, as the main mechanotransduction mechanism (see Kamm and Kaazempur-Mofrad, 2004 for a review). In particular, the hypothesis that links mechanotransduction phenomena to mechanically induced alterations in the molecular conformation of proteins has been gaining increased support. For example, certain proteins that reside in 'closed' conformation can be mechanically triggered to reveal cryptic binding sites. Similarly, small conformational changes may also change binding affinity or enzyme activity. For example, when protein binding occurs through hydrophobic site interactions, a conformational change could modify this function and potentially disrupt it totally. Force transmission from the extracellular matrix to the cell interior occurs through a chain of proteins, located in the focal adhesion sites, that are comprised of an integrin–extracellular matrix protein bond (primarily vitronectin and fibronectin), integrin-associated proteins on the intracellular side (paxillin, talin, vinculin, and others), and proteins linking the focal adhesion complex to the cytoskeleton. Stresses transmitted through adhesion receptors and distributed throughout the cell could cause conformational changes in individual force-transmitting proteins, any of which would be a candidate for force transduction into a biochemical signal. The process by which changes in protein conformation give rise to protein clustering at a focal adhesion or initiate intracellular signaling, however, remains largely unknown (Geiger et al., 2001).

External stresses imposed on the cell are transmitted through the cytoskeleton to remote locations within the cell. To understand these stress distributions requires knowledge of cytoskeletal rheology, as governed by the structural proteins, actin filaments, microtubules, and intermediate filaments. For example, a simplified picture can be painted of the cytoskeletal rheology that is limited to actin filaments and actin cross-linking proteins living in a dynamic equilibrium. These cross-links constantly form and unbind at rates that are largely influenced by the forces borne by the individual molecules. Cytoskeletal rheology would then be determined at the molecular scale by the mechanics and binding kinetics of the actin cross-linking proteins, as well as by the actin matrix itself (Gardel et al., 2004). To understand the phenomena related to mechanotransduction in living cells and their cytoskeletal rheology, the mechanics and chemistry of single molecules that form the biological signaling pathways that act in concert with the mechanics must be examined.

Another largely open question in the field of cytoskeletal mechanics is related to the cell migration and motility that is essential in a variety of biological processes in health (such as embryonic development, angiogenesis, and wound healing) or disease (as in cancer metastasis). As discussed in Chapter 9 and 10, the process of cell motility or

Summary

migration consists of several steps involving multiple mechanobiological signals and events starting with the leading edge protrusion, formation of new adhesion plaques at the front edge, followed by contraction of the cell and the release of adhesions at the rear (see Li et al., 2005 for a recent review). A host of mechanical and biochemical factors, namely extracellular matrix cues, chemoattractant concentration gradients, substrate rigidity, and other mechanical signals, influence these processes. Many unanswered questions remain in understanding the signaling molecules that play a key role in cell migration, and how they are regulated both in time and 3D space. It is largely unknown how a cell actively controls the traction force at a focal adhesion or how this force varies with time during the cell migration.

To understand the mechanobiology of the cell requires a multiscale/multiphysics view of how externally applied stresses or traction forces are transmitted through focal adhesion receptors and distributed throughout the cell, leading subsequently to conformational changes that occur in individual mechanosensing proteins that in turn lead to increased enzymatic activity or altered binding affinities. This presents both a challenge and an opportunity for further research into the intrinsically coupled mechanobiological phenomena that eventually determine the macroscopic behavior and function of the cell.

Because no one method has emerged as clearly superior in describing the mechanics and biology of the cell across all cell types and physical conditions, this might reflect the need for new approaches and ideas. We hope that this monograph has inspired new researchers with fresh ideas directed toward that goal. Perhaps the biggest question that still remains is whether it is at all possible to construct a single model that is universally applicable and can be used to describe all types of cell mechanical behavior.

References

Bursac P, Lenormand G, Fabry B, Oliver M, Weitz DA, Viasnoff V, Butler JP, Fredberg JJ. "Cytoskeletal remodelling and slow dynamics in the living cell." *Nat. Mater.*, 4(7):557–61, 2005.

Gardel ML, Shin JH, MacKintosh FC, Mahadevan L, Matsudaira PA, Weitz DA. "Scaling of F-actin network rheology to probe single filament elasticity and dynamics." *Phys. Rev. Lett.*, 93(18):188102, 2004.

Geiger B, Bershadsky A, Pankov R, Yamada KM. "Transmembrane crosstalk between the extracellular matrix–cytoskeleton crosstalk." *Nat. Rev. Mol. Cell Biol.*, 2(11): 793–805, 2001.

Haidekker MA, L'Heureux N, Frangos JA. "Fluid shear stress increases membrane fluidity in endothelial cells: a study with DCVJ fluorescence." *Am. J. Physiol. Heart Circ. Physiol.*, 278(4):H1401–6, 2000.

Hamill OP, Martinac C. "Molecular basis of mechanotransduction in living cells." *Physiol. Rev.*, 81: 685–740, 2001.

Kamm RD and Kaazempur-Mofrad MR. "On the molecular basis for mechanotransduction," Mech. Chem. Biosystems, Vol. 1(3):201–210, 2004.

Li S, Guan JL, Chien S. "Biochemistry and biomechanics of cell motility." *Annu. Rev. Biomed. Eng.*, 7:105–50, 2005.

Maniotis AJ, Chen CS, Ingber DE. "Demonstration of mechanical connections between integrins, cytoskeletal filaments and nucleoplasm that stabilize nuclear structure." *Proc. Natl. Acad. Sci. USA*, 94:849–54, 1997.

Marquez JP, Genin GM, Zahalak GI, Elson EL. "The relationship between cell and tissue strain in three-dimensional bio-artificial tissues." *Biophys. J.*, 88(2):778–89, 2005.

Odde DJ, Ma L, Briggs AH, DeMarco A, Kirschner MW. "Microtubule bending and breaking in living fibroblast cells." *J. Cell Sci.*, 112:3283–8, 1999.

Tseng Y, Lee J S, Kole, T P, Jiang, I, and Wirtz, D. "Micro-organization and visco-elasticity of the interphase nucleus revealed by particle nanotracking." *J. Cell Sci.*, 117, 2159–67, 2004.

Tschumperlin DJ, Dai G, Maly IV, Kikuchi T, Laiho LH, McVittie AK, Haley KJ, Lilly CM, So PT, Lauffenburger DA, Kamm RD, Drazen JM. "Mechanotransduction through growth-factor shedding into the extracellular space." *Nature*, 429: 83–6, 2004.

Index

Acetylated low-density lipoprotein, 58
Acoustic microscopy, 42–43
Actin
   cortical, 174
   depolymerization (*See* depolymerization of actin)
   diffusion coefficient of, 192
   dynamics, 180–188
   elastic modulus of, 178
   mechanics of, 177–180
   role of, 170–176
   swelling stress values, 215
F-Actin
   about, 153, 154, 172
   acoustic signals in, 42–43
   architecture, regulation of, 175
   and Arp 2/3, 186
   ATP hydrolysis rate, 182
   in cytosol conversion, 206
   discovery of, 181
   persistence length of, 155, 167
   shear response in, 163
   stiffness measurements of, 161, 165
G-Actin
   about, 153, 154
   assembly of, 182, 183–185
   in cell protrusion, 210
   cellular concentration of, 192
   discovery of, 181
   and profilin, 185, 188
   recycling of, 192, 193
Actin binding proteins (ABPs)
   about, 11, 14
   in actin dynamics regulation, 183–188
Actin cortex, 153
Actin filaments
   about, 12, 13, 73, 107, 153, 154, 171, 175
   in AFM, 25
   assembly of, 185–187 (*See also* polymerization of actin)
   bending of, 161, 178
   in cell migration, 75, 108, 210
   and cell stiffness, 10, 225
   crosslinking in, 166, 228
   and cytochalisin-D, 57, 111
   DLS measurement of, 40
   in the ECM, 173
   in load transfer, 114
   in muscle contraction, 108, 142–143, 145–146
   myosin, interactions with, 117, 124
   persistence length of, 38, 107, 178
   polymerization (*See* polymerization of actin)
   power-law relationships, 58
   and prestress, 104
   structure of, 179
   in tensegrity models, 106, 107, 117
   Young's modulus, 172
α-actinin
   about, 11
   in focal adhesions, 14
Actin monomers
   about, 171
   in cell protrusion, 210
   diffusion coefficient of, 171, 191
   recycling of, 187, 191–193
Actin networks
   cell membranes, dynamics near, 212
   compression in, 179
   crosslinking in, 166, 171, 228
   cytochalisin-D and, 57, 111, 181, 204
   differential modulus in, 166
   elastic moduli in, 167, 178, 227
   shear stress in, 179
   stiffness in, 178
   swelling stress and, 179, 215
   volume equations, 211
   Young's modulus in, 179
Action at a distance effect, 106, 110–112, 114
Actoclampin model, 193–194
ADF/cofilin, 183–185, 191

Index

Adherent cells
   cortical membrane model for, 115, 118
   protrusion models for, 218–222
Adhesion defined, 4
Affine network model
   about, 116, 164
   shear modulus of, 164
   uniform strain in, 166
Affine strain approximation, 117
Affinity in phase transitions, 138
AFM. *See* atomic force microscopy (AFM)
Aggregate modulus in osteoarthritis, 95
Airways, pulmonary, 8
Alginate capsules, acoustic signals in, 42–43
Amoeba, 75, 204
Anisotropy in cell behavior, 98
Apoptosis, 188
Arc length, short filaments, 157
Area expansion modulus defined, 10
Arp 2/3 complex
   in actin filament assembly, 185–187
   in cell locomotion, 189
   in the ECM, 173
Arterial wall cells, 3
Atherosclerosis, 171
Atomic force microscopy (AFM)
   and continuum models, 81
   in deformation measurement, 25–26, 43, 72, 73
   shear/loss moduli and, 58
   strain distributions, 74
Autocrine signaling, 228

Bead diameter, 116
Bead displacements, measurement of, 112
Bead pulling, cell response to, 118
Bead rotation and prestress, 115. *See also* rotation
Bending
   3D, 156
   of actin filaments, 161, 178
   energy in force extension relationships, 157
   moduli, 13, 155
   stiffness (*See* stiffness, bending)
   temperature and, 154–156
   and wavelength, 161
Binding affinity, changes in, 6, 228, 229
Biopolymers. *See also* polymers
   about, 154–162
   modeling, 226
   persistence lengths of, 155
   stiffness measurements of, 161, 165
Biphasic elastic formulations and creep response, 90
Biphasic models
   of cell mechanics, 85–88
   continuum viscoelastic, 84
   linear isotropic, equations for, 87
   time dependencies and, 88
   of viscoelasticity, 87–88, 98

Biphasic properties, measurement of, 92–94, 95, 96
Boltzmann's constant, 155, 210, 214
Bone
   deformation measurement in, 28–29
   mechanosensing in, 19
   repair of, 28
   stiffness in, 8
   stress, response to, 6
Bouchard's theory of glasses, 63
Boundary conditions
   in cell protrusion, 208
   in continuum models, 75, 76
   in force extension relationships, 157
Bowen's theory of incompressible mixtures, 86
Boyle van't Hoff relation and osmolality, 89
Brownian motion
   in cytoskeletal filaments, 160–162, 167, 227
   and mean square displacements, 161
   measurement of, 37, 81, 154, 155
Brownian ratchet model, 193–194, 209, 213
Buckling. *See* microtubules, buckling in

Cable-and-strut tensegrity model
   about, 106, 118, 121
   tension-compression synergy in, 108
   of viscoelasticity, 123
Cadherins, binding affinity in, 9, 228
Calcium, release of, 113, 138
Cancer cells
   melanoma, deformation measurement in, 20
   power-law relationships, 58, 61
Capping protein, 187
Cardiac tissue
   contraction in, 9
Cartilage
   behavior of, describing, 87–88
   deformation measurement in, 93, 96–98
   examination of, 171
   mechanical properties, measurement of, 85, 93
   mechanosensing in, 19
   momentum balance equations in, 86
   stiffness in, 8
   in vivo state, characterization of, 94
Caspases, 188
Cauchy-Green tensor in viscoelasticity measurements, 88
Cauchy stress tensors in momentum balance laws, 86
Caulerpa spp., 131
Cell division, 25
Cell indentation. *See* indentation; microindentation
Cell locomotion. *See also* migration of cells
   about, 152, 153
   actin's role in, 171, 193–195
   Arp 2/3 complex in, 189
   leukocytes, 189, 195

Index

Cell mechanics. *See also* cells, dynamics
   about, 1, 84
   biphasic models of, 85–88
   changes, measurements of, 59
   measurements of, 52–54, 56–57
   role of, 2
Cell membranes. *See also* lipid bilayer membrane
   about, 170, 172
   actin network dynamics near, 212
   in diffusion dynamics, 130–132
   disruption of, 130, 131
   fluidity changes, response to, 228
   force per unit area equation, 211
   force transmission in, 80
   and magnetic field gradients, 34–35
   mechanical tension applied to, 26
   modeling, 77, 80, 173, 227
   protrusion of, 189, 205–209
Cell pokers in deformation measurement, 22–23, 118
Cells. *See also individual cell type by name*
   about, 129, 170
   adherent, 115, 118, 218–222
   anatomy, structural, 7–9
   arterial wall, 3
   behavior
      anisotropy in, 98
      describing, 84–85, 95
      diffusion in, 129
      regulation of, 104
      stress/strain, 108–109
   biphasic properties, measurement of, 95
   cancer, power-law relationships, 58, 61
   crawling, actin dynamics in, 188–195
   dynamics (*See also* cell mechanics)
      about, 135–138
      modeling, 227
      tensegrity and, 121–124
   energy potential, 135
   environment
      interactions with, 152
      prediction of, 98
      sensing, 5–6
   function, 124, 134–135
   as gels, 132–135
   incompressible, Poisson ratio in, 92
   loading, response to, 87–89, 98
   mammalian, formin in, 186
   mechanical properties
      measurement of, 50, 51–62, 84, 92, 115
      regulation of, 98, 105, 117, 124, 152
      representation of, 206
   mechanosensing in, 5–6, 14, 19, 28
   membrane-wounded state in, 131
   migration of (*See* cell locomotion; migration of cells)
   modeling of, 71–72, 76–78, 81
   multiphasic properties, measurement of, 89, 94, 96
   network elasticity in, 163–165
   power-law relationships, 58
   secretion in, 139–140
   shape
      and behavior, 104
      deformation of, 74, 183
      maintenance of, 2–3, 171
      and prestress, 109, 118, 121, 124
      and stress/strain, 98
   spreading of (*See* cell spreading)
   stress/ion concentration, characterization of, 89
   stress/strain behavior in, 109
   structures in, 10–15, 152
   surface indentation in deformation measurement, 22–26
   suspended, modeling, 118
   temperature fluctuations in, 8, 36, 123, 165
   water retention in, 132–134
Cell spreading
   deformity and, 119
   and prestress, 113, 114, 121, 124
   and stiffness, 120
Cellular tensegrity model
   about, 104, 106
   experimental data, 107
   principles of, 104, 114, 129
Chains, force extension in, 156–159
Chondrocytes
   biphasic properties, modeling, 96
   creep response in, 90
   examination of, 171
   mechanical properties, measurement of, 85, 93
   modeling, 77, 87, 90
   volume changes, calculation of, 92
Chondrons, 93
Cilia, 11
Colchicines, 113, 119
Collagen
   binding affinity in, 9, 228
   in the ECM, 173
   in the PCM, 93
   power-law relationships, 58
   stiffness measurements of, 165
Compression
   in actin networks, 179
   in cytoskeletal filaments, 164, 227
   in microplates, 28
   in microtubules, 107, 112–113, 114, 118
   and prestress, 104, 118
   and tensed cables, 116
Computation domains, geometry of, 77–78
Conductance, changes in, 6
Cones, indentation depth in, 24
Connectin, 141
Connecting filaments in muscle contraction, 142, 146
Constant phase model. *See* structural damping equation

Constitutive laws
  in continuum models, 227
  Darcy's law in, 87
  hyperelastic, 98
  Lamé coefficients in, 87
  multiphasic about, 84, 85, 98
  in protrusion, 209, 217–218
  and stress-strain relationships, 72
  and structural behavior, 226
Continuum mechanics in cell representation, 206
Continuum viscoelastic models
  about, 71–72, 81, 227
  anisotropy in, 98
  biphasic/triphasic, 84
  limitations of, 81, 98
  principles of, 76, 78
  purpose of, 72–75
  stress in, 104
Contraction of muscles
  about, 4, 7, 8, 140–147
  actin filaments in, 108, 142–143, 145–146
  airways, pulmonary, 8
  cardiac tissue, 9
  in cell migration, 108
  connecting filaments in, 142, 146
  connectin in, 141
  cross-bridges in, 110, 145–146
  in the cytoskeleton, 14
  epithelial cells, 108
  fibroblasts, 14, 107, 113
  and force/stiffness, 66, 68
  force transmission and, 66, 68
  in HASM, 65, 109, 110
  immunoglobulins in, 141
  inchworm mechanism of, 143–147
  in lamellipodia, 108
  microtubules and, 108, 112
  modeling, 81, 227
  modulation of, 122
  myofibrils in, 144
  myosin-based, 191
  myosin in, 141
  phase transitions in, 129, 146
  power-law relationships, 123
  power-law rheology, 123
  and prestress, 104, 112, 114
  prestress in, 104, 112, 114
  sarcomeres in, 7, 8, 144
  and stiffness, 66, 68, 104, 111, 114
  stiffness and, 66, 68, 104, 111, 114
  temperature and, 158–159
  thermal fluctuations in, 158–159
Convergence-elongation theory, 191
Cortex
  about, 11
  modeling, 78, 80, 227
  shear moduli in, 35, 153
Cortical layer, tensile force in, 116

Cortical membrane, 205, 210. *See also* cell membranes
Cortical membrane model, 115–116, 117
Courant condition, 217–218
Creep response
  modeling, 87, 90, 123
  in the PCM, 93
  and shear moduli, 35
  in vascular endothelium, 172
Crk-associated substrate, 14
Cross-bridges in muscle contraction, 110, 145–146
Crosslinking
  in actin networks, 166, 171, 228
  and effective modulus, 180
  in polymers, 165, 180
  in proteins, 153, 163–165
  shear modulus of, 164–165, 180
  in smooth muscle, 110
  in tensegrity models, 106, 124
Crossover frequency. *See* frequency
Curvature, measurement of, 155, 156, 157
Cytochalisin-D
  and the actin network, 57, 111, 181, 204
  and cell migration, 34
  shear/loss moduli and, 56–57
Cytoindentation in cell modulus measurement, 94–96
Cytometry techniques, 109. *See also individual technique by name*
Cytoplasm
  about, 85, 170, 205
  dynamics, 205
  energy potential, 135
  as gel, 132
  tearing and strain, 114
  viscosity, measurement of, 20–21
  volume flow, 206
Cytoskeleton
  about, 62, 153, 170, 206
  deformability, mechanisms of, 117
  density, 204–205, 209
  environment, interactions with, 172–174, 176
  filaments
    about, 154, 167
    assembly of, 124
    behaviors, 159, 164
    Brownian motion in, 160–162, 167, 227
    compression in, 164
    dynamics of, 160–162, 167
    elastic properties of, 13
    imaging, 161
    loading, response to, 156–159
    nonlinear responses in, 167
    orientation, calculation of, 157
    relaxation rates of, 161
    rotation in, 164
    stretching of, 164
  force transmission in, 9–10, 14, 67, 111, 116
  mechanics, field of, 1–2

Index

mechanotransduction in, 227, 228
modeling of, 71, 77, 80, 81, 114–121, 124
momentum equations for, 208
particle movement in, 34
prestress in, 107–108, 114, 117, 124 (*See also* prestress)
stiffness values, 226
structure of, 176, 183
velocity field in, 222
volume equations for, 219
Cytosol, 206, 207, 210. *See also* cytoplasm

Darcy's law in constitutive law, 87
DBcAMP, 56–57, 61
Deformation measurement. *See* rheology
Deformations
   cell shape, 74, 183
   non-affine, 164
Depolymerization of actin, 177, 181–183, 189, 192
Dictyostelium, deformation measurement in, 20
Differential modulus in actin networks, 166
Diffusing wave spectroscopy (DWS), 40–41
Diffusion
   in cell behavior, 129, 140
   coefficients
      actin monomers, 171, 191
      extracting, 41–42, 171
   in deformation measurement, 21
   force-driven, 213
   and mean square displacements, 161
   membrane pumps in, 130
   paradigm, problems with, 130–132
   rate, changes in, 6
Discrete models, stress in, 105, 109
Disease
   mechanotransduction and, 228
   PCM response to, 93
   stress response to, 6
DNA
   force extension relationships in, 157
   persistence length of, 155
   stiffness measurements of, 161
Donnan osmotic pressure relation defined, 89
Drug treatment
   and actin polymerization, 110
   and cell migration, 34
   shear/loss moduli and, 57
Dynamic light scattering (DLS), 39–40
Dynamic moduli, calculation of, 95

Effective modulus and crosslinking, 180
Elastic energy and microtubule disruption, 112
Elasticity. *See also* viscoelasticity
   biphasic elastic formulations and creep response, 90
   of cytoskeletal filaments, 13
   Hookean, shear modulus of, 116, 123
   network, in cells, 163–165
   solutions, response to, 162
   stress/strain displacement, linear models of, 80
   Young's modulus of. *See* Young's modulus of elasticity
Elastic moduli
   in actin networks, 167, 178, 227
   in cell dynamics, 122, 166
   and frequency, 180
   in MTC, 54–55, 57, 58
   power-law relationships, 59–60
Elastic propulsion theory, 194, 195
Elastic ratchet model, 193–194
Electroporation, 130
Endothelial cells
   actin cytoskeleton in, 107, 117
   action at a distance effect in, 110–112
   deformation measurement in, 28
   examination of, 171, 172
   fluid shear stress in, 177
   focal adhesions in, 107
   mechanical properties of, 172, 173
   microtubules in, 108, 119
   modeling, 78, 80–81, 90
   power-law relationships, 58
   shear/loss moduli in, 58
   stiffness in, 119, 120
   stress response in, 19, 116, 120
Endothelial monolayer, 171
Energy
   bending, in force extension relationships, 157
   elastic and microtubule disruption, 112
   hyperelastic strain energy function, 98
   potential in cells, 135
   temperature and, 226
   wells in SGR, 64
Entanglement length in semi-flexible polymers, 162
Entropy in cell response, 157, 163, 165, 226
Environment
   and cell locomotion, 152
   cytoskeletal interactions with, 172–174, 176
   and phase transitions, 137, 139
   prediction of in cells, 98
   sensing by cells, 5–6, 14, 19, 28
Enzyme activity, changes in, 228, 229
Epithelial cells
   about, 3
   and cellular insult, 131
   contraction in, 108
   membrane-wounded state in, 131
   modeling, 79–81
   power-law relationships, 58, 61
   shear/loss moduli in, 25, 58
   structure of, 11
Equilibrium moduli
   calculation of, 95
   in endothelial cells, 171
   in protrusion, 219

Erythrocytes
    about, 2
    deformation measurement in, 26, 40, 73, 74
    microcirculation, 74
    modeling of, 71, 75, 78–79, 90, 118
    movement of, 34
    optical trapping of, 32
    rupture strength in, 10
    stiffness of, 11
    structure of, 11, 174
Euler-Lagrange scheme, 217–218
Extracellular matrix (ECM)
    about, 9, 93, 173
    behavior of, describing, 87–88, 98
    chondrocyte properties in, 92, 96
    interactions with, 84, 96–98, 124
    mechanotransduction in, 228
    prestress in, 104, 107, 116
    production of, 28

Falling sphere method, 20–21
Ferri/ferromagnetic particles, deformation measurements in, 34, 35, 51–52, 112, 115
Fibrin, stiffness measurements of, 165
Fibrinogen in endothelial cells, 172
Fibroblasts
    about, 3, 181
    actin dynamics in, 189, 190
    action at a distance effect in, 111
    contraction in, 14, 107, 113
    force transmission in, 80
    gelsolin-null, 187
    microtubules in, 108
    migration in, 75, 191
    modeling, 79–81, 90, 227
    power-law relationships, 58
    shear/loss moduli in, 25, 35, 58
    viscosity measurement in, 28
Fibronectin
    action at a distance effect, modeling, 110–112
    binding affinity in, 9, 228
    power-law relationships, 58
    viscosity measurement in, 28
Fick's law, 192
Ficoll, 192
Filamin-A, role of, 175–176
Filaments, cytoskeletal. See cytoskeleton, filaments
Filamin, 11
Filopodia, 4, 189, 191
Fimbrin, 11
Finite element methods (FEM)
    in cell behavior modeling, 90, 92, 93, 98
    in microcirculation studies, 74
    in multiphase models, 96–98
    in protrusion modeling, 217–218
    in stress/strain evaluation, 80–81
    in viscoelasticity measurements, 87, 88

FLMP, power-law relationships, 61
Fluctuation dissipation theorem in SGR, 65
Fluid flow
    and cell shape, 98
    deformation measurement, methods in, 28–29
    multiphasic/triphasic models, 88–89, 96
    and stress, 67 (See also shear stress)
    velocity in constitutive law, 87
Fluid mosaic model, 10
Fluorescence correlation spectroscopy (FCS), 41–42
Fluorescence microscopy, 154, 174
Focal adhesion complex (FAC)
    about, 14, 173
    in force transmission, 80, 114, 229
    formation of, 112
    mechanotransduction in, 227, 228
    prestress in, 107
Focal adhesion kinase (FAK), 14
Force balance in cell protrusion, 205
Force extension relationships
    about, 157, 159
    bending energy in, 157
    boundary conditions in, 157
    in chains, 156–159
    in DNA, 157
Force transmission
    in cell membranes, 80
    in cell migration, 75, 205
    cellular exposure to, 19, 228
    in chains, 156–159
    and contraction, 66, 68
    and crosslinking, 110, 111
    to the cytoskeleton, 9–10, 14, 67, 111, 116
    focal adhesions in, 80, 114, 229
    in fibroblasts, 80, 107
    and focal adhesion sites, 80
    in the glycocalyx, 3, 7
    and hysteresivity, 66, 68
    modeling, 72, 80, 82, 227
    in the nucleus, 111
    in particles, 30–31
    small particles, 30–31
    in solvent displacement, 132, 134
    and strain, 166
    in tensegrity models, 106, 113
    time dependencies and, 67
    vs. stiffness, 120
Formins in actin filament assembly, 185–186
Frequency
    and elastic moduli, 180
    in indentation studies, 95
    in MTC measurements, 53, 56
    in polymer solutions, 163, 166
    in power-law relationships, 61, 124
    and shear/loss moduli, 54–55, 57, 58, 163
    and stiffness, 163
    and viscoelasticity, 36, 123–124

Index

Galerkin finite element scheme, 217–218
Gel dynamics
    motion and, 138–147
    principles of, 129
    shear modulus of, 180
Gelsolin, 143, 187–188
Gene expression
    and cellular insult, 131
    changes in, 6, 152
    mechanotransduction and, 227
    regulation of, 124
Glass microneedles in deformation measurement, 22
Glycocalyx, 3, 7
Goblet cells, 139
GP1bα, 176
GTPase, 176

Hair cells, mechanosensing in, 5
Heart disease, 171
Hertz relation, 24
Histamine
    power-law relationships, 61
    and prestress, 110, 111, 119, 122
    shear/loss moduli of, 57, 58
Hookean elasticity, shear modulus of, 116, 123
Hooke's Law, 22
Human airway smooth muscle (HASM)
    action at a distance effect in, 112
    cell dynamics in, 115, 122–123
    contraction in, 65, 109, 110
    malleability of, 65
    MTC measurements of, 54–57
    power-law relationships, 60, 61
    stress/strain behavior in, 109
    work of traction in, 110, 111
Hydrostatic pressure protrusion, 215–216
Hyperelastic strain energy function and anisotropy, 98
Hypertonic permeability model of protrusion, 215, 216
Hysteresivity
    and force transmission, 66, 68
    in power-law relationships, 60, 61
    time dependencies and, 67
Hysteretic damping law. *See* structural damping equation

Immunoglobulins, 9
Inchworm mechanism of muscle contraction, 143–147
Indentation. *See also* microindentation
    in cell modulus measurement, 94–96
    cones, depth of, 24
    frequency in, 95
    spheres, depth in, 24
    and stiffness, 24
    surface, in deformation measurement, 22–26, 72
Inner ear cells, mechanosensing in, 5

Integral membrane proteins, binding affinity in, 10
Integrins
    in actin binding, 171, 176
    action at a distance effect, modeling, 110–112, 114
    binding affinity in, 9, 228
    in focal adhesions, 14
    modeling, 79, 115, 118
    and prestress, 128
Intermediate filaments
    about, 13
    and prestress, 104, 109, 114
    role of, 121, 171
    and stiffness, 114
    stiffness measurements of, 165
    in tensegrity models, 106, 113–114
Intervertebral discs, osmotic loading in, 98
Intracellular tomography technique
    action at a distance effect in, 112
    shear disturbance, modeling, 111
Ions
    channels, 228 (*See also* calcium, release of; potassium pump activity; sodium pump activity)
    concentration, characterization of, 89
    multiphasic/triphasic models, 88–89
    partitioning in cell function, 134–135
Isoproterenol and cell stiffness, 111, 122
Isotropy
    in modeling, 71, 88, 227
    in polymer-gel phase transitions, 138, 147

Kelvin model, 90
Keratocyte lamellipodia, 189, 191, 192

Lamé coefficients in constitutive law,
Lamellipodia
    about, 189
    actin depolymerization in, 192
    contraction in, 108
    modeling, 191, 227
    protrusion of, 221
Laminin, binding affinity in, 9, 228
Langevin equation, 160
Latrunculin A, 59
Leukocytes
    about, 2–3
    deformation measurements in, 43
    locomotion of, 189, 195
    modeling, 78–79, 227
    stiffness in, 11
    structure of, 11
Linear damping in cell dynamics, 123
Linear elastic models of stress/strain displacement, 80
Linear isotropic biphasic models, equations for, 87
Linear momentum conservation in continuum models, 75, 205
Linear solid model, 123

Lipid bilayer membrane, 9–10, 89. *See also* cell membranes
Lipid vesicles, rupture strength in, 10
Listeria monocytogenes, 193, 212
Loads
    and cell stiffness, 10, 108–109
    in discrete models, 109
    filament response to, 156–159, 165
    long-distance transfer of, 113, 114
    in tensegrity models, 106, 113
    work of traction transfer and, 113
Loss moduli
    in cell dynamics, 122
    in MTC, 54–55, 56
    power-law relationships, 58, 59–60

Macrophages, 58, 61
Magnetic methods in deformation measurement, 32–36
Magnetic tweezers, 34
Magnetic twisting cytometry (MTC)
    about, 51–52
    in the cortical membrane model, 115, 116
    in stiffness measurement, 120
Magnetocytometry, 72, 79, 111
Mass conservation in protrusion, 206
Mast cells, 139
Maxwell model, 123
Maxwell viscoelastic fluids, modeling, 78–79
Mean square displacements, 65, 161
Mechanosensing in cells, 5–6, 14, 19, 28. *See also* environment
Mechanosensitive channel of large conductance (MscL) defined, 5
Mechanotransduction, 227
Melanoma cells, deformation measurement in, 20
Membrane pumps in diffusion dynamics, 130
Mesh boundary in protrusion modeling, 217–218
Mesh size in semi-flexible polymers, 162
Mesoscopic model, 193
MG63 osteoblast cells, modeling of, 95
Mica surfaces, solvent displacement on, 132, 134
Microbeads
    force transmission in, 72, 80
    stress/strain displacements in, 80
Microcirculation and cell shape, 74
Microelectrodes, 130
Microindentation
    cell multiphasic properties modeling, 94–96
    in deformation measurement, 72, 73
Micropipette aspiration
    about, 90
    of chondrocytes, 77, 87, 93
    and continuum models, 73, 76–78
    in deformation measurement, 26, 43, 71, 72, 73
    in erythrocyte modeling, 118
    in viscoelasticity modeling, 78–79

Micropipettes in cell modeling, 90, 110–112, 131
Microplates, shearing/compression method in deformation measurement, 28
Microscopy. *See individual technique by name*
Microtubules
    about, 12, 13, 107, 154
    in AFM, 25
    buckling in, 107, 108, 112–114, 118
    disruption, cellular response to, 228
    in endothelial cells, 108
    and G-actin, 192, 193
    in load transfer, 113, 114
    modeling, 155, 167, 227
    persistence length of, 107, 155
    and prestress, 109, 113, 119, 121
    role of, 171
    in tensegrity models, 106
Migration of cells. *See also* cell locomotion
    about, 4, 7
    actin filaments in, 75, 108, 210
    anatomical structures in, 11
    contraction in, 108
    and drug treatment, 34
    force transmission in, 75, 205
    modeling, 75, 81, 228
Mitosis, measurement of, 25
Mixture momentum equation in triphasic models defined, 89
Molecular ratchet model, 194–195, 213
Momentum balance laws, solid/fluid phases, 86
Momentum conservation in cell protrusion, 205, 207–208, 217–218
Momentum exchange vector defined, 86
Motion
    3D, 156
    and gel dynamics, 138–147
    longitudinal dynamics, calculation of, 162
    planar, 155–156
    subdiffusive, 161
    transverse equations of, 160, 161
MTC. *See* magnetic twisting cytometry (MTC)
Multiphasic constitutive laws. *See* constitutive laws, multiphasic
Multiphasic/triphasic models
    about, 88–89
    applications of, 96
    cell environment prediction, constitutive models in, 98
    continuum viscoelastic, 84
    solids, 88–89
    triphasic continuum mixture models, 88, 89
Muscle contraction. *See* contraction of muscles
Muscle fibers, deformation measurement in, 20. *See also* human airway smooth muscle (HASM); myocytes, cardiac; smooth muscle cells; striated muscle

Index

Mutual compliance defined, 38
Myocytes, cardiac
    about, 3
    contraction in, 7, 9
    membrane-wounded state in, 131
    patch removal from, 131
    shear/loss moduli in, 25, 58
Myofibrils in muscle contraction, 144
Myosin
    actin filaments, interactions with, 117, 124
    in cell protrusion, 204, 209, 215, 217
    inhibitors and G-actin, 192, 193
    in muscle contraction, 141
    and reptation, 143
Myosin light chain kinase
    binding dynamics of, 66
    cross-bridge kinetics in, 110
    phosphorylation of, 113

Nematocyst vesicles, 140
Network elasticity in cells, 163–165
Network-membrane interaction theories of protrusion, 209–213
Network-network interaction theories of protrusion, 209, 213–215
Network phase, momentum equations for, 207–208
Neurites, extension of, 108
Neuronal cells
    about, 3
    action at a distance effect in, 111
    membrane disruption in, 131
Neutrophils
    about, 3
    gelsolin in, 187
    microcirculation in, 74
    modeling of, 71, 73, 78–79, 90
    power-law relationships, 58, 61
    protrusion in, 218
    stiffness in, 10
    stress response in, 19
Newtonian viscosity
    modeling, 78–79, 123
    in MTC measurements, 54, 56, 57
Nocodazole and cell migration, 34
Noise, 53, 63
Noise temperature
    about, 50
    in power-law relationships, 61, 68
    in SGR theory, 64, 67
    time dependencies and, 67
Non-affine deformations, 164
Nucleic acids, 132
Nucleus
    about, 11
    in cytosol conversion, 206
    disruption, cellular response to, 228
    force transmission in, 111
    and intermediate fibers, 113
    modeling, 77, 112, 227
    prestress in, 117
    stiffness in, 28

One-particle method, 37–38
Optical bead pulling, 75
Optical microscopy, 141, 174
Optical stretcher, 42
Optical traps in deformation measurement, 30–31
Optical tweezers, 32, 74
Oscillatory responses, modeling, 123
Osmolality
    cell response to, 98
    modeling, 88, 89
Osmometers, 89
Osteoarthritis, PCM measurements in, 93
Osteoblast cells, modeling of, 95
Osteocytes, 28

Particles
    ferri/ferromagnetic, deformation measurements in, 34, 35, 51–52, 112, 115
    force transmission, 30–31
    methods, 37–39
    movement in the cytoskeleton, 34
    sedimentation of in rheology, 20–21
    superparamagnetic, deformation measurements in, 34
    twisting of by magnetic forces, 35–36
Passive microrheology, 36
Patch-clamp method, 130
Paxilin, 14, 228
PCM. See pericellular matrix (PCM)
Pericellular matrix (PCM)
    biphasic properties of, 92–94
    role of, 93
    in stress/strain patterns, 96
    viscoelastic response, modeling, 93
Persistence lengths
    of actin, 38, 107, 178
    and bending stiffness, 13, 154, 157, 158
    of biopolymers, 155
    and filament behavior, 159, 167
    of filamentous proteins, 156
    of microtubules, 107, 155
    in semi-flexible polymers, 162
Phagocytosis, 34
Phase transitions
    about, 129, 140
    affinity in, 138
    in cell dynamics, 135–138
    environment and, 137, 139
    isotropy in, 138, 147
    motion and, 138–147
    in muscle contraction, 146
    in polymers, 137, 138

Phospholipids, binding affinity in, 10, 228
Phosphorylation of myosin light chain kinase, 113
Plasma membrane. *See* Cell membranes
Platelets, gelsolin in, 187
Poisson's ratio
   in chondrocytes, 92
   in constitutive law, 87
   in incompressible cells, 92
   in osteoarthritis, 94, 96
Polyacrylamide gel substrate, prestress transmission in, 107, 110
Polymerization of actin
   about, 175, 177, 206
   and cell rigidity, 180, 204–205
   in cytoskeleton production, 207
   discovery of, 181–183
   by drug treatment, 110
   in fibroblasts, 107, 189
   in protrusion, 193–195, 209–217
   regulation of, 187
   stiffness and, 180, 204–205
Polymerization zone, 211
Polymers. *See also* Biopolymers
   crosslinking in, 165, 180
   deformation measurements in, 40, 51–62
   hydrogel dynamics, 137
   modeling, 155–156, 227
   nonlinear responses in, 167
   phase transitions in, 129, 137, 138
   semi-flexible
      entanglement length in, 162
      solutions of, 163, 164, 167
   solutions, frequency in, 163, 166
   stiffness, mechanical in, 154, 156
Polysaccharides, 132
Porous solid model of actin filament structure, 179
Post-buckling equilibrium theory of Euler, 113
Potassium pump activity, 134. *See also* ions
Power-law rheology
   acetylated low-density lipoprotein, 58
   actin filaments, 58
   cancer cells, 58, 61
   collagen, 58
   in continuum modeling, 81
   contraction, 123
   data normalization method, 60–62
   DBcAMP, 61
   elastic moduli, 59–60
   endothelial cells, 58
   epithelial cells, 58, 61
   fibroblasts, 58, 107
   fibronectin, 58
   FLMP, 61
   frequency, 123–124
   HASMs, 61, 60
   histamine in, 61
   hysteresivity in, 60, 61
   macrophages, 58, 61
   myosin light chain kinase, 58, 61
   neutrophils, 58, 61
   noise temperature in, 61, 68
   relaxation rates, 226
   RGD peptide, 58
   shear/loss moduli, 58, 59–60
   in smooth muscle cells, 123
   stiffness, normalized, in, 60, 61
   structural damping equation in, 61
   urokinase, 58
   vitronectin, 58
Pressure in momentum conservation, 207
Prestress
   about, 104
   actin filaments and, 104
   balancing of, 110, 111, 113, 116, 118, 121
   and bead rotation, 115
   cell dynamics and, 122–123
   and cell shape, 109, 118, 121, 124
   cell spreading and, 113, 114, 121, 124
   colchicine and, 119
   compression and, 104, 118
   contractile, 104, 112, 113
   differential modulus in, 166
   in the ECM, 104, 107, 116
   in focal adhesions, 107
   histamine and, 110, 111, 119, 122
   integrins and, 107
   and intermediate filaments, 104, 109, 114
   measurement of, 109, 111, 113, 118
   microtubules and, 109, 113, 119, 121
   and model response, 118
   in the nucleus, 117
   and stiffness, 9, 105, 109–111, 118, 124
   structures, 105
   transmission in polyacrylamide gel substrates, 107, 110
   and viscoelasticity, 123
Probes, 34, 37
Profilin, 185, 188
Proteins
   about, 132
   crosslinking in, 153
   filamentous, 156
   interactions, dynamics of, 50
   mechanotransduction in, 227, 228
   modeling, 71, 227
Proteoglycans, binding affinity in, 9, 228
Protopod dynamics, measurement of, 75
Protrusion
   about, 4, 204–205, 218–222
   actin-based, 193–195, 209–217
   in amoeba, 204
   boundary conditions in, 208
   cell, viscosity in, 212, 219
   of cell membranes, 189, 205–209
   constitutive equations in, 209
   constitutive laws in, 209, 217–218

Index

cytoskeletal theories of, 209–217
hydrostatic pressure, 215–216
hypertonic permeability model of, 215–216
mass conservation in, 206
momentum conservation in, 205, 207–208, 217–218
myosin in, 141, 204, 209, 215, 217
repulsion-driven, 219–220, 222
shearing motor model of, 215, 217
swelling-driven, 220–221
swelling model of, 220–221
Pseudopodia, 189, 205, 216, 218–222

Quartz and water absorption, 132

Radial strain distributions via AFM, 74
Ratchet model, 193
Rate of deformation and viscoelasticity, 123–124
Reactive interpenetrative flows (RIF)
    about, 204
    numerical implementation of, 217–218
    principles of, 129, 205–209
Red blood cells. *See* erythrocytes
Relaxation rates
    of cytoskeletal filaments, 161
    power-law relationships, 226
Release defined, 4
Reptation, 142, 143, 144
RGD peptide, power-law relationships, 58
Rheology. *See also* passive microrheology; power-law rheology; soft glassy rheology (SGR)
    about, 18, 19–22, 74
    atomic force microscopy in, 25–26, 43, 72, 73
    in bone, 28–29
    in cartilage, 93, 96–98
    cell pokers in, 22–23, 118
    Dictyostelium, 20
    diffusion in, 21
    in endothelial cells, 28
    in erythrocytes, 26, 40, 73, 74
    in ferri/ferromagnetic particles, 34, 35, 51–52, 112, 115
    in fluid flow, 28–29
    glass microneedles in, 22
    in leukocytes, 43
    magnetic methods in, 32–36
    magnetocytometry in, 72, 79, 111
    in melanoma, 20
    in melanoma cells, 20
    microindentation in, 73
    micropipette aspiration, 26, 43, 71, 73
    microplates, shearing/compression method, 28
    multiphasic models of, 96
    in muscle fibers, 20 (*See also* human airway smooth muscle (HASM); smooth muscle cells; striated muscle)
    optical traps in, 30–31
    optical tweezers in, 74
    in polymers, 40, 51–62
    probe motion in, 34
    sedimentation of particles in, 20–21
    spectrin in, 32, 73
    in superparamagnetic particles, 34
    whole cell aggregates in, 20
RIF. *See* reactive interpenetrative flows (RIF)
Rotation
    in cytoskeletal filaments, 164, 227
    and prestress, 115
    and torque, 36, 115
Round configuration defined, 120
Rupture strength in cells, 10

Sarcomeres
    about, 141
    in contraction, 7, 8, 144
    experimental data, 144
Scanning probe microscopy, 94–96
Secretion, 139–140
Sedimentation of particles in deformation measurement, 20–21
Selectins, binding affinity in, 9, 228
Semi-flexible chains. *See* biopolymers; polymers
SGR. *See* soft glassy rheology (SGR)
Shear disturbance, modeling, 111
Shearing motor model of protrusion, 215, 217
Shear moduli
    actin cortex, 153
    cable-and-strut model, 118
    creep response and, 35
    in crosslinking, 164–165
    cytoskeletal, 225
    defined, 11
    fibroblasts, 80, 107
    gels, 180
    Hookean elasticity, 116
    in MTC, 54–55, 57, 58
    in the one-particle method, 37–38
    power-law relationships, 58–60
    in semi-flexible polymers, 163, 164
    tensed cable model, 117
Shear stress. *See also* stress
    about, 116, 165
    in actin networks, 179
    fluid, 172, 176
    nonlinear, 167
Shigella spp., 193–194
Signals, collecting, 41–42
Silicon rubber substrate, prestress transmission in, 107
Six-strut tensegrity model, 115, 119
Skeletal muscle cells, contraction in, 7
Smooth muscle cells
    about, 3
    actin polymerization in, 110
    contraction in, 7
    cross-bridges in, 110

Smooth muscle cells (*Contd.*)
  HASM (*See* human airway smooth muscle (HASM))
  membrane disruption in, 131
  membrane-wounded state in, 131
  microtubules in, 118
  MTC measurement of, 51–52, 56–57
  power-law relationships in, 123
  shear/loss moduli in, 25, 58
  time dependencies in, 67
Sodium pump activity, 134. *See also* ions
Softening in tensegrity architecture, 109
Soft glassy materials (SGMs), 62
Soft glassy rheology (SGR)
  biological insights from, 65, 226
  energy wells in, 64
  fluctuation dissipation theorem in, 65
  principles of, 51, 62–65, 129
  structural damping equation in, 64
Soft tissue behavior, modeling, 98
Solids
  modeling, 155, 226, 227
  multiphasic/triphasic models, 88–89
  stress/strain behavior in, 109
  viscoelastic interactions of, 87, 90
Sollich's Theory of SGMs, 63
Solutions, 162, 163. *See also* biopolymers; polymers
Solvent-network drag, 207
Solvent phase, momentum equations for, 207–208
Solvents
  cell response to, 139
  displacement of, 132, 134
  in mass conservation, 207
  momentum conservation in, 205, 207
Speckle microscopy, 192
Spectrin
  and cell stiffness, 11
  in deformation measurement, 32, 72, 73
  lattice, modeling, 118
Spheres, 24, 74
Spread configuration defined, 120
Stereocilia, 5
Stiffness
  and actin polymerization, 180, 204–205
  bending
    in actin networks, 178
    in the cortical layer, 116
    in microtubules, 13, 154
    and persistence length, 13, 154, 157, 158
    relations of, 13, 155–156
    Young's modulus, 13, 155–156
  in bone, 8
  cable-and-strut model, 118
  and cell load, 108–109
  and cell spreading, 120
  contraction and, 66, 68, 104, 111, 114
  and crosslinking, 110, 166
  frequency and, 163
  indentation and, 24
  and intermediate filaments, 114
  isoproterenol and, 111
  measurement of, 161
  mechanical in polymers, 154, 156
  and microtubule disruption, 119
  modeling, 77, 227
  in neutrophils, 10
  normalized, in power-law relationships, 60, 61
  and prestress, 105, 109–111, 119, 124
  shear, 116
  simulation of, 225
  static model of, 64
  structural, 120
  and temperature, 177
  tensed cable model, 117
  tensile, 107, 116
  and tension, 158
  time dependencies and, 67
  and viscosity, 226
  vs. applied stress, 120, 167, 227
Storage modulus. *See* elastic moduli
Strain. *See also* stress
  in actin networks, 179
  AFM indentation distributions, 74
  and bead displacement, 80
  in continuum models, 104
  cytoplasmic tearing and, 114
  evaluation of, 74, 76–78, 80–81
  multiphasic models of, 96
  network response to, 164, 166
  patterns
    identification of, 81
    prediction of, 81
Stress. *See also* strain
  in actin networks, 179, 215
  applied, *vs.* stiffness, 120, 167, 227
  and bead displacement, 80
  cell response to, 19, 67, 104, 116, 156–159
  and constitutive law, 72
  differential modulus in, 167
  fibers about, 11
  fluids, shear, 75
  hardening, prediction of, 120
  multiphasic models of, 96
  network response to, 164, 166
  patterns
    evaluation of, 72, 74, 76–78, 80–81
    identification of, 81, 229
    prediction of, 81
    time dependence of, 96
  response in disease, 6
  restoring, mechanism of, 105
  and stiffness, 114
  and strain field, 29, 166
  swelling, 164, 166, 213, 215
  and tension, 165

Index

Stress fibers
  about, 174, 181
  generation of, 186
Stress-supported structures, 105
Stretching of cytoskeletal filaments, 164
Striated muscle, contraction in, 65
Structural damping equation
  in MTC, 55–56
  in power-law relationships, 61
  and soft glassy rheology, 64
Substrates and prestress, 113, 118
Superparamagnetic particles, deformation measurements in, 34
Surface indentation in deformation measurement, 22–26
Suspended cells, modeling, 118
Swelling
  model of protrusion, 220–221
  stress and network dynamics, 179, 215 (*See also* stress, swelling)
Swinging cross-bridge mechanism model, 141

Talin
  in actin binding, 171
  in the ECM, 173
  in FACs, 14
Tangential strain distributions via AFM, 74
Tangent vectors, 155, 156
Temperature
  and contraction, 158–159
  and energy, 226
  filament bending and, 154–156
  fluctuations
    in cells, 8, 36, 123, 165
    imaging, 161
  and motion, measurement of, 18, 81
  noise (*See* noise temperature)
  and SGMs, 63
  shear/loss moduli and, 54–55
  and stiffness, 177
  viscosity and, 21
  and volume in actin networks, 211
Tensed cable nets model, 116
Tensegrity
  and cellular dynamics, 121–124
  defined, 105–106, 226
Tensegrity architecture, 104, 105, 109
Tensegrity methods. *See also* cellular tensegrity model
  about, 85, 98, 104, 124, 125
  mathematical models of, 114–121
Tensile stiffness, 107, 116
Tension
  cortical, 205, 216, 218
  and shear stress, 165, 166
  short filaments, 159
Tension-compression synergy in the cable-and-strut tensegrity model, 108
Tethered ratchet model, 194

Thermal concepts. *See* temperature
Thick filaments in muscle contraction, 142, 145–146
Thin filaments in muscle contraction, 142, 144, 146
Thrombin in microtubule buckling, 108
B-Thymosin, 188
Time dependencies
  and biphasic models, 88
  in cell behavior, 82
  fibroblasts, 80, 107
  modeling, 67, 88, 227
  smooth muscle, 67
  and stress patterns, 96
  viscoelasticity, 123–124
Tip link, 5
Titin in muscle contraction, 141
Torque
  measurements of, 52–54
  and rotation, 36, 115
Traction, work of. *See* work of traction
Traction cytometry technique in prestress measurements, 109, 111, 113, 118
Traction microscopy and elastic energy, 112
Treadmilling, 182, 204
Triphasic models
  about, 88–89
  applications of, 96
  cell environment prediction, constitutive models in, 98
  continuum mixture models, 88, 89
  continuum viscoelastic, 84
  solids, 88–89
Tropomyosin, 191
Trypsin and FACs, 107
Two-particle methods, 38–39

Upper convected Maxwell model, 88
Urokinase power-law relationships, 58

Vector tangents, 155, 156
Vertical displacement, calculation of, 120
Vertical strain distributions, 74, 75
Vesicles, 10, 139, 147
Vimentin in stress mapping, 29, 113
Vinculin, 228
Viscoelasticity
  biphasic models of, 87–88, 98
  cable-and-strut models of, 123
  Cauchy-Green tensor, 88
  and creep response, 90, 93
  finite element methods, 87, 88
  and frequency, 36, 123–124
  measurement of, 18, 28, 33, 36, 78–79, 90, 205
  mechanical basis of, 85
  modeling, 78–79, 124, 227
  prestress and, 123
  rate of deformation and, 123–124
  time dependencies of, 123–124

Viscosity
    in cell protrusion, 212, 219
    and cytoskeletal density, 209
    of cytosol, 207
    measurement of, 20–21, 26, 28
    modeling, 77, 227
    Newtonian (*See* Newtonian viscosity)
    and stiffness, 226
    temperature and, 21
Viscous modulus. *See* loss moduli
Vitronectin
    binding affinity in, 9, 228
    power-law relationships, 58
Voigt model, 123

WASp/Scar family proteins, 186, 189
Water
    bonding in, 134
    diffusion coefficient of, 171
    in polymer-gel phase transitions, 138
    retention in cells, 132–134

Wavelength
    drag coefficient and, 160
    and thermal properties, 158, 161
Whole cell aggregates in deformation measurement, 20
Work of traction
    in HASMs, 110, 111
    transfer of, 113, 229
Worm-like chain model, 155–156

Yeast and Arp 2/3, 186
Young's modulus
    actin filaments, 172
    in actin networks, 179
    and bending stiffness, 13, 155–156
    calculation of, 90
    in constitutive law, 87
    in elastic body cell model, 26
    in osteoarthritis, 93
    in the PCM, 93